A GEOGRAPHICAL ATLAS
OF WORLD WEEDS

A Geographical Atlas
of World Weeds

LEROY HOLM
JUAN V. PANCHO
JAMES P. HERBERGER
DONALD L. PLUCKNETT

A WILEY-INTERSCIENCE PUBLICATION

JOHN WILEY & SONS, New York · Chichester · Brisbane · Toronto

Library of Congress Cataloging in Publication Data:

Main entry under title:

A Geographical atlas of world weeds.

"A Wiley-Interscience publication."
English, Arabic, Chinese, French, German, Hindi,
Indonesian, Japanese, Russian, and Spanish.
1. Weeds—Geographical distribution. 2. Weeds—
Nomenclature.   I. Holm, LeRoy G.

SB611. G38   632'. 58'09   78–24280
ISBN 0–471–04393–1

Printed in the United States of America

10 9 8 7 6 5 4 3 2 1

To the great host of friends across the world, whose generous gifts of time and talent have made this work possible. We are grateful to the East-West Center in Hawaii for bringing us together in a common workplace to build this book from information gathered over more than a decade of searching for weeds in the world's fields, pastures, and waterways.

Introduction

This atlas of world weeds and their geographical distribution has been prepared with the hope that it will help people to realize that the same weed species may be common to a number of different regions of the world. A worker who needs help with a species that is a serious problem in his or her own country may thus be able to discover in what areas the weed is troublesome and turn directly to weed scientists in other parts of the world who have information on the biology and control of that species.

The literature of agriculture contains suggestions that as many as 50,000 species of plants may behave as weeds, but the authors now know that such estimates are unrealistic. During the past decade, in preparing a two-volume work on the biology and distribution of the world's worst weeds, we have visited most of the countries of the world and searched much of the world's literature on weeds and weed science. We recorded all of the species that came to our attention and found that fewer than 8000 are said to behave as weeds in agriculture. More important, we know from our work that only about 250 of these are important for *world* agriculture. This book does not include all of the species for any geographical region, nor does it present a complete list of all of the weed species of the world. It is simply a record of the weed species we have found in ten years of searching. Surely there are more.

The initial efforts of soil scientists to describe and classify the soils of the United States and its territories were evaluated and summarized as working documents in a series of "approximations," the first of which was issued in 1951. These approximations were statements—agreements—about soil classification that could be accepted at that time, with full understanding that further refinements were being determined. The "seventh approximation" was published in 1960. As a result of these studies there is now a highly sophisticated working system available to soil scientists; it was published in final form in 1975 as "The United States Soil Taxonomy." We believe the present atlas, the most comprehensive weed list ever assembled, is a "first approximation" of the weed species of the world and their distribution. The authors welcome all corrections and additions in order that publication of a "second approximation" may be hastened. For this purpose, a self-addressed page is provided at the end of the book that may serve as a model for the reader's response or may be torn out and mailed in. We ask the reader's help in this effort.

Because Latin is the universal language of taxonomy, the scientific names of the weed species are used here. Nomenclature has been brought into agreement with present-day usage. However, for the convenience of the reader who may be more familiar with an older name, up to three of the most common synonyms are given.

The synonyms are themselves listed in their proper place in alphabetical order and referred back to the correct taxon designation. A country in which a certain weed is found is recorded opposite the species name that is presently in use in that area—whether it is a synonym or the currently accepted taxon designation.

RANKING OF IMPORTANCE

Opposite each species name are recorded the symbols (abbreviations) representing the countries or areas in which it occurs, ranked according to the importance of the weed in each area. Thus each country symbol listed for a species is placed under one of the following headings:

S   for Serious weed
P   for Principal weed
C   for Common weed
X   for Present as a weed.(the species is present and behaves as a weed, but its rank of importance is unknown)
F   for Flora (the species is known to be present in the flora of the country, but confirming evidence is needed that the plant behaves as a weed)

Every ranking of Serious, Principal, or Common was provided by a weed scientist for his own country. A country symbol, such as IDO (Indonesia) placed under "P" (Principal), for example, means that a weed scientist has given such a ranking for the species in some crop in some area of Indonesia. It does not mean that a total evaluation of every weed species for every crop has been completed on a countrywide scale. There are perhaps only two or three countries of the world for which even estimates can be made at this level.

As we studied the ways in which workers across the world expressed their estimates of weed importance, our experience grew, and we began to perceive that we were learning from those with whom we visited, worked, and corresponded and that, although it goes unrecorded, there is some commonality of understanding about what is meant by "serious," "principal," and "common" weeds in a crop. When a worker states that a certain species ranks as one of the two or three most "serious" weeds in a crop, there is no problem of meaning. It is our experience that when the worker speaks of "principal" weeds he or she is usually referring to about the five most troublesome species for the crop. Rarely are ten weeds named, for example. A "common" weed refers to one that is very widespread in many of his crops or regions of that country, requiring constant effort and expense to hold at bay but never seriously threatening a crop.

THE LANGUAGE TABLE OF COUNTRY NAMES AND SYMBOLS

Because of space requirements the 124 country names used in the study are expressed with two- or three-letter symbols (abbreviations). The symbol and the full English name of the country represented by that symbol are shown in the first two columns of the "Language Table of Country Names and Symbols" just preceding the text of the book. In the remaining columns, the name of the country is given in nine other languages. This provision, together with the custom of using Latin for the names of plant species across the world, makes it likely that all weed scientists, everywhere, can use this geographical atlas of world weeds.

المقدمة

هذا الدليل لحشائش العالم وتوزيعها الجغرافي قد تم اعداده املا في ان يساعد القارئ على ادراك ان نوعا واحدا من الحشائش يمكن ان يسود عدة مناطق مختلفة من العالم . ولذلك فان اي دارس محتاج لتعريف نوع من الحشائش ضار بقطره يمكنه ان يحدد اضرار هذا الحشيش ثم يلجأ الى علماء الحشائش في مناطق أخرى مَن العالم ليعدوه بمعلومات عن حياته ومقاومته .

هناك تقديرات بأن حوالي ٥٠،٠٠٠ نوع من النباتات يمكنها ان تقوم بدور الحشائش ، ولكننا وجدنا أن هذه تقديراته غير واقعية . ففي العشر سنوات الماضية التي أعددنا فيها كتابا (في جزئين) عن حيوية وتوزيع أكثر حشائش العالم ضررا ، قمنا بزيارة معظم أقطار العالم ودراسة مخطوطات كثيرة عن الحشائش والعلوم الخاصة بها . وبعد تسجيل كل أنواع الحشائش التي أثارت انتباهنا ، وجدنا أن أقل من ٨٠٠ نوع يمكن اعتبارها ذات أضرار زراعية ، وأهم من ذلك فأننا نعلم من دراساتنا حوالي ٢٥٠ فقط من هذا العدد لها أهمية زراعية عالمية . هذا الكتاب لا يحتوي على جميع أنواع الحشائش بكل منطقة جغرافية ، كما انه لا يحتوي على قائمة بأنواع جميع حشائش العالم ، ولكنه مجرد سجل لأنواع الحشائش التي وجدناها في بحثنا الذي دام عشر سنوات ، ولا شك ان هناك انواعا أخرى .

مجهودات علماء الأراضي الأولى لوصف وتقسيم أراضي الولايات المتحدة ومحمياتها قد درست ولخصت كرئائث عطية في سلسلة من «التقريبات» أولها نشر في عام ١٩٥١ م . هذه «التقريبات» عن علم تقسيم الأراضي كانت تحتوي على ما انفق عليه العلماء في ذلك الوقت . «التقريب السابع» في تلك السلسلة تم نشره في عام ١٩٦٠ م . ونتيجة هذه الدراسات فانه يوجد لدى علماء الأراضي نظام عمل رائع سوف ينشر في صيغته النهائية في عام ١٩٧٥ تحت اسم «تقسيم أراضي الولايات المتحدة» . واننا نؤمن ان هذا الدليل الذي يعتبر أضخم قائمة حشائش جمعت حتى الآن ، يمثل «التقريب الأول» لأنواع حشائش العالم وتوزيعها . اننا نرحب بكل توجيه وأي اضافة تمكننا من نشر «التقريب الثاني» بسرعة . ومن اجل هذا الغرض أرفقنا في نهاية هذا الكتاب صفحة عنواننا يمكن قطعها وارسالها الينا مع ابداء رأيك فيها ايها القارئ راجين منك المساعدة في ذلك .

وحيث ان اللغة اللاتينية هي اللغة العالمية المستعملة في علم تقسيم النباتات ، فاننا قد استعملنا الأسماء العلمية الحديثة لأنواع الحشائش وبالاضافة الى هذا فان الدليل يحتوي على عدة أسماء عامة لمساعدة القارئ على معرفة الحشيش . وهذه الأسماء مرتبة ترتيبا «ابجديا» وكل منها يرمز الى التقسيم العلمي السليم . اسم القطر الذي وجد فيه الحشيش مسجل أمام الاسم المستعمل في تلك المنطقة سواء كان اسما عاما أوعليها .

الترتيب من حيث الأهمية

أمام اسم كل نوع من أنواع الحشائش توجد رموز (اختصارات) تشير الى الأقطار أو المناطق التي يوجد بها ذلك النوع ، مرتبة حسب أهمية الحشيش في كل منطقة . وعلى ذلك فرمز كل قطر مسجل لنوع معين يأتي تحت أحد الرتب الآتية :

IX

S	للحشيش الخطير .
P	للحشيش الرئيسي .
C	للحشيش العام .
X	لنبات يعتبر حشيشا ولكن أهميته غير معروفة .
F	لنبات موجود ضمن الكائنات النباتية في القطر ولكن كه كحشيش لم يثبت بعد .

أن أي رتبة من الرتب الثلاث الأولى قد حددها عالم من علماء الحشائش . في قطره . مثلا عندما يوضع رمز قطر مثل (اندونيسيا) تحت P (حشيش رئيسي) فذلك يعني أن عالم حشائش قد أعطى هذه الرتبة لنوع من الحشائش من محصول ما في منطقة ما من اندونيسيا . وهذا لا يعني أن تقييما كاملا لكل أنواع الحشائش من كل محصول قد أنجز على نطاق القطر ، فان هناك على الأكثر قطرين أو ثلاثة في العالم كله يمكن أن يعمل لها مجرد تقديرات للتقييم على هذا المستوى .

وفي خلال دراستنا للطرق التي يعبر بها علماء الحشائش حول العالم عن تقديراتهم لأهمية الحشائش فأن خبرتنا زادت وبدأنا نفهم من خلال احتكاكنا بهم أن هناك مفهوما مشتركا بخصوص الرتب «خطير – رئيسي – عام» من محصول ما . عندما يذكر أحد علماء الحشائش أن حشيشا ما يعتبر أخطر اثنين او ثلاثة حشائش من محصول ما من منطقة معينة ، فمن الواضح أنه يتحدث عن الحشائش S (الخطيرة) . أما عندما يتحدث عن الحشائش P (الرئيسية) فانه يعني حوالي الخمسة حشائش الأكثر ضررا لمحصوله ، ونادرا ما يذكر عشرة حشائش . أما C «الحشيش العام» فهو الحشيش المنتشر في عدة مناطق ومحاصيل ، والحاجة تدعو الى بذل جهود متواصلة لمنعه من الانتشار، ولكنه لم يهدد أي محصول بصفة جدية .

جدول اللغات لأسماء ورموز الأقطار

أسماء ال ١٢٤ قطرا المستعملة في هذا البحث مختصرة لحرفين أو ثلاثة بسبب ضيق المجال . الرمز والاسم الانجليزي الكامل لكل قطر موجودان في العمودين الأولين في «جدول اللغات لأسماء ورموز الأقطار» الذي تجده في بدء هذا الكتاب . أسماء الأقطار مكتوبة بتسع لغات أخرى في الأعمدة الباقية . استعمال هذا النمط مع استعمال الأسماء اللاتينية لأنواع النباتات سوف يسهل على علماء الحشائش حول العالم استخدام هذا الدليل الجغرافي لحشائش العالم .

緒　言

　　本**世界雜草種類及其分佈索引**編輯之目的乃希望它能幫助人類了解同一種雜草可能爲世界上幾個不同地區的人所熟悉。如此一個工作者在其國內若某種雜草對他來說是個束手無措的嚴重問題而需求助於外界時，就會發現在世界其他什麼地方這種雜草也是一個嚴重的問題而能直接求敎於世界其他地區有生物學及防治此類雜草知識的專家來幫助他。

　　在農業文獻裏包含約有五萬種植物爲雜草，然本書作者們現已發現此估計數字實爲不確。在過去十年中，我們於準備出版二册關於世界最嚴重的雜草的生物學及其分佈時，曾走遍世界上許多國家並搜集研究世界各地有關雜草以及雜草科學的著作。我們記摘下來所研究過的雜草，發現到約有八千種爲農業方面的雜草。更重要的是,研究的結果我們發現在這八千種裏,只有二百五十種對世界農作物是極其重要的。本書沒有包括任何一地區裏所有的雜草種類，亦非代表世界所有雜草一個完整的一覽表。它只代表我們在過去這十年中所研究及發現的雜草種類的記錄。世界雜草種類當不只此數。

　　美國土壤科學家們對其國內及領地的記紋及分類所作的最初努力已被評估並摘要爲工作報告於一九五一年起以**概略** (Approximation)爲名連續發表。這些**概略**是記載當時所認爲可採用的對土壤分類的協議，且在當時更精細的區別已進行決斷中。該**概略**第七次於一九六○年發表。那次土壤分類研究的結果促使今日土壤科學家們得以有了一套發展得完整的處理系統。這套處理系統

於一九七五年出版爲**美國土壤分類學**。我們相信本索引是
前所未有的最充實而完整的雜草目錄，爲世界雜草種類
及其分佈的首次「概略」。本書作者們歡迎讀者的改正意
見及增益以促使第二次「概略」得以早日出版。爲此目的
本書後供有幾頁可撕下來的自備地址通信紙，希望讀者
們多提供意見指教。

　　因爲拉丁文是世界植物分類學上的共同語言，本書
裏所採用的乃是雜草種類的科學學名，且所用的科學學
名爲最新所通用的。爲使一些比較慣用較老學名之讀者
方便起見，本書作者們決定包括多至三種最普遍的科學
別名。這些科學別名是按照英文字母次序列出。同時也
提及它們正式的類名。國名是記錄在那類雜草名稱的右
邊，而這名稱乃是那一國所通用的，不論它是科學別名
或是目前所正式採用的雜草類名。

重要性的分類

　　每一雜草種類名稱的右邊有代表國家或地區的英文
字母符號（簡稱）。根據那類雜草在那個地區的重要性
將國名的英文字母符號列在底下幾個標題的其中之一

　　　S　　表示嚴重的雜草
　　　P　　表示主要的雜草
　　　C　　表示普通的雜草
　　　X　　表示出現爲雜草（其重要性不詳）
　　　F　　表示植物（表示那一類出現於該國的植
　　　　　　物裏，但需有更多的證明來確定它確爲雜
　　　　　　草）

　　每一嚴重、主要或普通的重要性的排列乃是根據該
國的雜草科學家所提供的資料。例如：一個國家的簡稱
符號若是IDO（代表印尼）列在「P」（主要的）底下即表
示在印尼的某些農作物或某些地區，這一雜草經該國
的科學家們鑒定爲主要的雜草。它並不代表每一雜草種

類在每一農作物裏的全國性完整估計。在目前全世界能做如此完整估計的國家大概只有二、三個。

　　當我們研究世界各地的工作人員對他們雜草重要性鑒定的種種方法時，使我們增長了不少經驗。並開始覺察到由於我們從以前訪問過的、共同工作過的和通信過的那些人學到了不少，雖然沒有記摘下來，對於他們所謂「嚴重的」、「主要的」、和「普通的」雜草在農作物裏是何等程度或意義却有共同的了解。當一個工作人員指出某種雜草在某農作物裏被列爲該國幾種最嚴重的雜草之一時，其意所指甚爲明白。經驗告訴我們當他說「主要」的雜草時所指的是，在他的農作物裏有五種最使他束手無措的雜草，通常很少提到十種。「普通」的雜草指的是，在很多農作物裏或在該國很多地區裏很普遍，這一種雜草需得經常的努力及經費來防治，但並不嚴重地爲害農作物。

國名及其簡稱符號的九種語言一覽表

　　由於篇幅有限，本書裏所用的一百廿四個國家的名字是用二個到三個英文字母符號做代表。這代表符號及其整個英文國名出現在書前的**國名及其簡稱符號的九種語言一覽表**裏的頭二欄。其餘幾欄表示國名的九種世界其他語言。本一覽表以及世界通用植物類名用拉丁文表示的慣例將能使世界各地雜草科學家們使用本**世界雜草種類及其分佈索引**。

Introduction

Cet Atlas des mauvaises herbes et de leur répartition géographique dans le monde a été réalisé dans le but de faire prendre conscience que les mêmes espèces d'herbes se retrouvent souvent dans des régions différentes du globe. Ainsi, toute personne qui aurait des difficultés avec une espèce d'herbe dans un pays donné, peut utiliser cet index pour déterminer d'abord, dans quelle autre partie du monde cette herbe pose un problème identique, et faire ensuite appel à des spécialistes travaillant dans d'autres centres de recherche pour obtenir les informations relatives à la biologie et au contrôle chimique de cette espèce particulière.

Les publications ayant trait à l'agriculture semblent indiquer qu'environ 50 mille espèces de plantes s'apparentent aux herbes, mais les auteurs de cet ouvrage sont d'avis que ces chiffres sont grandement exagérés. Au cours de la préparation de ces deux volumes portant sur la biologie et la répartition des pires espèces de mauvaises herbes—réalisation qui nous a pris presque 10 ans—nous avons visité de nombreux pays du monde et pris connaissance avec presque tout ce qui a été publié sur ces herbes et leur contrôle. Nous avons trouvé qu'un peu moins de 8 mille espèces de plantes présentent les caractéristiques attribuées aux mauvaises herbes à l'échelle agricole mondiale. De plus il ressort de notre travail qu'environ seulement 250 jouent un rôle vraiment significatif au point de vue agricole. Ce livre ne mentionne pas toutes les espèces d'herbes que l'on trouve dans une région géographique donnée. Il ne constitue pas non plus un catalogue exhaustif de toutes les espèces que l'on trouve dans le monde. Il représente seulement la somme de nos dix années de recherche et d'expérience dans ce domaine. Il y en a surement bien plus.

Les premiers efforts faits par des spécialistes pour classifier et décrire les différents types de sol que l'on trouve aux Etats-Unis et dans les territoires américains, ont été évalués et publiés sous forme de documents de travail dans une série de fascicules appelés : "Approximation." Le premier numéro de cette série a été publié en 1951. Ces approximations ne constituent en fait, que des déclarations—des points communs—relatifs à la classification des sols. A cette époque ces approximations étaient acceptées comme telles, bien que tous étaient conscients que la matière demanderait à être approfondie ultérieurement. La "Septième Approximation" a été publiée en 1960. A la suite de ces travaux, il existe maintenant, pour les spécialistes des sols, une méthode de travail et une documentation extrêmement specialisée. Elle á été publiée sous sa forme définitive en 1975 comme "La taxonomie des sols des Etats-Unis."

Nous souhaitons que cet ouvrage, qui constitue la documentation la plus complète jamais rassemblée dans ce domaine, représente au même titre, la première "Approximation" de la classification et de la répartition géographique des mauvaises herbes dans le monde. Les auteurs seraient reconnaissants aux lecteurs pour toute correction ou suggestion qui pourrait contribuer à hâter la publication de leur "Seconde Approximation." Ils comptent naturellement sur la collaboration de leurs lecteurs dans ce domaine, et à cet effet, des feuilles portant l'adresse du destinataire ont été incluses à la fin de l'ouvrage.

Le Latin étant la langue universelle de la taxonomie, nous avons utilisé les noms scientifiques des différentes espèces d'herbe. La nomenclature de l'ouvrage est conforme à l'usage moderne. Toutefois, pour le lecteur qui serait plus familier avec les noms communs de certaines espèces, nous avons inclus à cet effet, jusqu'à trois des appellations les plus répandues. Elles figurent par ordre alphabétique dans l'index, avec références appropriées à leur nom scientifique. Ainsi, en face de chaque espèce d'herbe—qu'il s'agisse du nom "local" ou du nom scientifique—on trouvera le nom du pays où cette espèce particulière se rencontre.

LA CLASSIFICATION SELON L'IMPORTANCE

En face de chaque nom de plante on trouvera des symboles (ou abréviations) indiquant le nom des pays ou des régions où cette espèce particulière se rencontre. Ces symboles sont classés en fonction de l'importance de cette herbe dans le pays en question. Ainsi, chaque symbole de pays est accompagné par un de ces principaux signes :

S   pour Herbe posant des problèmes *serieux*
P   pour Herbe largement répandue et constituant le *principal* problème
C   pour Herbe *commune*
X   pour existant en tant qu'herbe (cette espèce particulière se comportant comme une herbe mais son importance dans le pays en question n'ayant pas été établie)
F   pour *Flore* (cette espèce particulière étant présente dans la flore, mais il n'a pas été établi de façon définitive si la plante en question se comporte comme une herbe)

Le choix de ces trois signes—Sérieuse, Principale et Commune—pour une herbe donnée, aura généralement été établi par un spécialiste travaillant dans le pays même. Ainsi, quand le symbole d'un pays, tel IDO (Indonésie) figure à côté de P (Principale) cela veut dire que cette classification a été attribuée, par un spécialiste des mauvaises herbes travaillant localement, à cette espèce qui se rencontre dans telle culture et dans telle région d'Indonésie. Mais cela ne veut pas dire qu'une évaluation globale de chaque espèce de mauvaise herbe pour chaque culture a été établie à l'échelle du pays tout entier. Ce genre de documentation ne pourrait exister à l'heure actuelle, que pour deux ou trois pays tout au plus.

Notre expérience n'a cessé de croître au fur et à mesure que nous avons pris contact avec les chercheurs du monde entier et avec leur méthode de classification des mauvaises herbes selon leur importance. Les personnes avec lesquelles nous avons travaillé ou entretenu une correspondance suivie nous ont tout d'abord beaucoup appris. Et le temps aidant, nous avons fini par découvrir en quelque sorte, un "critère commun" pour la classification des mauvaises herbes. Ainsi, quand nous parlons d'une herbe qualifiée de "Sérieuse," "Principale" ou "Commune," dans une culture il existe peu de chances de confusion. Quand un cultivateur dit qu'une certaine espèce de mauvaise herbe figure parmi les deux ou trois espèces les plus "sérieuses" d'une plantation, nous savons immédiatement ce qu'il veut dire. Et quand il parle de ses "principales" mauvaises herbes, il se réfère généralement aux espèces les plus nuisibles de son champ—au nombre de cinq et très rarement au nombre de dix au grand maximum. La classification appelée "commune" désigne généralement une espèce très répandue dans une culture ou dans une région, dont le contrôle nécessite des efforts sérieux et coûteux, mais l'espèce en question ne pose que rarement une menace grave à la culture.

CODE DES NOMS DE PAYS ET DE SYMBOLES

Pour des raisons purement pratiques, les noms des 124 pays mentionnés dans notre ouvrage sont désignés par un code de 3 lettres (abréviations). Les deux premières colonnes intitulées "The Language Table of Country Names and Symbols" (Code des Noms de Pays et des Symboles) figurant au début de l'ouvrage, donneront le symbole et en entier le nom anglais de ces pays. Dans les colonnes suivantes ces noms de pays sont traduits dans neuf autres langues. Nous espérons que ces précautions, ainsi que l'attribution des noms latins pour désigner les différentes espèces de plantes, permettront aux chercheurs du monde entier, quelque soit le pays où ils se trouvent, d'utiliser ce catalogue des mauvaises herbes et de leur répartition géographique dans le monde.

Einführung

Der vorliegende Atlas aller Unkräuter und ihrer geographischen Verbreitung soll zeigen, dass dieselben Unkrautarten in verschiedenen Regionen der Welt vorkommen können. Jemand, der sich in seinem eigenen Land mit einem ernsthaften Unkrautproblem konfrontiert sieht, wird durch diesen Atlas informiert, in welchen Gebieten dieses Unkraut vorkommt, und er kann sich dann direkt an die Unkrautwissenschaftler in anderen Teilen der Welt wenden. Sie besitzen höchstwahrscheinlich die notwendige Information, uber Biologie und Bekampfung dieser Art.

In der Fachliteratur kann man lesen, dass bis zu 50000 Pflanzenarten als Unkräuter vorkommen können. Diese Schätzung hat sich nun als übertrieben erwiesen. Während der letzten zehn Jahre haben wir ein zweibändiges Werk über Biologie und Verbreitungsareale der schlimmsten Unkräuter der Welt vorbereitet. Wir haben fast alle Länder der Welt besucht und haben die Weltliteratur über Unkräuter und Unkräutwissenschaft durchforscht. Und auf Grund unsere Forschungen können wir sagen, dass weniger als 8000 Pflanzenarten als Unkräuter in der Landwirtschaft vorkommen. Vielleicht noch wichtiger ist die Tatsache, dass nur ungefähr 250 davon für die Weltlandwirtschaft wichtig sind. In dem vorliegenden Buch sind nicht alle Arten einer gegebenen geographischen Region aufgeführt. Es bietet auch nicht eine vollständige Liste aller Unkrautarten der Welt. Es ist ganz einfach ein Atlas der Unkrautarten, die wir nach zehnjähriger Suche gefunden haben. Zweifellos gibt es noch mehr.

Die ursprünglichen Bemühungen der Bodenkundler, die Bodenarten der Vereinigten Staaten und ihrer Territorien zu beschreiben und zu klassifizieren wurden ausgewertet und als Arbeitsdokumente in einer Serie sogenannter "Schätzungen" zusammengefasst, von denen die erste in 1951 herausgegeben wurde. Diese "Schätzungen" waren Darstellungen über die Bodenklassifikation, die zu jener Zeit für allgemein annehmbar galten. Jederman war klar, dass spätere Korrekturen dieser "Schätzungen" notwendig sein würden. Die siebte "Schätzung" wurde 1960 veröffentlicht. Auf Grund dieser Studien haben Bodenkundler nun ein höchst ausgeklügeltes Arbeitssystem zur Verfügung. Es wurde als "Bodentaxonomie der Vereinigten Staaten" 1975 veröffentlicht. Wir glauben, dass unser Atlas als die bis jetzt vollständigste Unkrautliste eine erste "Schätzung" der Unkrautarten der Welt und ihrer Verteilung ist. Die Autoren sind für alle Korrekturen und Ergänzungen dankbar, damit eine zweite "Schätzung" bald veröffentlicht werden kann. Wir bitten unsere Leser um ihre Kooperation. Am Ende des Buches finden sie ein adressiertes Blatt, das zum Zwecke der Korrespondenz herausgerissen werden kann.

Da Latein die Universalsprache der Taxonomie ist werden hier die wissenschaftlichen Namen der Unkrautarten gebraucht. Die Terminologie is auf den letzten Stand der Wissenschaft gebracht worden. Für solche Leser, die mit den älteren Namen bekannt sind, haben wir drei der gebräuchlichsten Synonyme angeführt. Die Synonyme sind amgebotenen Platz in alphabetischer Ordnung angeführt. und dort wird auch auf die richtige taxonomische Bezeichnung verwiesen. Der Name eines Landes in dem ein gewisses Unkraut vorkommt, ist gegenüber der Artbezeichnung, die gegenwärtig dort in Gebrauch ist, angeführt, ganz gleich, ob es sich um ein Synonym oder um die gegenwärtig akzeptierte taxonomische Bezeichnung handelt.

GLIEDERUNG NACH WICHTIGKEIT

Jeder Artbezeichnung gegenüber sind die Symbole (Abkürzungen) der Länder oder Gebiete, in denen die Art vorkommt, aufgeführt, je nach dem wie wichtig ein Unkraut in einem gegebenen Gebiet ist. Das bedeutet, dass jedes Ländersymbol unter den folgenden Kategorien aufgeführt ist:

S   gefährliches Unkraut
P   Hauptunkraut
C   allgemeines Unkraut
X   kommt als Unkraut vor (die Art kommt als Unkraut vor, aber seine Wichtigkeit ist unbekannt)
F   Flora (die Art kommt in der Flora des Landes vor, aber Beweise, dass die Pflanze als Unkraut vorkommt, müssen noch erbracht werden)

Ob ein Unkraut als gefährlich, als Haupt oder als allgemeines Unkraut kategorisiert werden sollte, wurde von den Unkrautwissenschaftlern eines jeden Landes bestimmt. Wo ein Landsymbol wie IDO (= Indonesien) unter P (Haupt) eingeordnet ist, bedeutet, zum Beispiel, dass ein Unkrautwissenspezialist der Unkrautart diesen Rang für ein Gebiet Indonesiens zugesprochen hat. Es bedeutet nicht, dass eine totale Auswertung aller Unkrautarten jeder Kultur auf landweiter Basis vorgenommen worden ist. Es gibt höchstens zwei oder drei Länder in der Welt, für die Schätzungen auf dieser Ebene gemacht werden könnten.

Wir haben untersucht, wie Fachleute in der ganzen Welt ihre Schätzungen über die Wichtigkeit eines Unkrautes ausdrückten. Wir haben so Erfahrungen gesammelt und erkannten bald, dass es ein allgemeines Verständnis, eine weltweite Übereinstimmung, gibt über das, was man "gefährliche," "Haupt," "allgemeine" Unkräuter in einer Kultur nennt. Wenn ein Fachleute sagt, dass eine gewisse Art als eins der zwei oder drei gefährlichsten Unkräuter in einer kultur angesehen wird, dann gibt es kein Verstandnisproblem. Unsere Erfahrung hat uns gelehrt, dass, wenn er von einem Hauptunkraut spricht, er sich normalerweise auf die fünf lästigsten Arten bezieht. Es ist sehr selten, dass zehn Unkräuter in derselben Kategorie genannt werden. Ein "allgemeines" Unkraut ist eins, das sehr weit verbreitet ist, in vielen Kulturen oder Gebieten eines Landes vorkommt, das dauernde Bemühungen und Ausgaben zu seiner Bekampfung erfordert, das aber niemals eine Kultur ernsthaft gefährdet.

DIE SPRACHENTABELLE DER LÄNDERNAMEN UND SYMBOLE

Aus Raumgründen sind die 124 Ländernamen, die in der Studie vorkommen, auf zwei oder drei Buchstaben abgekürzt. Das Symbol und der volle englische Name des Landes, das durch das Symbol repräsentiert ist, sind in den ersten beiden Reihen der "Sprachentabelle der Ländernamen und Symbole" im Anfang des Buches aufgeführt. In den restlichen Reihen ist der Name des Landes in neun anderen Sprachen gegeben. Diese Landes-Methode, wie auch die Sitte lateinische Namen für die Pflanzenarten in aller Welt zu verwenden, wird Unkrautwissenspezialisten in aller Welt den Gebrauch dieses geographischen Atlas der Weltunkräuter ermöglichen.

प्रस्तावना

विश्व में पाए जाने वाले अपतृणों (घास पात, खरपतवार) का यह सूचीपत्र और इनका भौगोलिक-वितरण इस आशय से संकलित किया गया है कि लोगों को यह ज्ञात हो सके कि एक ही जाति के अपतृण विश्व के विभिन्न भागों में पाए जाते हैं। कोई भी अपतृण वैज्ञानिक, जो किसी एक जाति के अपतृण, जो उसके अपने देश में एक गम्भीर समस्या बनी हुई है, इस सूचीपत्र से जान सकता है कि किस क्षेत्र में यह अपतृण आपातीय है और वह विश्व के दूसरे भागों के अपतृण-वैज्ञानिकों से जिन को इसके जैव-चक्र तथा नियंत्रण का ज्ञान है, सीधा सम्पर्क स्थापित कर सकता है।

कृषि-साहित्य के अध्ययन से यह ज्ञात होता है कि पौधों की लगभग ५०,००० जातियां ऐसी हैं जो अपतृणों के अनुरूप हैं, परन्तु अब वैज्ञानिकों के मतानुसार यह अनुमान अवास्तविक है। पिछले दशक में विश्व में पाए जाने वाले अपतृणों के जैव-चक्र तथा नियंत्रण से सम्बन्धित, दो खण्डों में, इस पुस्तक के निर्माणार्थ हम लोग विश्व के लगभग समस्त देशों में गए और वहां के सभी अपतृण साहित्य तथा विज्ञान की खोज की। अपतृणों की सभी जातियों को, जो हमारी दृष्टि में आ सकीं, अभिलिखित किया और यह पाया कि ८,००० से कुछ कम ही जातियां कृषि में खरपतवार के रूप में महत्वपूर्ण हैं। इससे भी महत्वपूर्ण तथ्य हमको अपनी इस खोज से ज्ञात हुआ कि इनमें से लगभग २५० जातियां ही विश्व स्तर पर कृषि के महत्व की हैं। इस पुस्तक में किसी एक भौगोलिक खण्ड की समस्त अपतृण जातियां सम्मिलित नहीं हैं और न ही यह विश्व की अपतृण जातियों की पूर्ण सूची है। यह तो विश्व की अपतृण जातियों का केवल एक अभिलेख है जो हमारे दस वर्षों के प्रयास का परिणाम है। इसके अतिरिक्त अन्य जातियां भी हैं।

मृदा-वैज्ञानिकों का अमरीका तथा उसके प्रदेशों की मिट्टी के वर्णन तथा वर्गीकरण के प्रारम्भिक प्रयास का मूल्यांकन तथा संक्षेपण कार्य-योग प्रलेख के रूप में "समीपीकरण" की एक सीरीज में प्रकाशित हुआ है। इसका प्रथम प्रकाशन सन् १८५१ ई० में हुआ। यह समीपीकरण मिट्टी के वर्गीकरण से सम्बन्धित उस समय के स्वीकृत कथन तथा इकरारनामें थे जिनका आगे और भी शुद्धिकरण होना शेष था। इसका "सातवां समीपीकरण" सन् १९६० ई० में प्रकाशित हुआ। इस समस्त अध्ययन के फलस्वरूप अब एक उच्च कोटि की विशिष्ट कार्य-योग पद्धति मृदा वैज्ञानिकों के लिए उपलब्ध है जो पूर्ण रूप से सन् १९७५ ई० में "The United States Soil Taxonomy" के रूप में प्रकाशित हुआ। हमारा विश्वास है कि हमारी यह प्रस्तुत सूची अब तक की अपतृणों की सभी संग्रहित सूचियों से अत्यन्त विस्तृत और विश्वव्यापी अपतृण जातियों का "प्रथम समीपीकरण" तथा वितरण है। लेखकगण सभी सुझाए गए परिवर्तनों तथा परिवर्धनों का स्वागत करते हैं जिससे "दूसरा समीपीकरण" शीघ्र ही प्रकाशित हो सके। इस उद्देश्य से एक स्वयं-प्रेषित पत्र पुस्तक के अन्त में उद्धृत है। हम पाठकों से अपने इस प्रयास में उनकी सहायता के लिए निवेदन करते हैं।

क्योंकि लेटिन वर्गीकरण की सर्वलौकिक भाषा है, अतः अपतृणों के वैज्ञानिक नाम इस पुस्तक में दिये गये हैं। इनकी नाम-पद्धति आजकल की व्यवहारिक पद्धति के अनुकूल है। तथापि उन पाठकों की सुविधा के लिए जो पुरानी नामावली से अधिक परिचित हैं––अपतृणों के तीन सबसे अधिक प्रचलित पर्याय दिए गए हैं। यह प्रायः अपने उचित स्थान पर वर्णमाला के अनुसार सूचीबद्ध किए गए हैं। और उचित वर्गसूची में पुनः

प्रेषित हैं । जिस देश में अमुक अपतृण उपपादित है, उस देश का नाम उस देश में अमुक अपतृण के प्रचलित नाम के सम्मुख अभिलिखित है । यह प्रायः या तो उसके पर्याय हैं या उसका तत्कालीन स्वीकृत वर्ग-अभिधान है ।

महत्ता का वर्गीकरण

प्रत्येक अपतृण जाति के नाम के सम्मुख अभिलिखित चिन्ह (संक्षिप्तकरण) उन देशों या क्षेत्रों का प्रति-निधित्व करते हैं, जहां यह अपतृण जातियां पाई जाती हैं । यह उस क्षेत्र में उस अपतृण की महत्ता के अनुसार वर्गीकृत है । इस प्रकार एक जाति के सूचीगत देशों के चिन्ह, निम्नलिखित शीर्षकों के अन्तर्गत लिखे हुए हैं :

S समस्यापूर्ण अपतृण के लिए

P प्रमुख अपतृण के लिए

C सर्वनिष्ठ अपतृण के लिए

X विद्यमान अपतृण के लिए (यह जाति अपतृण के रूप में विद्यमान है परन्तु इसकी महत्ता अज्ञात है।)

F उद्भिज्जात के लिए (यह जाति उस देश के उद्भिज्जात में समाविष्ट समझी जाती है परन्तु इसके अपतृण होने के प्रमाण की पुष्टि होनी है।)

अपने देश के समस्यापूर्ण, प्रमुख अथवा सर्वनिष्ठ अपतृण का प्रत्येक वर्गीकरण वहां के अपतृण-वैज्ञानिकों द्वारा किया गया है । जब कि एक देश का चिन्ह जैसे कि IDO (इण्डोनेशिया), "P" (प्रमुख अपतृण) के अन्तर्गत है यह प्रकट करता है कि अपतृण वैज्ञानिकों ने इन जातियों का, जो इण्डोनेशिया के कुछ क्षेत्रों की कुछ फसलों में उपलब्ध है, यह वर्गीकरण किया है । इसका तात्पर्य यह नहीं है कि प्रत्येक जाति के अपतृण का मूल्यांकन प्रत्येक फसलों के अनुसार एक देश-व्यापी स्तर पर परिपूर्ण है । सम्भवतः विश्व में केवल दो या तीन देश ही ऐसे हैं जिनके विषय में इस स्तर का समग्र मूल्यांकन किया जा सकता है ।

जब हमने उन विधियों का अध्ययन किया जिसके द्वारा विश्व के वैज्ञानिक अपतृण की महत्ता की प्रांगणना व्यक्त करते हैं, तो हमारा अनुभव बढ़ा और हम यह समझने लगे कि हमने उनसे कुछ ज्ञान प्राप्त किया था जिनके यहां हम कभी गये थे, जिनके साथ हम काम कर चुके हैं तथा जिनसे हम पत्र-व्यवहार कर चुके हैं । यद्यपि यह अभिलिखित नहीं है, फिर भी एक फसल के "समस्यापूर्ण", "प्रमुख" व "सर्वनिष्ठ" अपतृण के अवबोध में कुछ सामान्यता है । यदि कोई वैज्ञानिक यह बताए कि अमुक जाति किसी एक फसल की उन दो या तीन "समस्यापूर्ण" अपतृण में से एक है तो इस बात को समझने में कोई कठिनाई नहीं होगी । यह हमारा अनुभव है कि जब कोई एक वैज्ञानिक अपने किसी "प्रमुख" अपतृण के विषय में चर्चा करता है तो वह साधारणतया अपने फसल की लगभग पांच अत्यन्त आपातीय अपतृण जातियों का उल्लेख करता है । कदाचित् ऐसे दस अपतृण ही उदाहरणार्थ हैं । एक "सर्वनिष्ठ" अपतृण अपनी अनेक फसलों या अपने देश के अनेक प्रदेशों में अत्यन्त विस्तृत रूप में वितरित है । उसके नियंत्रण के लिए सतत् प्रयत्नों तथा व्यय की आवश्यकता है । यद्यपि यह कभी भी फसलों को गंभीर रूप से आतंकित नहीं कर सकता ।

देश के नामों तथा चिन्हों की भाषा सारणी

स्थानाभाव के कारण, इस अध्ययन में चर्चित १२४ देशों के नाम दो या तीन शब्दों के संकेत (शब्द-संक्षेप) से अभिव्यक्त हैं । देशों के पूरे अंग्रेजी नाम तथा उनका प्रतिनिधित्व करते हुए उनके संकेत-शब्द "The Language Table of Country Names and Symbols" के प्रथम दो स्तम्भों में पुस्तक के प्रारम्भ में दिए हुए हैं । शेष स्तम्भों में देशों के नाम नौ भाषाओं में दिए गए हैं । इस व्यवस्था तथा सम्पूर्ण विश्व में पौधों की जातियों के नामकरण के लिए लेटिन प्रणाली के परम्परागत प्रयोग के फलस्वरूप समस्त अपतृण-वैज्ञानिक सर्वत्र ही विश्व की इस भौगोलिक अपतृण-सूची को उपयोग में ला सकेंगे ।

Pendahuluan

Pedoman tentang tumbuhan pengganggu didunia ini dan penyebaran geografinya dimaksudkan untuk dapat membantu kita guna mengetahui bahwa jenis tumbuhan pengganggu yang sama mungkin juga terdapat didaerah-daerah lain didunia. Seseorang yang membutuhkan pertolongan mengenai salah satu jenis tumbuhan yang merupakan masalah yang serius didaerahnya akan mendapatkan keterangan mengenai tempat-tempat dimana tumbuhan itu menimbulkan kerugian dan dapat langsung menghubungi akhli-akhli tumbuhan pengganggu didaerah lain yang mempunyai keterangan-keterangan tentang kehidupan dan cara mengendalikan jenis tumbuhan tersebut.

Dari pustaka pertanian dapat diperkirakan bahwa lebih dari 50.000 jenis tumbuhan dapat merupakan tumbuhan pengganggu tetapi sekarang kami menyadari bahwa perkiraan itu tidak tepat. Dalam 10 tahun terakhir ini, dalam mempersiapkan 2 jilid karya tentang kehidupan dan penyebaran dari tumbuhan pengganggu yang menimbulkan kerugian besar didunia, kami telah mengunjungi hampir semua negara serta menyelidiki banyak pustaka tentang tumbuhan pengganggu dan ilmu tumbuhan pengganggu. Kami mencatat semua jenis yang menarik perhatian kami dan mendapatkan kenyataan bahwa hanya kurang dari 8.000 jenis tumbuhan yang dapat disebut sebagai tumbuhan pengganggu pertanian. Yang lebih penting lagi dari penyelidikan kami tersebut ialah bahwa ternyata dari jumlah itu hanya kira-kira 250 jenis saja yang sebenarnya merupakan pengganggu terpenting terhadap pertanian didunia.

Buku ini tidak mencakup semua jenis dari semua daerah, tidak juga memuat daftar dari semua jenis tumbuhan pengganggu didunia. Buku ini hanya merupakan catatan dari jenis-jenis tumbuhan pengganggu yang telah kami ketemukan selama 10 tahun penyelidikan itu.

Sudah barang tentu masih banyak lagi yang belum kami catat. Usaha-usaha permulaan dari akhli-akhli ilmu tanah untuk menerangkan dan menggolongkan jenis-jenis tanah di Amerika dan penyebarannya telah dievaluasi dan diringkaskan menjadi karya-karya dalam sebuah seri 'perkiraan' dan diterbitkan pertama kali ditahun 1950.

Perkiraan-perkiraan ini merupakan pernyataan-pernyataan dan persetujuan-persetujuan tentang penggolongan tanah yang dapat diterima pada waktu itu, dengan pengertian bahwa penyempurnaan-penyempurnaan selanjutnya akan diusahakan. "Perkiraan yang ketujuh" diterbitkan pada tahun 1960. Sebagai hasil dari penyelidikan-penyelidikan ini, suatu sistim kerja yang baik sekali sekarang telah dapat dipergunakan oleh para akhli-akhli tanah; sistim tersebut diterbitkan dalam bentuknya yang terakhir ditahun 1975 sebagai "Taksonomi Tanah Amerika Serikat" ("The United States Soil Taxonomy"). Kami percaya bahwa daftar yang sekarang ini, suatu daftar tumbuhan pengganggu yang paling mudah dimengerti, merupakan "perkiraan yang pertama" dari jenis dan penyebaran tumbuhan pengganggu didunia.

Para pengarang menerima baik semua saran-saran untuk pembetulan dan tambahan-tambahan dengan maksud agar penerbitan "perkiraan kedua" dapat dipercepat penyelesaiannya. Untuk keperluan ini, sebuah lampiran yang dapat dirobek dan dialamatkan kepada pengarang telah disertakan pada akhir buku ini. Kami mengharapkan bantuan para pembaca dalam usaha ini.

Karena bahasa Latin adalah bahasa yang universil bagi taksonomi, maka nama ilmiah dari jenis-jenis tumbuhan pengganggu tersebut digunakan disini. Didalam cara penamaan (nomenklatur) telah dipergunakan sistim yang umum digunakan pada waktu sekarang. Akan tetapi untuk kepentingan para pembaca yang barangkali lebih terbiasa dengan nama yang lama, 1–3 nama lain (sinonim) yang umum juga disertakan. Sinonim tersebut disusun dengan urutan abjad darimana nama Latin-nya yang benar dapat diketemukan kembali. Untuk negara dimana suatu tumbuhan pengganggu tertentu terdapat, akan dicatat pula nama daerah yang umum dipakai diwilayah tersebut; baik itu sinonim ataupun nama umum yang resmi diterima.

TINGKAT KEPENTINGAN

Dibelakang tiap nama jenis dituliskan lambang (singkatan) yang menunjukkan negara atau wilayah dimana jenis itu terdapat, yang disusun menurut tingkat kepentingan dari tumbuhan pengganggu itu ditiap daerah. Jadi tiap lambang negara yang disusun untuk suatu jenis, ditempatkan dibawah salah satu dari lambang-lambang berikut:

S   untuk tumbuhan pengganggu serius
P   untuk tumbuhan pengganggu penting
C   untuk tumbuhan pengganggu umum
X   untuk tumbuhan yang dikenal dan berkelakuan sebagai tumbuhan pengganggu, tetapi tingkat kepentingannya tidak diketahui
F   untuk jenis tumbuhan (yang dikenal terdapat dinegeri yang bersangkutan tetapi bukti-bukti masih diperlukan untuk meyakinkan bahwa tumbuhan tersebut berlaku sebagai tumbuhan pengganggu)

Tingkatan serius, penting atau umum ini diberikan oleh seorang akhli tumbuhan pengganggu untuk negerinya sendiri. Dimana lambang negara seperti IDO (Indonesia) diletakkan dibawah 'P' (penting), ini dapat diartikan bahwa seorang akhli tumbuhan pengganggu telah memberikan urutan untuk jenis itu dibeberapa tanaman pertanian dipelbagai daerah di Indonesia. Hal ini tidak berarti bahwa suatu evaluasi yang menyeluruh dari tiap jenis tumbuhan pengganggu ditiap tanaman pertanian telah diselésaikan diseluruh negeri tersebut. Barangkali hanya 2 atau 3 negara saja yang dapat dianggap telah mengevaluasikan dengan baik sampai tingkatan itu.

Ketika kami mempelajari cara-cara bagaimana akhli-akhli bekerja diseluruh dunia dalam menyatakan perkiraan-perkiraan mereka tentang kepentingan tumbuhan pengganggu, pengalaman kami bertambah, dan kami mulai memahami bahwa kami sedang belajar dari mereka-mereka yang kami kunjungi, yang kami ajak bekerjasama dan kami ajak berkoresponden, bahwa walaupun tidak dicatat, ada suatu keseragaman pendapat tentang apa yang dimaksud dengan "serius", "penting", dan "unum" dari tumbuhan pengganggu pada tanaman pertanian. Apabila seorang akhli mengatakan bahwa suatu jenis menduduki tempat sebagai salah satu dari 2 atau 3 tumbuhan pengganggu yang paling serius pada sesuatu tanaman pertanian, hal ini berarti tidak ada persoalan pengertian. Adalah merupakan pengalaman kami, bahwa apabila ia berkata tentang tumbuhan pengganggu 'penting' nya, ia biasanya menunjuk kira-kira 5 jenis yang paling menyulitkan bagi sesuatu jenis tanaman pertanian. Jarang sekali 10 tumbuhan disebut sebagai contohnya. Suatu tumbuhan pengganggu "umum" menunjukkan suatu tumbuhan pengganggu yang tersebar luas sekali pada tanaman pertanian atau dibeberapa wilayah dari negaranya, dimana diperlukan suatu usaha dan biaya yang terus menerus untuk menjaganya pada suatu keadaan yang tidak merugikan tanaman pertanian tersebut secara serius.

DAFTAR BAHASA DARI NAMA NEGARA DAN LAMBANG

Karena kebutuhan ruang, nama-nama 124 negara yang digunakan dalam penyelidikan ini dinyatakan dengan 2 atau 3 huruf saja (singkatan). Lambang dan nama lengkap dalam bahasa Inggris dari negara yang diwakili oleh lambang tersebut terlihat pada 2 kolom pertama dari "Daftar Bahasa dari Negara dan Lambang" (The Language Table of Country Names and Symbols) dibagian muka buku ini. Pada kolom-kolom lainnya, nama negara-negara ditulis dalam 9 bahasa lain. Daftar ini, bersama dengan kebiasaan menggunakan bahasa Latin untuk nama-nama jenis tumbuhan diseluruh dunia, diharapkan akan mengajak semua akhli tumbuhan pengganggu, dimanapun ia berada untuk dapat memanfaatkan daftar geografi dari tumbuhan pengganggu didunia.

緒　言

　　本書は世界の雑草とその地理的分布に関するもの
であり、読まれれば同じ種類の雑草が世界のどの地域
にもみられることがよくおわかりと思う。ある国であ
る雑草が非常に問題である時、世界の他のどの地域で
その雑草が問題になっているかもわかるし、またその
地域でその雑草の生物学やコントロールのしかたにく
わしい雑草学者に問いあわせることもできる。
　　農業の文献によるとおよそ50,000種の植物は雑草
たりうるとのことであるが、著者らはその見積りはあ
まり現実的でないと思っている。ここ10年の間、世界
最悪の雑草の生物学と分布に関する二巻からなるこの
本書を準備中、著者らは世界のほとんどの国をおとず
れたし、また雑草やその科学に関するたいていの世界
の文献も調査した。著者らの注意をひいたすべての雑
草を記録したが、8,000 種以上は農業上雑草たりうるこ
ともわかった。この著者らの仕事からわかったのであ
るが、これら雑草のうちただ250種ほどが**世界の**農業に
とって重要なのである。本書はいかなる地域のすべて
の雑草種を記載したものでもないし、全世界の雑草雑
の完全なリストをのせたものでもない。ただ著者らの
10年間の調査中みっけたものを記録したものにすぎな
い。それ以上あることはいうまでもない。
　　土壌学者のアメリカ及びその領土の土壌やその分
類に関する初期の仕事はそれなりに評価され、"Approximations" シリーズの一環として記載されたが、その初

版は1951年に発行された。これら Approximations は当時受入れられた土壌の分類についてのべたものであり、将来より完全なかたちにしていこうという前提があった。この "Approximation 第7版" は1960年に発刊された。これらの集大成として現在土壌学者が利用することの出来る非常にすぐれた研究の体系がある。それは1975年に最終版として "アメリカ合衆国の土壌分類学" の名で発刊されるはずである。

　著者らの本書はかってないほど広汎な雑草のリストであり、いわば世界の雑草とその分布に関する "Approximation 第一版" であるとおもっている。"Approximation 第2版" を迅速な発行のため本書のあろゆる訂正やまた追加のご指摘を歓迎する。そのためにも本書の終りのページに質問書がっけてある。読者のご協力をお願いしたい。

　分類学ではラテン語が広く用いられているので、本書でも雑草種の科学名にそれを用いた。命名法は現在使われている様式にしたがった。しかし古い名や習慣的な名になじんでいる読者のため、もっとも一般的な名を3つまであげた。これら一般名は ABC 順に記載し、本来の分類学上の名前とてらしあわせてある。ある雑草がみられる国の名前は、その国で使われているその雑草の名——一般的であれ分類学上の名であれ一の反対側に記載してある。

重要性の順位について

　各々の種名の反対側にその種がみられる国又は地域の記号（略号）が記載してあるが、その順位はその地域におけるその雑草の重要度にしたがってならべた。また国名記号は次の記号の下に付加した。

S　　最害を及ぼす雑草
P　　基本的な雑草

C　一般的な雑草
X　雑草性を有するもの（すなわち雑草とし
　　て知られているが、その重要性について
　　はわからないもの）
F　フローラ（その国の植物相中に見い出し
　　うるが、雑草たりうることの証明が必要
　　なもの）

　　最害を及ぼす雑草、基本的な雑草、あるいは一般
的な雑草といつた各順位づけは各々の国の雑草学者に
よって決められたものである。ひとつの例としてある
国、たとえばIDO（インドネシア）の項に"P"（基本的）
とあれば、これはインドネシアのある地域のある作物
の雑草に対して雑草学者の決めたものである。このこ
とはすべての作物に対するすべての雑草種の全体的な
評価がその国全般にわたって完成していることを意味
するものではない。そのようなことのあをはまるのは
世界中でもほんの二、三の国にすぎないであろう。
　　世界の研究者がどのように雑草の重要性をきめる
のかを研究するにつれて著者らの経験も豊富になり、
会ったり、共に研究したり、通信し合った人々をとお
して、記録にはなされなかったがある作物の雑草
に対する"最害"、"基本的"あるいは"一般的"と
いう評価基準に共通性カーあるンとを知った。たと
えばある人がある雑草をさしてその作物に対して
二、三の"最害の"雑草のひとっであるといえば、そ
のことはただちにどこでも通じるのである。また我々
の経験からしてある人がある雑草が"基本的"雑草で
あるといえば、一般にその人はえの作物に対する5つ
のもつともめんどうな種のことをいっているのである。
とれが5つでなく10もあるようなことはめったにない。
"一般的な"雑草の意味するところは、それがある国の作
物中にあるいは地域に広く分布しているものであり、
常時防衛を必要とはするが決して甚大なひがいは及ぼ
さない雑草のことである。

国名の記号表について

　　スペースの関係から研究に用いた124ケ国の名は2
または3字の記号（略号）で表わしてある。記号とそ
の意味するその国の英語名は本書のはじめにある国名
の記号表についての最初の二段に記載してある。残り
の段にはその国の名を9つの他の言葉で記してある。
このことは、世界中の植物種の名にラテン語を使う習
慣とともに、世界中のすべての雑草学者に世界の雑草
の地誌たる本書を使用する便宜を与えるものである。

Введение

Настоящий атлас сорняков мира и их географического распространения составлен в надежде, что он поможет людям понять, что одни и те-же сорные виды могут быть общими для различных частей мира. Работник нуждающийся в помощи в отношении вида представляющего серьёзную проблему в его стране сможет обратиться непосредственно к учёным других частей света знакомым с биологией данного сорняка и мерами борьбы с ним.

В сельскохозяйственной литературе можно найти указания, что до 50,000 видов растений могут вести себя как сорняки, но авторы этого атласа выяснили теперь, что эта цифра не реальна. В течение последнего десятилетия, приготовляя двухтомник по биологии и распространению наиболее злостных сорняков мира, мы побывали в большинстве стран света и просмотрели значительную часть литературы о сорняках. Мы учли все те виды, которые привлекли наше внимание и нашли, что менее 8,000 ведут себя, как нам сказали, как сорные растения в земледелии. Ещё более важно, что на основании наших работ мы выяснили, что только около 250 из них играют важную роль в **мировом** сельском хозяйстве. Эта книга не включает все сорняки каждого географического района и не предлагает полного списка всех сорных растений мира. Это всего только список сорных видов, которые мы обнаружили за десять лет поисков. На самом деле их, несомненно, больше.

Первоначальные усилия почвоведов описать и классифицировать почвы Соединенных Штатов и их территорий были суммированы в качестве рабочего документа в серии "приближений", первое из которых было выпущено в 1951 году. Эти приближения представляли собой предложения-согласования о классификации, приемлемые в то время, с полным пониманием, что дальнейшие уточнения уже разрабатывались. Седьмое приближение было опубликовано в 1960 г. В результате этих усилий в настоящее время имеется основательно разработанная рабочая система доступная почвоведам и опубликованная в окончательной форме в 1975 году как **Классификация почв Соединённых Штатов.** Мы считаем, что настоящий атлас, наиболее полный список сорняков когда либо составленный, является "первым приближением" видов сорных растений мира и их распространения. Авторы будут приветствовать все исправления и дополнения с тем, чтобы выпуск "второго приближения" мог быть ускорен. Для этой цели, в конце книги предлагается адресованная страница, которая может послужить моделью реакции и и ответов читателя или же вырвана и послана автором. В этом мы просим помощи наших читателей.

Латынь является универсальным языком классификации и для сорных видов растений иы употребляем научные названия, номенклатура которых согласована с её современным употреблением. Однако для удобства читателей более знакомых с прежними названиями даны до трёх наиболее обычных синонимов. Синонимы даны в алфавитном порядке и отсылают читателя к современному названию таксона. Старана, в которой данный вид был найден указывается против видового названия употребляемого в данной области, будь то синоним или современное название.

РАЗРЯД СТЕПЕНИ ЗНАЧИМОСТИ

Против названия каждого вида указываются символы,/сокращения,/ указывающие страны или области в которых данный вид встречается и его значение как сорняка в каждой области. Таким образом символы стран для данного вида даются под одним из следующих заголовков:

S   для Серьёзного сорняка
P   для Главного сорняка
C   для Обычного сорняка
X   для присутствующего сорняка/вид присутствует и ведет себя как сорняк, но степень его значения неизвестна/
F   для флоры/вид известен для флоры страны, но требует подтверждения, что он ведёт себя как сорняк/

Каждая степень значения сорняка, Серьёзный, Главный или Обычный, определялась иестным специалистом для своей страны. Символ страны, такой как например IDO /Индонезия/ помещённый под P /Главный/ значит, что местный специалист по сорнякам дал этому виду сгепень его значимости для определённой культуры в некоторой области Индонезии. Это не значит, однако, что общая оценка каждого сорного вида для каждой культуры была закончена в масштабе всей страны. Есть, быть может, только две или три страны во всём мире, для которых предварительная оценка сорняков может быть сделана на этом уровне.

При изучении способов, которыми работники всего мира выражали оценку значимости сорняков наш опыт обогащался и мы начали замечать, что учимся у тех кого мы встречали, с кем сотрудничали или вели переписку, и хотя мы не фиксировали нашего опыта мы замечали некоторую общность в понимании того какие сорняки считать "серьёзными", "главными" или "обычными" в посевах разных культур. Если местный работник считает, что известный вид входит в разряд одного из двух или трёх наиболее "серьёзных" сорняков, затруднений в понимании не возникает. Наш опыт показал, что когда специалист говорит о "главных" сорняках он имеет в виду всего около пяти вредоносных видов для определённой культуры. Только изредка число их может достигнуть до десяти. "Обычный" сорняк пониматся как широко-распространённый в ряде культур или районов страны, требующий постоянных усилий и расходов чтобы держать его в границах, но никогда серьёзно не угрожающий культурным растениям.

ЯЗЫКОВАЯ ТАБЛИНА НАЗВАНИЙ СТРАН И СИМВОЛОВ

Из соображений экономии, места, названия 124-Х стран упоминаемых в книге выражены символами /сокращениями/ из двух-трёх букв. Символ и полное английское название страны представленной этим символом показаны в первых двух столбцах языковой таблицы названий стран и символов помещённой перед самым началом текста. В оставшемся столбце названия стран даны на десяти других языках. Такое расположение, вместе с обычным во всём мире употреблением латинского языка для названий растений даст возможность всем специалистам по сорнякам повсеместно употреблять настоящий атлас сорняков мира.

Introducción

Este Atlas de malezas mundiales y su distribución geográfica ha sido preparado con la esperanza de que pueda ayudar a la gente a darse cuenta que las mismas especies de malezas pueden ser comunes en muchas regiones del mundo. Un investigador que necesita ayuda con una especie que representa un serio problema en su propio país, puede de esta manera, aprender cuales son las áreas en las que esa misma maleza es un problema, y dirigirse directamente, entonces, a otros científicos en otras partes del mundo que ya tienen la información acerca de la biología y el control de aquella especie.

Según la literatura agrícola existen hasta 50,000 especies de plantas que pueden comportarse como malezas, pero ahora sabemos que tal estimación escapa a la realidad. Durante la década pasada, cuando estaba en preparación la obra de dos volúmenes sobre la biología y la distribución de las peores malezas del mundo, nosotros visitamos la mayoría de los países del mundo e investigamos casi toda la literatura mundial en malezas y la ciencia de las malezas. Apuntamos todas las especies que llamaron nuestra atención y encontramos que alrededor de 8,000, nada más, se comportan como malezas en la agricultura. Lo que es más importante, durante nuestro estudio encontramos que solamente alrededor de 250 de aquellas son relevantes para la agricultura mundial.

Este libro no incluye todas las especies de todas las regiones geográficas, ni presenta una lista completa de todas las especies de malezas del mundo. Es, sencillamente, una compilación de las especies de malezas que hemos encontrado en diez años de continua búsqueda. Seguramente hay más.

Los esfuerzos iniciales de científicos de suelos para describir y clasificar los suelos de los Estados Unidos y sus territorios fueron evaluados y sumarizados como trabajos documentados en una serie de "aproximaciones," la primera de las cuales fue publicada en 1951. Estas aproximaciones fueron definiciones—contratos acerca de la clasificación del suelo, que podría haber sido aceptada en aquel momento, con completo entendimiento de que ulteriores modificaciones podría ser hechas. La "séptima aproximación" fue publicada en 1960. Como resultado de estos estudios ahora existe un sistema muy sofisticado disponible a los científicos del suelo; fue publicado en forma final en 1975 bajo el título "The United States Soil Taxonomy" ("Taxonomía de Suelos de los Estados Unidos"). Creemos que el presente Atlas, el más extenso de los publicados hasta la fecha, es la "primera aproximación" de la distribución de las malezas en el mundo. Los autores aceptan complacidos todas las correcciones y adiciones para que la publicación de una "segunda aproximación" pueda llevarse a cabo rápidamente.

Con este propósito, hemos incluído al final del libro, hojas con la dirección del autor. Estas hojas pueden ser arrancadas y enviadas con las correcciones necesarias. Mucho agredecemos a los lectores la ayuda que nos presten.

Como el latín es la lengua universal de taxonomía, los nombres científicos de las malezas son usados aquí. Hemos usado una nomenclatura que esté de acuerdo con el uso actual. Sin embargo, algunos de los lectores pueden estar más familiarizados con los nombres antiguos. Para ellos, hemos dado hasta tres de los sinónimos más comunes. Cada sinónimo figura en la lista por sí mismo y en orden alfabético. Además estos sinónimos tienen directa referencia al nombre taxonómico correcto. Un país en el cual se encuentra determinada maleza, está anotado al lado del nombre de la especie que está en uso en esa área—ya sea sinónimo o la designación taxonómica aceptada.

ORDEN DE IMPORTANCIA

Junto a cada especie hemos anotado abreviaciones que representan los países, o las áreas en los que esas malezas se dan más comúnmente. Estos han sido ordenados de acuerdo a la abundancia con que la maleza crece en cada país. De manera que, cada abreviatura para cada país está encasillada debajo de uno de los siguientes símbolos:

S   por "maleza muy seria" (grave, peligro)
P   por "maleza principal"
C   por "maleza común"
X   por "maleza en este caso" (Quiere decir que se comporta como maleza, pero su rango de importancia es aún desconocido.)
F   por "flora" (Se sabe que la especie está presente en la flora del país, pero se necesita más evidencia de que la planta se comporta como maleza.)

Esta clasificación de S, P, o C, nos ha sido proporcionada por un investigador del país a que cada especie corresponde.

Cuando se encuentre un simbolo como IDO (Indonesia), debajo de una P (Principal) por ejemplo, quiere decir que un científico en malezas de ese país ha otorgado ese rango para las malezas de determinado cultivo en determinada área de Indonesia.

Esto no significa que hayamos completado un estudio tan amplio que abarque todos los cultivos Y todos los paises del mundo. Tal vez hay solamente dos o tres países del mundo para los que las estimaciones pueden ser hechas a este nivel.

A medida que nos familiarizábamos con la forma en que los diferentes investigadores clasificaban las malezas por su importancia, nuestra experiencia crecía y nos era más fácil la comunicación de ideas. Es así que hay un entendimiento absoluto en lo que "seria," "principal," y "común" significan en un cultivo. En el caso de que un investigador clasifique una maleza como una de las dos o tres malezas más "serias" para un cultivo, no hay problema de lo que eso significa.

También entendemos que cuando ese investigador habla de malezas "principales," se está refiriendo a, más o menos, cinco de las especies más problemáticas para ese cultivo. Casi nunca va a haber diez malezas nombradas allí, por ejemplo. Cuando se habla de maleza "común," se refiere generalmente a una especie muy difundida en muchos cultivos o regiones de su país, y que requiere un esfuerzo constante para mantenerla bajo control. Esta maleza no representa en ningún caso una amenaza para el cultivo.

IDIOMA USADO EN LAS TABLAS PARA LOS NOMBRES DE LOS PAISES Y LOS SIMBOLOS

A causa de falta de espacio, hemos expresado los nombres de los 124 países presentes por medio de símbolos de dos o tres letras (abreviaciones). El símbolo y el correspondiente nombre completo en inglés, aparecen en las dos primeras columnas de "Idioma usado en las tablas para los nombres de los países y los símbolos ("The Language Table of Country Names and Symbols"), en el comienzo del libro. En las restante columnas se dan los nombres de los países en los nueve otros idiomas. Esta medida, junto con la costumbre de usar el latín para el nombre de las especies de plantas en todo el mundo, hace posible que todos los científicos, especialistas en malezas, en todas partes, puedan usar este Atlas.

Language Table
of
Country Names and Symbols

	ENGLISH	ARABIC	CHINESE	FRENCH	GERMAN
AFG	Afghanistan	افغانستان	阿富汗	l'Afghanistan	Afghanistan
ALK	Alaska	الاسكا	阿拉斯加	l'Alaska	Alaska
ALG	Algeria	الجزائر	阿爾及利亞	l'Algerie	Algerien
ANG	Angola	انجولا	安哥拉	l'Angola	Angola
ANT	Antilles, Lesser	انتيلس لسر	小安地利斯羣島	les Petites Antilles	Antillen
ARB	Arabian Peninsula	الجزيرة العربية	阿拉伯半島	la Peninsule Arabique	Arabien
ARG	Argentina	الأرجنتين	阿根廷	l'Argentine	Argentinien
AST	Austria	النمسا	奧地利	l'Autriche	Osterreich
AUS	Australia	استراليا	奧大利亞	l'Australie	Australien
BND	Bangladesh	بنغلا دش	班哥拉答斯	le Bangladesh	Bangladesch
BEL	Belgium	بلجيكا	比利時	la Belgique	Belgien
	Belize *See* Honduras				
BER	Bermuda	برمودا	百慕達	les Bermudes	Bermuda
BOL	Bolivia	بوليفيا	玻利維亞	la Bolivie	Bolivien
BOR	Borneo	بورنيو	婆羅洲	le Borneo	Borneo
BOT	Botswana	بوتسوانا	玻專納	le Botswana	Botswana
BRA	Brazil	البرازيل	巴西	le Bresil	Brasilien
BUL	Bulgaria	بلغاريا	保加利亞	la Bulgarie	Bulgarien
BUR	Burma	بورما	緬甸	la Birmanie	Burma
CAB	Cambodia	كمبوديا	束埔寨	le Cambodge	Kambodscha
CAM	Cameroon	كاميرون	喀麥隆	le Cameroun	Kamerun
CAN	Canada	كندا	加拿大	le Canada	Kanada
CEL	Celebes Island Region	جزيرة سيليبس	西里伯島	Ile Celebes	Celebes
CEY	Ceylon (Sri Lanka)	سيلان	錫蘭	le Ceylan	Ceylon
CHA	Chad	شاد	乍得	le Tchad	Tschad
CHL	Chile	شيلي	智利	le Chili	Chile
CHN	China	الصين	中國	la Chine	China
COL	Colombia	كولومبيا	哥倫比亞	la Colombie	Kolumbien
CNK	Congo-Kinshasa (Zaire)	كونغو(كنشاشا)	剛果	le Zaire	Kongo
CR	Costa Rica	كوستريكا	哥斯達黎加	le Costa Rica	Costa Rica
CUB	Cuba	كوبا	古巴	la Cuba	Kuba
CZE	Czechoslovakia	تشيكوسلوفاكيا	捷克	la Tchecoslovaquie	Tschechoslowakei
DAH	Dahomey	داهومي	達荷美	le Dahomey	Dahomey
DEN	Denmark	الدنُمارك	丹麥	le Danemark	Danemark
DR	Dominican Republic	جمهورية الدومنيكان	多明尼加共和國	la Republique Dominicaine	Dominikanische Republik
ECU	Ecuador	اكوادو	厄瓜多爾	l'Equateur	Ekuador
EGY	Egypt	مصر	埃及	l'Egypte	Agypten
SAL	El Salvador	السلقادور	薩爾瓦多	El Salvador	El Salvador
ENG	England	انجلترا	英國	l'Angleterre	Gross Britannien
ETH	Ethiopia	الحبشة (ايتوبيا)	衣索匹亞	l'Ethiopie	Athiopien
FIJ	Fiji	فيجي	飛枝羣島	Isles Fidji	Fidschi

HINDI	INDONESIAN	JAPANESE	RUSSIAN	SPANISH
अफगानिस्तान	Afganistan	アフガニスタン	Афганистан	el Afganistan
अलास्का	Alaska	アラスカ	Аляска	Alaska
अल्जीरिया	Aljazair	アルジェリア	Алжир	Argelia
अंगोला	Angola	アンゴラ	Ангола	Angola
अंटिलिस	Antila Kecil	アンチル列島	Малые Антильские о-ва	Antillas Menores
अरब पेनिसुला	Jazirah Arab	アラビア半島	Аравийский полуостров	Peninsula Arabica
अर्जेंटिना	Argentina	アルゼンチン	Аргентина	la Argentina
आस्ट्रिया	Austria	オーストリア	Австрия	Austria
आस्ट्रेलिया	Australia	オーストラリア	Австралия	Australia
बंगला देश	Bangladesh	バングラティッシュ	Бангладеш	Bangladesh
बेल्जियम	Belgia	ベルギー	Бельгия	Belgica
बरंयूडा	Bermuda	バミューダ	Бермудские острова	Bermudas
बोलिविया	Bolivia	ボリビア	Боливия	Bolivia
बोर्नियो	Kalimantan	ボルネオ	Борнео	Borneo
वोट्सवाना	Botswana	ボックト	Ботсвана	Botswana
ब्राजील	Brazilia	ブラジル	Бразилия	el Brasil
बुल्गारिया	Bulgaria	ブルガリア	Болгария	Bulgaria
बर्मा	Burma	ビルマ	Бирма	Birmania
कम्बोडिया	Kamboja	カンボジア	Камбоджа	Camboya
केमरुन	Kanerun	カメルーン	Камерун	el Camerun
कनाडा	Kanada	カナダ	Канада	el Canada
सेलेबीज द्वीप	Sulawesi	セレビス島	Целебес	Islas Celibes
श्री लंका	Srilangka	セイロン	Цейлон	Ceilan
छाद	Chad	チャド	Чад	el Chad
चीली	Chili	チリ	Чили	Chile
चीन	Republik Rakyat Cina	中国	Китай	China
कोम्बोडिया	Kolombia	コロンビア	Колумбия	Colombia
कांगो	Kongo	コンゴ	Конго (Киншасса)	Congo
कोस्टारिका	Kosta Rika	コスタリカ	Коста-Рика	Costa Rica
क्यूबा	Kuba	キューバ	Куба	Cuba
चेकोस्लोवाकिया	Cekoslovakia	チェコスロバキア	Чехословакия	Checoslovaquia
दहोमी	Dahomey	ダホメ	Дагомея	el Dahamey
डेनमार्क	Denmark	デンマーク	Дания	Dinamarco
डोमिनियन गणराज्य	Republik Dominika	ドミニカ共和国	Доминиканская республика	la Republica Dominicana
इक्वेडर	Ekuador	エクアドル	Эквадор	el Ecuador
ईजप्ट	Mesir	エジフト	Египет	Egipto
एल सेल्वेडोर	El Salvador	エルサルバドル	Сальвадор	el Salvador
इंग्लैंड	Inggris	イギリス	Англия	Inglaterra
इथोपिया	Ethiopia	エチオピア	Эфиопия	Etiopia
फिजी	Fiji	フィジー諸島	Фиджи	Figi

	ENGLISH	ARABIC	CHINESE	FRENCH	GERMAN
FIN	Finland	فنلندا	芬蘭	la Finlande	Finnland
FRA	France	فرنسا	法國	la France	Frankreich
GAB	Gabon	جابون	加本	le Gabon	Gabun
GER	Germany	المانيا	德國	l'Allemagne	Deutschland
GHA	Ghana	غانا	迦納	le Ghana	Ghana
GRE	Greece	اليونان	西臘	la Grece	Griechenland
GUA	Guatemala	جواتيمالا	瓜地馬拉	le Guatemala	Guatemala
GUI	Guinea	غينيا	幾內亞	la Guinee	Guinea
GUY	Guyana	جيانا	蓋亞那	la Guyane	Guyana
HAW	Hawaii	هواي	夏威夷	Iles Hawaii	Hawaii
HON	Honduras (Belize)	هندوراس	宏都拉斯	Le Honduras	Honduras
HK	Hongkong	هونج كونج	香港	Hongkong	Hongkong
HUN	Hungary	هنغاريا	匈牙利	la Hongrie	Ungarn
ICE	Iceland	ايسلندا	冰島	l'Islande	Island
IND	India	الهند	印度	l'Inde	Indien
IDO	Indonesia	اندونيسيا	印尼	l'Indonesie	Indonesien
IRA	Iran	ايران	伊郎	l'Iran	Iran
IRQ	Iraq	العراق	伊拉克	l'Irak	Irak
IRE	Ireland	ايرلندا	愛爾蘭	l'Irlande	Irland
ISR	Israel	اسرائيل	以色列	l'Israel	Israel
ITA	Italy	ايطاليا	意大利	l'Italie	Italien
IVO	Ivory Coast	ساحل العاج	象牙海岸	la Cote-d'Ivoire	Elfenbeinkuste
JAM	Jamaica	جمايكا	牙買加	la Jamaique	Jamaica
JPN	Japan	اليابان	日本	le Japon	Japan
JOR	Jordan	الأردن	約旦	la Jordanie	Jordanien
KEN	Kenya	كينيا	肯牙	le Kenya	Kenia
KOR	Korea	كوريا	韓國	la Coree	Korea
LEB	Lebanon	لبنان	黎巴嫩	le Liban	Libanon
LIB	Liberia	ليبيريا	賴比瑞亞	le Liberia	Liberia
LBY	Libya	ليبيا	利比亞	la Libye	Libyen
MAD	Madagascar (Malagasy)	مدغشقر	馬達加斯加	le Madagascar	Madagaskar
MLI	Mali	مالي	馬利	le Mali	Mali
MAL	Malaysia	ملاسيا	馬來西亞	la Malaisie	Malaysia
MRE	Mauritania	موريتانيا	毛里塔尼亞	le Mauritanie	Mauretanien
MAU	Mauritius	موريتس	毛里斯	le Maurice	Mauritius
MEL	Melanesia	ميليسيا	美拉尼西亞	la Melanesie	Melanesien
MEX	Mexico	المكسيك	墨西哥	le Mexique	Mexiko

HINDI	INDONESIAN	JAPANESE	RUSSIAN	SPANISH
फिनलैंड	Finlandia	フィンランド	Финляндия	Finlandia
फ्रांस	Perancis	フランス	Франция	Francia
गैबोन	Gabon	ガボン	Габон	el Gabon
जर्मनी	Jerman	ドイツ	Гермения	Alemania
घाना	Gana	ガーナ	Гана	Ghana
ग्रीस	Yunani	ギリシャ	Греция	Grecia
गुआटेमाला	Guatemala	グァテマラ	Гватемала	Guatemala
गिनी	Guinea	ギィニア	Гвинея	Guinea
गायना	Guyana	ギャナ	Гвиана	Guyana
हवाई	Hawaii	ハワイ	Гавайи	Hawaii
होण्डुरास	Honduras	ホンジュラス	Гондурас	Honduras
हांगकांग	Hongkong	ホンコン	Гонконг	Hongkong
हंगरी	Hongaria	ハンガリー	Венгрия	Hungria
आइसलैंड	Islandia	アイスランド	Исландия	Islandia
भारत	India	インド	Индия	la India
इंडोनेशिया	Indonesia	インドネシア	Индонезия	Indonesia
ईरान	Iran	イラン	Иран	el Iran
ईराक	Irak	イラク	Ирак	el Irak
आयरलैंड	Irlandia	アイランド	Ирландия	Irlanda
इज़्राईल	Israel	イズラエル	Израиль	Israel
इटली	Italia	イタリア	Италия	Italia
आइबरी कोस्ट	Pantai Gading	ぞうげ海岸	Берег Слоновой Кости	la Costa de Marfil
जमाइका	Yamaika	ジャマイカ	Ямайка	Jamaica
जापान	Japang	日本	Япония	el Japon
जोर्डन	Yordania	ヨルダン	Иордания	Jordania
केन्या	Kenya	ケニヤ	Кения	Kenia
कोरिया	Korea	韓国	Корея	Corea
			Ливан	
लेबनान	Libanon	レバノン	Либерия	el Libano
लाइबेरिया	Liberia	リベリア	Ливия	Liberia
लीबिया	Libia	リビア		Libia
मदागास्कर	Madagaskar	マダガスカル	Мадагаскар	Madagascar
माली	Mali	マリ	Мали	Mali
मलेशिया	Malaysia	マレーシア	Малайзия	Malasia
मारीतानिया	Mauritania	モーレタニア	Мавритания	Mauritania
मारीशस	Mauritius	モーリシャス	Маврикий	Mauricio
मेलानेशिया	Melanesia	メラネシア	Меланезия	Melanesia
मैक्सिको	Mexiko	メキシコ	Мексика	Mexico

	ENGLISH	ARABIC	CHINESE	FRENCH	GERMAN
MIC	Micronesia	ميكرونيسيا	密克羅尼西亞	la Micronesie	Mikronesien
MOR	Morocco	مراكش	摩洛哥	le Maroc	Marokko
MOC	Mozambique	موزامبيق	莫三鼻克	le Mozambique	Mozambique
NEP	Nepal	نيبال	尼泊爾	le Nepal	Nepal
NET	Netherlands	هولندا	荷蘭	les Pays-Bas	Niederlande
NGI	New Guinea	غينيا الجديدة	新畿內亞	la Nouvelle-Guinee	Neuguinea
NZ	New Zealand	نيوزيلندا	紐西蘭	la Nouvelle-Zelande	Neuseeland
NIC	Nicaragua	نيكاراجوا	尼加拉瓜	le Nicaragua	Nicaragua
NGR	Niger	النيجر	尼日	le Niger	Niger
NIG	Nigeria	نيجيريا	尼日利亞	le Nigeria	Nigeria
NOR	Norway	النرويج	挪威	la Norvege	Norwegen
PAK	Pakistan	الباكستان	巴基斯坦	le Pakistan	Pakistan
PAN	Panama	بناما	巴拿馬	le Panama	Panama
PAR	Paraguay	براجواى	巴拉圭	le Paraguay	Paraguay
PER	Peru	بيرو	秘魯	le Perou	Peru
PHI	Philippines	الفلبين	菲律賓	les Philippines	Philippinen
POL	Poland	بولندا	波蘭	la Pologne	Polen
PLE	Polynesia, East	شرور بولينيسيا	東坡里尼西亞	la Polynesie Orientale	Ost Polynesien
PLW	Polynesia, West	غرب بولينيسيا	西坡里尼西亞	la Polynesie Occidentale	West Polynesien
POR	Portugal	البرتغال	葡萄牙	le Portugal	Portugal
PR	Puerto Rico	بورتوريكو	波多黎各	la Porto Rico	Puerto Rico
RHO	Rhodesia	روديسيا	羅得西亞	la Rhodesie	Rhodesien
ROM	Romania	رومانيا	羅馬尼亞	la Roumanie	Rumanien
SEG	Senegal	السنغال	塞內加爾	le Senegal	Senegal
SAF	South Africa	جنوب افريقيا	南非洲	l'Afrique du Sud	Sudafrika
SOV	Soviet Union	الاتحاد الوثياني	蘇聯	l'Union Sovietique	Sovietunion
SPA	Spain	اسبانيا	西班牙	l'Espagne	Spanien
	Sri Lanka *See* Ceylon				
SUD	Sudan	السودان	蘇丹	le Soudan	Sudan
SUR	Surinam	سيريتام	蘇利南	le Surinam	Surinam
SWZ	Swaziland	سوازيلاندا	斯威士蘭	le Swaziland	Swasiland
SWE	Sweden	السويد	瑞典	la Suede	Schweden
SWT	Switzerland	سويسرا	瑞士	la Suisse	Schweiz
TAI	Taiwan	الصين الوطنية	台灣	la Formose	Taiwan
TNZ	Tanzania	تانزانيا	坦桑尼亞	la Tanzanie	Tansania
TAS	Tasmania	تاسمانيا	塔斯馬尼亞	la Tasmanie	Tasmanien
THI	Thailand	تايلاند	泰國	la Thailande	Thailand
TRI	Trinidad	ترينداد	千里達	la Trinite	Trinidad
TUN	Tunisia	تونس	突尼西亞	la Tunisie	Tunesien
TUR	Turkey	تركيا	土耳其	la Turquie	Turkei

HINDI	INDONESIAN	JAPANESE	RUSSIAN	SPANISH
माइक्रोनेशिया	Mikronesia	マイクロネシア	Микронезия	Micronesia
मोरक्को	Maroko	モロッコ	Марокко	Marruecos
मोज़ाम्बिक	Mozambique	モザンビーク	Мозамбик	Mozambique
नेपाल	Nepal	ネパール	Непал	Nepal
हालैंड	Negeri Belanda	オランダ	Нидерланды (Голландия)	Los Paises Bajos
न्यू गिनी	Papua-Irian	ニューギニア	Новая Гвинея	Nueva Guinea
न्यूज़ीलैंड	Selandia Baru	ニュージーランド	Новая Зеландия	Nueva Zelandia
निकारागुआ	Nikaragua	ニカラグア	Никарагуа	Nicaragua
नाइगर	Niger	ニジェル	Нигер	el Niger
नाइजीरिया	Nigeria	ナイジェリア	Нигерия	Nigeria
नोर्वे	Norwegia	ノルウェー	Норвегия	Noruega
पाकिस्तान	Pakistan	パキスタン	Пакистан	el Paquistan
पनामा	Panama	パナマ	Панама	Panama
परागुबे	Paraguay	パラグァイ	Парагвай	el Paraguay
पेरु	Peru	ペルー	Перу	el Peru
फ़िलिपीन	Philipina	フィリピン	Филиппины	Filipinas
पोलेण्ड	Polandia	ポーランド	Польша	Polonia
पूर्वी पालीनेशिया	Polynesia Timur	ポリネシア(東)	Восточная Полинезия	Polinesia Oriental
पश्चिमी पालीनेशिया	Polynesia Barat	ポリネシア(西)	Западная Полинезия	Polinesia Occidental
पुर्तगाल	Portugal	ポルトガル	Португалия	Portugal
पोटोरिको	Porto Riko	プェルトリコ	Пуэрто-Рико	Puerto Rico
रोडेशिया	Rhodesia	ローデシア	Родезия	Rodesia
रोमानिया	Rumania	ローマニア	Румыния	Rumania
सेनेगाल	Senegal	セネガル	Сенегал	el Senegal
दक्षिणी अफ्रीका	Afrika Selatan	南アフリカ	Южно-Африканская республика	Sud Africa
सोवियत संघ	Sovyet Rusia	ソ連	Советский Союз	Union Sovietica
स्पेन	Spanyol	スペイン	Испания	Espana
सूडान	Sudan	スーダン	Судан	el Sudan
सुरिनाम	Suriname	スリナム	Суринам	Surinam
स्वाज़ीलैंड	Swaziland	スワジランド	Свазиленд	Suazilandia
स्वीडेन	Swedia	スウェーデン	Швеция	Suecia
स्वीटज़रलैंड	Swis	スイス	Швейцария	Suiza
तैवान	Taiwan	台湾	Тайван	Taiwan
तन्ज़ानियां	Tanzania	ダンザニア共和国	Танзания	Tanzania
तसमानिया	Tasmania	タアスマニア	Тасмания	Tasmania
थाईलैंड	Muangthai	タイ	Тайланд	Tailandia
ट्रिनिडाड	Trinidad	トリニダード	Тринидад	Trinidad
टूनीसिया	Tunisia	チユニジア	Тунис	Tunez
तर्की	Turki	トルコ	Турция	Turquia

	ENGLISH	ARABIC	CHINESE	FRENCH	GERMAN
UGA	Uganda	اوغندا	烏干達	l'Ouganda	Uganda
URU	Uruguay	اروجواي	烏拉圭	l'Uruguay	Uruguay
USA	United States	الولايات المتحدة	美國	les Etats-Unis d'Amerique	Vereinigte Staaten von Amerika
VEN	Venezuela	فنزويلا	委內瑞拉	le Venezuela	Venezuela
VIE	Vietnam	فيتنام	越南	le Viet-nam	Vietnam
VOL	Volta, Upper	تولتا العليا	上伏達	l'Haute-Volta	Obervolta
YUG	Yugoslavia	يوغسلافيا	南斯拉夫	la Yougoslavie	Jugoslawien
	Zaire. *See* Congo (Kinshasa)				
ZAM	Zambia	زامبيا	桑比亞	la Zambie	Zambia

HINDI	INDONESIAN	JAPANESE	RUSSIAN	SPANISH
युगाण्डा	Uganda	ウガンダ	Уганда	Uganda
युरुगुवे	Uruguay	ウルダァイ	Уругвай	Uruguay
अमेरिका	Amerika Serikat	アメリカ合衆国	Соединенные Штаты Америки	Los Estados Unidos de America
वैनेज़ुएला	Venezuela	ベネズエラ	Венесуэла	Venezuela
वितनाम	Vietnam	ベトナム	Вьетнам	Viet-Nam
उपरीवोल्टा	Volta Hulu	マッパボルタ	Верхняя Вольта	el Alto Volta
युगोस्लाबिया	Yugoslavia	ユーゴスラビア	Югославия	Yugoslavia
जाम्बिया	Zambia	ザンビア	Замбия	Zambia

ABELMOSCHUS Medic. **Malvaceae**	S	P	C	X	F
mindanaensis Warb.				PHI	
moschatus Medic. (*See* **Hibiscus abelmoschus** L.)			TRI		
ABIES Mill. **Pinaceae**					
balsamea (L.) Mill.				USA	
concolor (Gord. & Glend.) Lindl.				USA	
fraseri (Pursh) Poir.				USA	
grandis (Dougl.) Lindl.				USA	
lasiocarpa (Hook.) Nutt.				USA	
ABRUS L. **Papilionaceae** (*Leguminosae*)					
precatorius L.		IND PR		DR HON JAM PLW USA	IDO
ABUTILON Mill. **Malvaceae**					
asiaticum (L.) Sweet				MAU	
figarianum Webb. (*See* **A. graveolens** Wight & Arn.)				SUD	
fruticosum Guill. & Perr.				ANG	
glaucum G. Don (Syn. *A. pannosum* (Forsk. *f.*) Schlecht.)			SUD		
graveolens Wight & Arn. (Syn. *A. figarianum* Webb.)				THI	
guineense (Schumach.) E. G. Bak. & Exell			KEN	GHA	
indicum (L.) Sweet (Syn. *A. mauritianum* (Jacq.) Medic.)			IND	BOR CAB PHI PK THI VIE	IDO ISR
mauritianum (Jacq.) Medic. (*See* **A. indicum** (L.) Sweet)	GHA			KEN MOZ NIG	
molle Sweet			HAW	ARG USA	
oxycarpum F. Muell.				AUS	
pannosum (Forsk. *f.*) Schlecht. (*See* **A. glaucum** G. Don)			SUD		
striatum Dicks. *ex* Lindl.				ARG	
theophrasti Medic.			CAN	USA	AFG
trisulcatum Urb.				MEX	

ACACIA Mill. Mimosaceae (*Leguminosae*)	S	P	C	X	F
albicorticata Burkart				BOL	
arabica (Lam.) Willd. (Syn. *A. nilotica* (L.) Delile)	SAF			IND PK SUD	
armata R. Br.			AUS	NZ	
aroma Gill. *ex* Hook. & Arn.			ARG		
atramentaria Benth.				ARG	
berlandieri Benth.				USA	
bonariensis Gill.				ARG	
brevispica Harms			KEN		
catechu (L. *f.*) Willd. (Syn. *A. polyacantha* Willd.)					
concinna (Willd.) Dc.				MAU	
constricta Benth.				USA	
cyanophylla Lindl.		AUS	SAF		
cyclops A. Cunn. *ex* G. Don			SAF		
dealbata Link		AUS	SAF	NZ	
decurrens (Wendl.) Willd. (Syn. *A. mollissima* Willd.)			HAW IDO	COL NZ USA	
detinens Burch. (*See* **A. mellifera** (Vahl) Benth. *ssp.* detinens (Burch.) Brenan)	SAF				
drepanolobium (Harms) Sjostedt.		KEN			
erubescens Welw. *ex* Oliver				RHO	
farnesiana (L.) Willd. (Syn. *Vachellia farnesiana* Wight & Arn.)	IRQ	AUS FIJ IDO MEX PAR	HAW PHI	ARG CAM COL CR GUA HON IRA JOR MEL NIC PK PLW PR SAL SUD SUR THI USA VEN	AFG CAB CHN HK
georginae F. M. Bail.			AUS		
gerrardi Benth.		RHO		KEN	BND
glaucescens Willd.			AUS		
glomerosa Benth.				NIC	

ACACIA Mill. Mimosaceae *(Leguminosae) (Cont'd)*	S	P	C	X	F
glomerosa Benth. *(Cont'd)*				VEN	
greggii A. Gray				USA	
harpophylla F. Muell. *ex* Benth.		AUS			
hebecladoides Harms		TNZ	KEN		
heteracantha Burch.	SAF			RHO	
hindsii Benth.				GUA HON SAL	
hockii De Wild.	UGA		KEN		
intsia Willd.				IND	
ixiophylla Benth.				AUS	
karroo Hayne	SAF	RHO			
lahai Steud. & Hochst. *ex* Benth.			KEN		
longifolia (Andr.) Willd.			SAF	NZ	
macracantha Humb. & Bonpl. *ex* Willd.				PER	
mangium Willd.		AUS			
mearnsii De Wild.			SAF		
melanoxylon R. Br.		AUS	SAF	NZ	
mellifera (Vahl) Benth. *ssp.* detinens (Burch.) Brenan (Syn. *A. detinens* Burch.)					
mollissima Willd. *(See* **A. decurrens** (Wendl.) Willd.*)*		AUS			
nigrescens Oliver				RHO	
nilotica (L.) Delile *(See* **A. arabica** (Lam.) Willd.*)*	MOZ		KEN	AUS	SWZ
paniculata Willd.				BRA PER VEN	
pennata (L.) Willd.			KEN	IND	TNZ
polyacantha Willd. *(See* **A. catechu** (L. *f.*) Willd.*)*			KEN		
polyphylla Dc.				BRA	
pycnantha Benth.			SAF		
rehmanniana Schinz		RHO			
retinodes Schlecht.		AUS			
rigidula Benth.		MEX		USA	
robusta Burch.				RHO	SAF
seyal Delile			KEN	ANG	
stuhlmannii Taub.		TNZ			

ACACIA Mill. Mimosaceae (*Leguminosae*) (*Cont'd*)	S	P	C	X	F
subalata Vatke		RHO	KEN		
tortuosa Willd.				VEN	
westiana Dc. (*See* **Senegalia westiana** (Dc.) Britton & Rose)					PR
zanzibarica (S. Moore) Taub. (Syn. *Pithecellobium zanzibaricum* S. Moore)			KEN		
ACAENA L. Rosaceae					
anserinifolia (Forst.) Druce			AUS	NZ	
argentea Ruiz & Pav.				CHL	
ovalifolia Ruiz & Pav.				CHL	
ovina A. Cunn.				AUS	
pinnatifida Ruiz & Pav.			ARG	CHL	
sanguisorbae Vahl				USA	
splendens Hook. & Arn.			ARG		
ACALYPHA L. Euphorbiaceae					
alopecuroides Jacq.		DR	COL	MEX VEN	
arvensis Poepp. & Endl.				GUA TRI	
australis L.		JPN	TAI	KOR	
boehmerioides Miq.				PLW	IDO
ciliata Forsk.		GHA	IND NIG		
fallax Muell.-Arg.		MAL			
hispida Burm. *f.*				BRA	
indica L.		SUD	RHO	IND MAU PLW THI VIE	IDO
macrostachya Jacq.				VEN	
neomexicana Muell.-Arg.				USA	
ostryaefolia Riddell			USA		
poiretti Spreng.				ARG	
rhomboidea Rafin.				USA	
schiedeana Schlecht.				VEN	
segetalis Muell.-Arg.	MOZ		RHO		
setosa A. Rich.			SAL		
virginica L.				USA	

ACALYPHA L. Euphorbiaceae *(Cont'd)*	S	P	C	X	F
wilkesiana Muell.-Arg.				DR JAM	
ACANTHOCEREUS Britton & Rose **Cactaceae**					
pentagonus (L.) Britton & Rose				AUS	
ACANTHOSPERMUM Schrank **Asteraceae** *(Compositae)*					
australe (Loefl.) O. Ktze.			ARG HAW RHO SAF	BRA PER URU USA	
brasilum Schrank			SAF		
hispidum Dc.	ANG BRA CEY GHA IND MOZ RHO	IVO MAD NIG USA	ARG AUS CNK ETH HAW HON KEN PER SAF SEG	BOT CAN COL DAH MAU PR SAL THI TNZ	
ACANTHOSTYLES R.M. King & H. Robinson **Asteraceae** *(Compositae)*					
buniifolius (Hook. & Arn.) R. M. King & H. Robinson (Syn. *Eupatorium bunifolium* H. B. K.)					
ACANTHUS L. **Acanthaceae**					
caudatus Lindau				VIE	
ebracteatus Vahl				VIE	
ilicifolius L.				VIE	IDO
mollis L.				NZ	
montanus T. Anders.		GHA			
syriacus Boiss.			LEB		ISR
ACER L. **Aceraceae**					
barbatum Michx.				USA	
circinnatum Pursh				USA	
macrophyllum Pursh				USA	
negundo L.				USA	
pensylvanicum L.				USA	
platanoides L.				USA	
rubrum L.				USA	
saccharinum Wangenh. *(See* **A. saccharum** Marsh.)				TUR	
saccharum Marsh. (Syn. *A. saccharinum* Wangenh.)				USA	
spicatum Lam.				TUR	

	S	P	C	X	F
ACETOSELLA Fourr. **Polygonaceae**					
vulgaris Fourr.			AUS		
ACHETARIA Cham. & Schlechtend. **Scrophulariaceae**					
guianensis Pennell			TRI		
ACHILLEA L. **Asteraceae** *(Compositae)*					
borealis Bong.			CAN		
lanulosa Nutt.			CAN USA		
micrantha Willd.			SOV		
millefolium L.		FIN NOR NZ SWE	ARG AUS CAN ENG GER HAW IRA SOV SPA USA	CHL IND	AFG ALK POL
nobilis L.			SOV		
ptarmica L.		FIN			
santolina L.			IRA	PK TUR	AFG
ACHYRANTHES L. **Amaranthaceae**					
aspera L. (Syn. *A. indica* (L.) Mill.) (Syn. *A. obtusifolia* Lam.)	AFG CNK ECU HON	COL GHA IDO IND KEN	HAW ITA MAU PHI PK RHO SAF UGA	AUS BND BOR BOT CAB CAM CHN CR DAH EGY FIJ GUI IRA MAL MEL NEP NGI NIG PLW PR SAL SEG SUD SUR THI TNZ VIE ZAM	HK ISR
fauriei Lev. & Van.			JPN		
indica (L.) Mill. (*See* **A. aspera** L.)			HAW PR TRI	DR HON JAM	

	S	P	C	X	F
ACHYRANTHES L. **Amaranthaceae** *(Cont'd)*					
indica (L.) Mill. *(Cont'd)*				NIC	
japonica Nakai			JPN		KOR
longifolia Makino			JPN		
obtusifolia Lam. *(See* **A. aspera** L.)			TAI		
ACHYROCLINE Less. **Asteraceae** *(Compositae)*					
satureioides (Lam.) Dc.				BRA	
vargasiana Dc.				BRA	
ACICARPHA Juss. **Calyceraceae**					
tribuloides Juss.			ARG	URU	
ACIOTIS D. Don **Melastomataceae**					
caulialata (R. & P.) Triana				PER	
dichotoma Cogn.				TRI	
ACNIDA L. **Amaranthaceae**					
altissima Riddell				USA	
ACONITUM L. **Ranunculaceae**					
napellus L.			AUS		
ACORUS L. **Araceae**					
calamus L.			IRA	BEL GER	
ACRACHNE Wight & Arn. *ex* Chiov. **Poaceae** *(Gramineae)*					
verticillata (Roxb.) Lindl. *ex* Chiov.				ANG	
ACROCERAS Stapf **Poaceae** *(Gramineae)*					
zizanioides (H. B. K.) Dandy (Syn. *Panicum zizanioides* H. B. K.)					
ACROPTILON Cass. **Asteraceae** *(Compositae)* *(See* **CENTAUREA** L.)					
picris (L.) Dc. *(See* **Centaurea repens** L.)				CAN SOV TUR	
repens (L.) Dc. *(See* **Centaurea repens** L.)			AUS IRA		AFG
ACROSTICHUM L. **Pteridaceae**					
aureum L.				VIE	IDO
ACTAEA L. **Ranunculaceae**					
rubra (Ait.) Willd.				USA	
ACTINOSTEMMA Griff. **Cucurbitaceae**					
lobatum Maxim.			JPN		

	S	P	C	X	F
ADENIA Forsk. **Passifloraceae**					
gracilis Harms				CNK	
ADENOCALYMNA Mart. **Bignoniaceae**					
alliaceum Miers		BOL			
bracteatum Dc.				BRA	
ADENODOLICHOS Harms **Papilionaceae** *(Leguminosae)*					
rhomboideus (Hoffm.) Harms				ANG	
ADENOROPIUM Pohl **Euphorbiaceae** *(See* **JATROPHA** L.*)*					
gossypifolium (L.) Pohl *(See* **Jatropha gossypifolia** L.*)*		PR		DR	
ADENOSTEMMA Forst. **Asteraceae** *(Compositae)*					
lavenia (L.) O. Ktze.				PHI	
ADENOSTOMA Bl. **Scrophulariaceae**					
fasciculatum Hook. & Arn.				USA	
sparsifolium Torr.				USA	
ADIANTUM L. **Sinopteridaceae**					
capillus-veneris L.			EGY	TUR	
cristatum L.			PR		
cuneatum Langsd. & Fisch.			IDO		ARG
tenerum Sw.				JAM	
ADONIS L. **Ranunculaceae**					
aestivalis L.			GER SPA	MOR TUR	
annua L.			AUS	MOR USA	
autumnalis L.				SPA TUR	
baetica Coss.			POR		
dentata Delile				MOR	
flammea Jacq.				TUR	IRA
microcarpus Dc.				POR	ISR
vernalis L.				GER	
AEGILOPS L. **Poaceae** *(Gramineae)*					
cylindrica Host		TUR	USA		
ovata L.			MOR POR	JOR	ISR
squarrosa L. *(See* **A. tauschii** Cosson)			IRA		

AEGILOPS L. Poaceae *(Gramineae) (Cont'd)*	S	P	C	X	F
tauschii Cosson (Syn. *A. squarrosa* L.)					
triuncialis L.			MOR	TUR	
ventricosa Tausch				MOR	
AEGINETIA L. Orobanchaceae					
indica L.				JPN PHI THI	
AEGOPODIUM L. Apiaceae *(Umbelliferae)*					
podagraria L.			FIN GER SOV	ICE NZ USA	
AELUROPUS Trin. Poaceae *(Gramineae)*					
litoralis (Willd.) Parl.				SOV	
villosus Trin.				IRQ	
AERVA Forsk. Amaranthaceae					
javanica (Burm. *f.*) Juss. *ex* Schult. (Syn. *A. tomentosa* Lam.)				PK SUD	AFG BND
lanata (L.) Juss.				GHA IND	
tomentosa Lam. *(See* **A. javanica** (Burm. *f.*) Juss. *ex* Schult.)				IND PK SUD	
AESCHYNOMENE L. Papilionaceae *(Leguminosae)*					
americana L.			IND PR	MEX PER PHI SAL	IDO MAU TRI
aspera L.		THI		BND IND VIE	IDO
indica L.	AFG CAB ECU KOR THI	CEY IND SEG	JPN MAU PHI	BOT CHN CNK COL DAH FIJ GHA GUA HK IDO NEP NGI PLW PR RHO TAI VIE	USA
marginata Benth.				BRA	
portoricensis Urb.			PR		

AESCHYNOMENE L. Papilionaceae *(Leguminosae)* *(Cont'd)*	S	P	C	X	F
rudis Benth. *(See* **A. sensitiva** Sw. *var.* hispidula (H. B. K.) Rudd)		BRA			
sensitiva Sw.			PR	ANT COL SUR TRI	
sensitiva Sw. *var.* hispidula (H. B. K.) Rudd (Syn. *A. rudis* Benth.)					
AETHUSA L. Apiaceae *(Umbelliferae)*					
cynapium L.			ENG SOV	NZ	
AGALINIS Rafin. Scrophulariaceae					
fasciculata (Ell.) Raf.			PR		
AGAVE L. Agavaceae					
lecheguilla Torr.				USA	
sisalana Perrine				AUS	
AGERATINA Spach Asteraceae *(Compositae)*					
adenophora (Spreng.) R. M. King & H. Robinson (Syn. *Eupatorium adenophorum* Spreng.)	AUS	HAW	NZ	NGI PHI TAS THI TRI	
altissima (L.) R. M. King & H. Robinson (Syn. *Eupatorium rugosum* Hout.)					
riparia (Regel) R. M. King & H. Robinson (Syn. *Eupatorium riparium* Regel)					
AGERATUM L. Asteraceae *(Compositae)*					
conyzoides L.	ANG BOR CEY GHA MAL NIG PHI TAI	AUS BRA COL FIJ HAW IDO IND KEN MAU MEL NGI THI TNZ UGA	CHN EGY HK PLW RHO TRI	CAM CNK CR CUB DR GUI IVO LIB MLI NEP NIC PER PK SAF SAL SEG SUR VEN VIE	ETH SOV USA
houstonianum Mill. (Syn. *A. mexicanum* Sims)	AUS	TAI		CHN FIJ GER HAW IND	IDO

AGERATUM L. Asteraceae *(Compositae)* *(Cont'd)*	S	P	C	X	F
mexicanum Sims *(See* **A. houstonianum** Mill.)			IDO		
AGRIMONIA L. Rosaceae					
eupatoria L.			IRA TAI		
odorata Mill.			SAF		
pilosa Ledeb.			JPN		
AGROPYRON Gaertn. **Poaceae** *(Gramineae)*					
kamoji Ohwi		JPN			
ramosum (Trin.) Richt.				SOV	
repens (L.) Beauv. (Syn. *Triticum repens* L.)	ENG FIN GRE IRE SOV USA	ALK BEL CAN GER NGI NOR NZ SPA SWE	ARG ECU IRA	AFG AUS BND BUL CZE DEN FRA HUN ICE IND ITA JPN NET POL TUR YUG	HAW
semicostatum Nees		JPN			AFG
smithii Rydb.				CAN	
tsukushiense Ohwi			JPN		KOR
AGROSTEMMA L. Caryophyllaceae					
githago L.	ARG HUN	GRE ITA POL	AST AUS CAN ENG GER IRA ISR MOR POR SAF SOV SPA TUN TUR USA	CHL JOR JPN NOR SWE URU	KOR
linicola Terech.				SOV	
AGROSTIS L. Poaceae *(Gramineae)*					
alba L.			HAW JPN	CHL COL ITA TUR URU	AFG GER POR SAF

AGROSTIS L. Poaceae *(Gramineae) (Cont'd)*	S	P	C	X	F
canina L.			FIN SOV		HAW
castellana Boiss. & Reut.				POR	
clavata Trin.			JPN		
exarata Trin.			JPN		
filifolia Link			POR		
gigantea Roth			IRA	NZ	
salmantica (Lag.) Kunth			POR		
semiverticillata (Forsk.) Christ. (Syn. *A. verticillata* Vill.)			EGY IRQ POR	NZ	AFG HAW
spicaventi L.		TUN		ITA SOV	
stolonifera L.				AUS GER	
tenuis Sibth. (Syn. *A. vulgaris* With.)	NZ	FIN	CHL		HAW SAF
verticillata Vill. *(See* **A. semiverticillata** (Forsk.) Christ.*)*				EGY	AFG HAW POR
vulgaris With. *(See* **A. tenuis** Sibth.*)*		SOV			HAW
AILANTHUS Desf. **Simaroubaceae**					
altissima (Mill.) Swingle			AUS	CHL	USA
AIRA L. **Poaceae** *(Gramineae)*					
caespitosa L. *(See* **Deschampsia caespitosa** (L.) Beauv.*)*				TUR	
cappillaris Host			IRA		ISR
caryophyllea L.				CHL NZ	
AJUGA L. **Lamiaceae** *(Labiatae)*					
australis R. Br.				AUS	
bracteosa Benth.			TAI		
chamaepytis (L.) Schreb.				TUR	
chia Schreb.				TUR	
decumbens Thunb.			JPN	CHN	
remota Benth.				KEN	
reptans L.				GER	
ALBERSIA Kunth **Amaranthaceae** *(See* **AMARANTHUS** L.*)*					

ALBERSIA Kunth Amaranthaceae *(Cont'd)*	S	P	C	X	F
blitum Kunth			SOV		
(See A. lividus L.)					
ALBIZIA Durazz. **Mimosaceae** *(Leguminosae)*					
amara Boig.				RHO	
chinensis (Osb.) Merr.			IDO		
distichya Mcbride *(See A. lophantha (Willd.) Benth.)*					
falcata (L.) Back. *(See A. falcataria (L.) Fosb.)*				IDO	

falcataria (L.) Fosb.
   (Syn. *A. falcata* (L.) Back.)
   (Syn. *A. moluccana* Miq.)					

lebbek (L.) Benth.				FIJ	

lophantha (Willd.) Benth.
   (Syn. *A. distichya* Mcbride)
   (Syn. *A. montana* Benth.)
				AUS
				NZ	

moluccana Miq. *(See A. falcataria (L.) Fosb.)*			HAW		
montana Benth. *(See A. lophantha (Willd.) Benth.)*				IDO	
ALCHEMILLA L. **Rosaceae**					
arvensis Scop.				CHL GER	
gracilipes Engl.		ETH			
ALCHORNEA Sw. **Euphorbiaceae**					
cordifolia (Schu. & Thon.) Muell.-Arg.		GHA			
latifolia Sw.				JAM	
ALECTOROLOPHUS Hall **Scrophulariaceae**					
apterus (Fr.) Ostenf. *(See Rhinanthus apterus Ostenf.)*				SOV	
major Reichb. *(See Rhinanthus major Ehrh.)*				SOV	
ALECTRA Thunb. **Scrophulariaceae**					
kerkii Hemsl.			RHO		
senegalensis Benth.				NIG	
vogelii Benth.			RHO SAF		
ALHAGI Adans. **Papilionaceae** *(Leguminosae)*					
camelorum Fisch.	AUS TUR	SAF		IND IRA SOV	USA

ALHAGI Adans. Papilionaceae *(Leguminosae) (Cont'd)*	S	P	C	X	F
maurorum Medic.			EGY IRQ ISR	PK	
persarum Boiss. & Buhse			IRA		AFG
pseudalhagi Desv.			IND		
ALISMA L. **Alismataceae**					
canaliculatum A. Br. & Bouche			JPN		
gramineum K. C. Gmel.			USA	TUR	
lanceolatum With.	HUN ITA	POR			ISR
plantago-aquatica L.	ITA	POR SPA	IRA JPN	AUS BEL BND CHL ENG FRA GER MOR NET NZ TUR	AFG PK
triviale Pursh			ENG		USA
ALKANNA Tausch **Boraginaceae**					
orientalis Boiss.			TUR		
ALLIARIA B. Ehrh. **Brassicaceae** *(Cruciferae)*					
petiolata Cavara & Grande			ENG		
ALLIUM L. **Alliaceae**					
ampeloprasum L.			IRA POR	TUR	
angulosum L.			SOV		
cepa L.			YUG		
flavum L.				TUR	
macrostemon Bunge		JPN			
nigrum L.		TUN	MOR POR		
oleraceum L.			SOV		
roseum L.			POR	CHL	
rotundum L.			SOV		
rubellum Bieb.				PK	AFG
sativum L.				JAM	CAB
subhirsutum L.				TUR	
triquetrum L.			AUS	NZ	CAB

ALLIUM L. Alliaceae *(Cont'd)*	S	P	C	X	F
vineale L.		AUS TUR	GER IRA SPA USA	CHL ITA NZ	
ALLMANIA R. Br. *ex* Wight **Amaranthaceae**					
nodiflora (L.) R. Br. *ex* Wight		IND			
ALLOTEROPSIS C. Presl **Poaceae** *(Gramineae)*					
cimicina (L.) Stapf (Syn. *Axonopus cimicinus* (L.) Beauv.					
ALOCASIA (Schott) G. Don **Araceae**					
indica (Lour.) Koch				IDO	
macrorrhiza (L.) G. Don				MAU	
ALOE L. **Liliaceae**					
barbadensis Mill. (Syn. *A. vera* L.) (Syn. *A. vulgaris* L.)					
vera L. *(See* **A. barbadensis** Mill.)					
vulgaris L. *(See* **A. barbadensis** Mill.)		PR		JAM	
ALOPECURUS L. **Poaceae** *(Gramineae)*					
aequalis Sobol.	JPN KOR NEP TAI	PHI POL SOV	AFG CAN CHN FIN HK ITA TUR	EGY GRE NZ PK SUD YUG	
agrestis L.			IRA SPA	CHN ITA TUR	AFG AUS
amurensis Komarov	PHI			KOR	
anthoxantoides Boiss.			LEB		
arundinaceus Poir.				TUR	
fulvus J. E. Sm.		JPN			
geniculatus L.			CHN POR TAI	GER NZ USA	AFG AUS FIN
japonicus Steud.				JPN	
monspeliensis L. *(See* **Polypogon monspeliensis** (L.) Desf.)					HAW JPN POR
myosuroides Huds.	ENG GER HUN ITA	BEL IND IRA LEB	AST IRQ ISR NET	AFG FRA GRE JOR	AUS

ALOPECURUS L. Poaceae *(Gramineae) (Cont'd)*	S	P	C	X	F
myosuroides Huds. *(Cont'd)*		SWE	SOV TUR	NOR NZ PK USA	
pratensis L.			FIN JPN SOV	COL NZ URU USA	AFG AUS HAW
ALOYSIA Ort. & Palau *ex* L'herit. **Verbenceae**					
lycioides Cham.				USA	
ALSOPHILA R. Br. **Cyatheaceae**					
latebrosa Wall. (Syn. *Hemitelia latebrosa* Mett.)					
ALSTONIA R. Br. **Apocynaceae**					
constricta F. Muell.			AUS		
ALTERNANTHERA Forsk. **Amaranthaceae**					
achyrantha R. Br.				BRA	AUS
amabilis Lam.				BRA	
denticulata R. Br.				AUS NZ	
echinata Sm.				IND	
ficoidea (L.) R. Br. *ex* Griseb. *var.* Bettzickiana (Nicholson)Baker (Syn. *A. ficoides* Roem. & Schult. *non* Beauv.)			BRA	DR	
ficoides Roem. & Schult. *non* Beauv. *(See* **A. ficoidea** (L.) R. Br. *ex* Griseb. *var.* Bettzickiana (Nicholson)Baker			PHI TRI	ANT JAM SUR	
frutescens R. Br. *ex* Spreng.				THI	
halimifolia Standl. *ex* Pittier				PER	
nana R. Br.				AUS	
nodiflora R. Br.		TAI		AUS NIG	
philoxeroides (Mart.) Griseb.	ARG USA	NZ TAI	IDO IND	BRA COL HON MEX SUR THI URU	
polygonoides R. Br.				BRA VEN	
pugens H. B. K.				ARG AUS IND KEN SAF	

ALTERNANTHERA Forsk. Amaranthaceae *(Cont'd)*	S	P	C	X	F
repens (L.) Link *(See* **A. sessilis** (L.) Dc.	GHA		AUS HAW KEN RHO	ARG MAU THI USA VIE	IDO
sessilis (L.) Dc. (Syn. *A. triandra* Lam.) (Syn. *A. repens* (L.) Link)	IVO MOZ NIG PHI THI	IDO IND MAL TAI	COL EGY GHA HAW JPN NGI PR TRI	BND BOR BOT CAB CAM CHN CNK DAH DR ECU FIJ GUI IRA IRQ JAM JOR KEN MAU MIC NEP PK PLW RHO SAF SEG SUD SUR TNZ UGA VIE ZAM	HK ISR KOR MRE
triandra Lam. *(See* **A. sessilis** (L.) Dc.)		MAL		IND THI	
ALTHAEA L. Malvaceae					
acaulis Cav.			LEB		
longiflora Boiss. & Reut.			MOR		
ludwigii L.			PK		ISR
officinalis L.			POR SPA YUG		AFG AUS ISR
sulphurea Boiss. & Hohen.			IRA		
ALYSICARPUS Desv. Papilionaceae *(Leguminosae)*					
longifolius Wight & Arn.				IND	
monilifer Dc.				IND	
nummularifolius (L.) Dc.			TAI	PHI	IDO
ovalifolius (Schum.) F. Leon.				GHA	
rugosus (Willd.) Dc.			IND TAI	GHA PHI	IDO
vaginalis (L.) Dc.		THI	HAW	CEY	

ALYSICARPUS Desv. Papilionaceae *(Leguminosae) (Cont'd)*	S	P	C	X	F
vaginalis (L.) Dc. *(Cont'd)*			PHI TAI	FIJ IND NGI PLW TRI USA	
ALYSSUM L. Brassicaceae *(Cruciferae)*					
campestre L.			POR	TUR	
desertorum Stapf			SOV		AFG
incanum L.			SOV		
maritimum Lam.			SPA		
minimum Willd.			TUR		
AMARANTHUS L. Amaranthaceae *(Syn.ALBERSIA* Kunth)					
albus L.		MEX POR	ARG AUS CAN MOR USA	COL NZ SOV	CAB
angustifolius Lam.			EGY KEN LEB	MOR TNZ	
ascendens Lois. *(See* **A. viridis** L.)			EGY		
australis (Gray) J. Sauer				USA	
blitoides S. Wats. **blitum**			CAN LEB MOR POR USA	IRQ NZ	
blitum L. *(See* **A. lividus** L.)	JPN NGI SPA	GRE IDO ISR MOZ PK	KEN NZ	CHL CHN IND ITA	AFG ARG USA
caudatus L.	ANG	TUR	SAF	BRA MAU VEN	AFG AUS ISR
chlorostachys Willd.		PER	EGY		BRA
crassipes Schlecht.			PR		
crispus Terrac.			ARG		
cruentus L.			SAF	AUS	
deflexus L.			ARG SAF	CHL MOR NZ	AUS
dubius Mart.	COL	CUB PR TNZ	HAW NIC	DR HON KEN MEL	

AMARANTHUS L. Amaranthaceae *(Cont'd)*	S	P	C	X	F
dubius Mart. *(Cont'd)*				PER TRI VEN	
gangeticus L.				USA	
giganticus L.			CAB		
gracilis Desf. (*See* **A. viridis** L.)	MOZ	ARG IND PHI	IDO JPN LEB PR TRI	BRA MOR	HAW
graecizans L.	MEX	ISR KEN	LEB POR	GHA NZ SUD USA ZAM	
hybridus L.	MEX NZ RHO SAF USA ZAM	ARG BRA COL ETH KEN PER THI TNZ	AUS JPN LEB MOR SAL	CHL HAW IDO NIG PK	DR GUI HUN IND IVO MLI NIC SEG UGA
hypochondriacus L.			KEN		

lividus L.
   (Syn. *A. blitum* L.)
   (Syn. *Albersia blitum* Kunth)
   (Syn. *Euxolus blitum* Gren.)
		AST
		GRE
		ISR
		JOR
		JPN
			ARG
			GER
			HUN
			IDO
			KEN
			PER
			SEG
			SUD
			TUN
				CAN
				CHN
				CNK
				IRA
				IRQ
				MOZ
				NET
				NGI
				NZ
				PLW
				RHO
				SAF
				SPA
				THI
				TNZ
				UGA
				USA	

macrocarpus Benth.				AUS	
mangostanus L.			JPN TAI		
mitchelli Benth.				AUS	
muricatus (Jacq.) Gill. *ex* Moq.			ARG SAF	MOR URU	
oleraceus L.				CHL	
palmeri S. Wats.				USA	
paniculatus L.		SAF	CHN EGY		AFG SOV
patulus Bertol.			JPN		

AMARANTHUS L. Amaranthaceae (Cont'd)	S	P	C	X	F
polygamus L.			PR	IND PK	
powellii S. Wats.			CAN	USA	
quitensis H. B. K.	ARG			URU	
retroflexus L.	AFG BRA CAN FRA GER HUN ISR ITA KOR MEX MOZ POL SOV SPA TUR YUG	AST AUS BUL COL CZE ECU GUA IRA JOR JPN LEB NZ PER TNZ TUN USA	MOR POR	ARG CHL CHN CR EGY GRE IND NEP PR ROM SWE VEN	
spinosus L.	GHA HAW MEL MEX MOZ PHI TAI THI	BRA CAB CUB IDO IND JPN KEN MAL NIC NIG RHO SAF SAL USA	ANG ARG AUS CEY CHN COL HK PER PR	BND FIJ HON IVO JAM MAD MAU NEP PLW SOV TRI VEN VIE	ANT KOR TNZ UGA
standleyanus L. R. Parodi *ex* Covas.			ARG		
thunbergii Moq.		MOZ RHO SAF	KEN		
tricolor L.			HAW	IND KEN	IDO
tristis L.				ANT CHL JAM TRI	
tuberculatos (Moq.) J. Sauer				USA	
viridis L. (Syn. *A. ascendens* Lois.) (Syn. *A. gracilis* Desf.)	AFG BRA HUN IND ITA TAI TUR	ARB AST AUS CUB GUA IRA IVO JOR PHI PK THI	CAB CEY CHN EGY GHA HAW HK IRQ JAM JPN MIC NIG PER PR	ANT ARG BND CAN CHL DAH FIJ GRE HON KOR MAL MAU MEL NEP	

AMARANTHUS L. Amaranthaceae *(Cont'd)*	S	P	C	X	F
viridis L. *(Cont'd)*			SAF SEG SUD ZAM	NGI PLW RHO SAL USA VEN VIE	
AMASONIA L.f. **Verbenaceae**					
campestrus (Aubl.) Moldenke			TRI		
AMBROSIA L. **Asteraceae** *(Compositae)*					
artemisiifolia L. (Syn. *A. elatior* L.)		CAN GUA	AUS COL HAW JPN USA	BRA CHL JAM MAU	
bidentata Michx.				USA	
cumanensis H. B. K.				COL	VEN
elatior L. (*See* **A. artemisiifolia** L.)			ARG CAN JPN	CHL	AUS
maritima L.			EGY		AUS ISR
paniculata Michx.				JAM MAU	
peruviana Willd.				PER	
polystachya Dc.				BRA	
psilostachya Dc.		AUS MAU	CAN USA		
tarapacana Phil.				CHL	
tenuifolia Spreng.		ARG	AUS	PER URU	BRA
trifida L.			CAN JPN USA		
AMELANCHIER Medic. **Rosaceae**					
alnifolia Nutt.				USA	
arborea (Michx. *f.*) Fern.				USA	
canadensis (L.) Medic.				USA	
florida Lindl.				USA	
laevis Wieg.				USA	
sanguinea (Pursh) Dc.				USA	
AMETHYSTEA L. **Lamiaceae** *(Labiatae)*					
caerulea L.				SOV	

AMMANNIA L. Lythraceae	S	P	C	X	F
auriculata Willd.		NIG		USA	AFG
baccifera L.	IND		TAI	CEY IDO VIE	AFG
coccinea Rottb.	ECU USA	COL POR SPA	ARB HAW PER TUR	CAB IRA JAM MEX MOZ PR TRI VIE	JPN
latifolia L.				TRI	
multiflora Roxb.			JPN		AFG IDO
octandra L. _f._				BND IND	IDO
pentandra Roxb.		IND			
peploides Spreng.		IND	JPN		
prieuriana Guill. & Perr.		NIG			
rotundifolia Buch.-Ham.				BND	
senegalensis Lam.				PK	
AMMI L. **Apiaceae** _(Umbelliferae)_					
majus L.			ARG EGY IRQ ISR LEB MOR POR	URU USA	AUS
visnaga (L.) Lam.	ISR		ARG EGY MOR POR	CHL IRA PER TUR URU USA	
AMMOPHILA Host **Poaceae** _(Gramineae)_					
arenaria (L.) Link				USA	
breviligulata Fern.				USA	
AMORPHOPHALLUS Bl. **Araceae**					
campanulatus (Roxb.) Bl. _ex_ Decne.				FIJ	
AMPELANUS Rafin. **Asclepiadaceae**					
albidus (Nutt.) Britt.				USA	
AMPELOPSIS Rich. **Vitaceae**					
arborea (L.) Koehne				USA	

AMPHIBROMUS Nees Poaceae *(Gramineae)*	S	P	C	X	F
neesii Steud.				NZ	AUS
AMPHICARPAEA Ell. Papilionaceae *(Leguminosae)*					
edgeworthii Benth.			JPN		
AMPHILOPHIS Nash Poaceae *(Gramineae)*					
glabra (Roxb.) Stapf				VIE	
pertusa (L.) Nash *ex* Stapf (*See* **Bothriochloa pertusa** (L.) A. Camus)				VIE	
AMPHILOPHIUM Kunth Bignoniaceae					
oxylophium Donn. Sm.				PER	
AMSINCKIA Lehm. Boraginaceae					
angustifolia Lehm.				CHL	
douglasiana A. Dc.				USA	
hispida (Ruiz & Pav.) I. M. Johnston		AUS		CHL NZ	ARG
intermedia Fisch. & Mey.		AUS	ENG	USA	
lycopsioides Lehm.		AUS		USA	
tessellata Gray				USA	
ANABASIS L. Chenopodiaceae					
aphylla L.			MOR		SOV
ANACARDIUM L. Anacardiaceae					
occidentale L.				CAB JAM	PR
ANACYCLUS L. Asteraceae *(Compositae)*					
clavatus (Desf.) Pers.			LEB	MOR	
radiatus Loisel.			MOR		
valentinus L.				MOR	
ANADENDRUM Schott Araceae					
montanum Schott				IDO	
ANAGALLIS L. Primulaceae					
arvensis L.	IND SEG	CHN ETH FRA GRE IRA MEX NZ PK POR TUN USA	ARG AUS EGY ENG GER HAW IRQ ISR LEB MOR POL SAF SOV SPA	ARB BEL BRA CAN CHL CNK CZE HUN ITA KEN MAU PER TUR	AFG KOR

ANAGALLIS L. Primulaceae (Cont'd)	S	P	C	X	F
arvensis L. (Cont'd)			TAI YUG		
coerulea Schreb.			LEB	CHL GER ISR TUR	AUS POL
femina Mill.				AUS TUR	LEB
ANANAS Mill. Bromeliaceae					
microstachys Lindm.			BRA		
ANAPHALIS Dc. Asteraceae (Compositae)					
margaritacea (L.) C. B. Clarke				USA	
ANASTROPHUS Schlecht. Poaceae (Gramineae)					
compressus (Sw.) Schlecht. (See Axonopus compressus (Sw.) Beauv.)		IDO			
ANCHUSA L. Boraginaceae					
arvensis Bieb. (See Lycopsis arvensis L.)			ENG FIN	ICE SOV	
azurea Mill. (See A. italica Retz.)		LEB	MOR	JOR POR	ISR
hybrida Ten.			ISR		
italica Retz. (Syn. A. azurea Mill.)		LEB	IRA	GRE TUR	AFG
officinalis L.			SOV	TUR USA	AUS POL
ANDIRA A. L. Juss. Papilionaceae (Leguminosae)					
humilis Mart. ex Benth.			BRA		
inermis H. B. K.			JAM		
ANDRACHNE L. Euphorbiaceae					
telephioides L.				TUR	
ANDROGRAPHIS Nees Acanthaceae					
echioides Nees				IND	
paniculata Nees				JAM	
ANDROPOGON L. Poaceae (Gramineae)					
aciculatus Retz. (See Chrysopogon aciculatus (Retz.) Trin.)		IDO			HAW
annulatus Forsk. (See Dichanthium annulatum (Forsk.) Stapf)		IND		PR TRI	AFG AUS HAW ISR
arundinacea Scop. (See Sorghum halepense (L.) Pers.)					IDO

ANDROPOGON L. Poaceae *(Gramineae) (Cont'd)*	S	P	C	X	F
barbinodis Lag.				USA	HAW
bicornis L.		TRI	HAW	DR JAM NIC PER USA VEN	
brevifolius Sw. *(See* **Schizachyrium brevifolium** (Sw.) Nees *ex* Buse)				JPN	
caricosus L. *(See* **Dichanthium caricosum** (L.) A. Camus)					IDO
condensatus H. B. K.			TRI		
contortus L. *(See* **Heteropogon contortus** (L.) Beauv. *ex* Roem. & Schult.)					HAW IDO
filiculmis Hook. *f.* *(See* **Capillipedium filiculme** (Hook. *f.*) Stapf)				VIE	
gerardi Vitman				USA	HAW
glomeratus (Walt.) B. S. P.				USA	HAW PR
halepensis (L.) Brot. *(See* **Sorghum halepense** (L.) Pers.)		PHI		BRA ITA SOV TUR	HAW
hirtus L. *(See* **Hyparrhenia hirta** (L.) Stapf)			SPA		
intermedius R. Br. *(See* **Bothriochloa intermedia** (R. Br.) A. Camus)				VIE	HAW IDO
ischaemum L. *(See* **Bothriochloa ischaemum** (L.) Keng)			IRQ		AFG HAW
lateralis Nees			BRA		
leucostachyus H. B. K.				GUA	
nodosus (Willem.) Nash *(See* **Dichanthium aristatum** (Poir.) C. E. Hubb.)			HAW		
pertusus (L.) Willd. *(See* **Bothriochloa pertusa** (L.) A. Camus)	JAM	CUB	HAW	DR	AFG AUS IDO TRI
saccharoides Sw.				CHL USA	ARG HAW PER PR
scoparius Michx. *(See* **Schizachyrium scoparium** (Michx.) Nash)				USA	
serratus Thunb. *non* Miq. *(See* **Capillipedium parviflorum** (R. Br.) Stapf)				PHI	
sorghum (L.) Brot. *(See* **Sorghum bicolor** (L.) Moench)					IDO
tectorum Schum. & Thonn.				NIG	

ANDROPOGON L. Poaceae *(Gramineae) (Cont'd)*	S	P	C	X	F
ternarius Michx.				USA	
virginicus L.				HAW USA	AUS PR
ANDROSACE L. **Primulaceae**					
filiformis Retz.			SOV		
maxima L.				TUR	
septentrionalis L.				USA	
umbellata Merr.				CHN	
ANEILEMA R. Br. **Commelinaceae**					
beninense Kunth				CNK NIG	
japonica (Thunb.) Kunth		JPN KOR			
keisak Hassk.		JPN			
nudiflorum (L.) Wall. (*See* **Murdannia nudiflora** (L.) Brenan)			IDO	IND THI TRI VIE	
sinicum Lindl.			TAI		
spiratum R. Br.				IND	
umbrosum (Vahl) Kunth				NIG	
vitiense Seem.				PHI	
ANEMONE L. **Ranunculaceae**					
cernua Thunb.				CHN	
coronaria L.				TUR	
nemorosa L.			AUS		
pulsatilla L.			AUS		
ANEMOPSIS Hook. & Arn. **Saururaceae**					
californica Hook. & Arn.				USA	
ANETHUM L. **Apiaceae** *(Umbelliferae)*					
graveolens L.			LEB POR SOV SPA	MOR TUR	AUS ISR
ANGELICA L. **Apiaceae** *(Umbelliferae)*					
archangelica L.				SPA	
atropurpurea L.				USA	
sylvestris L.			FIN		

	S	P	C	X	F
ANGELONIA Humb. & Bonpl. **Scrophulariaceae**					
salicariaefolia Humb. & Bonpl.			PR		
ANISOMELES R. Br. **Lamiaceae** *(Labiatae)*					
indica (L.) O. Ktze.			TAI	IND PLW TRI	IDO JAM
ovata R. Br.				IND	
ANNONA L. **Annonaceae**					
coriacea Mart.				BRA	
glabra L.				VIE	
muricata L.				BRA CAB JAM	
purpurea Moc. & Sesse *ex* Dunal				BRA	
reticulata L.				BRA CAB JAM	
senegalensis Pers.		TNZ			
squamosa L.				CAB JAM	
ANODA Cav. **Malvaceae**					
acerifolia Cav.			PR	COL	
cristata (L.) Schlecht.	MEX			ARG AUS CHL GUA PER SAL USA	
hastata Cav.				CHL COL	
ANOTIS Dc. **Rubiaceae**					
spermacoce L.	IND				
ANREDERA Juss. **Basellaceae**					
cordifolia (Ten.) Steen.			AUS		
ANTENNARIA Gaertn. **Asteraceae** *(Compositae)*					
neglecta Greene				USA	
neodioca Greene				USA	
plantaginea R. Br. (Syn. *A. plantaginifolia* (L.) Hook.)					
plantaginifolia (L.) Hook. *(See* **A. plantaginea** R. Br.)				VOL	
ANTHEMIS L. **Asteraceae** *(Compositae)*					
arvensis L.				CAN	CHL	AUS

ANTHEMIS L. Asteraceae *(Compositae) (Cont'd)*	S	P	C	X	F
arvensis L. *(Cont'd)*			ENG FIN GER SOV SPA	ICE NZ TUR URU USA	POL POR
austriaca Jacq.				GER	
chia L.				GRE JOR	
cotula L.	ARG TUR	ENG ITA NZ POL	AST CAN GER GRE IRA IRQ ISR LEB POR SAF SOV SPA TUN USA YUG	AUS BRA CHL EGY JOR NOR URU	
fuscata Brot.				POR	
graveolens Boiss.				SPA	
hyalina Dc.				LEB	ISR
melanolepis Boiss.				LEB	
mixta L.				POR URU	BRA
montana L.				TUR	
nobilis L.				BRA NZ PER VEN	AUS POR
pseudocotula Boiss.				EGY ISR	
ruthenica Bieb.				SOV	POL
tinctoria L.				FIN SOV YUG	TUR USA
ANTHOXANTHUM L. Poaceae *(Gramineae)*					
aristatum Boiss.				GER	
odoratum L.				COL JPN	CHL ITA NZ USA	AUS FIN HAW SAF
ANTHRISCUS Pers. Emend. Hoffm. Apiaceae *(Umbelliferae)*					
neglecta Boiss. & Reut. *ex* Lange				POR	
scandicina (Web.) Mansf.				USA	

ANTHRISCUS Pers. Emend. Hoffm. **Apiaceae** *(Umbelliferae) (Cont'd)*	S	P	C	X	F
sylvestris (L.) Hoffm.			ENG GER SPA YUG	FIN NZ USA	
vulgaris Bernh.				CHL	POR
ANTHURIUM Schott **Araceae**					
acaule Schott			PR	DR	
ANTIGONON Endl. **Polygonaceae**					
leptopus Hook. & Arn.				HON	PR
ANTIRRHINUM L. **Scrophulariaceae**					
majus L.			SPA YUG		
orontium L.		POR	EGY MOR SPA	NZ TUR	ISR
APARGIA Scop. **Asteraceae** *(Compositae)*					
autumnale (L.) Hoffm.				CAN	
APERA Adans. **Poaceae** *(Gramineae)*					
spica-venti (L.) Beauv.		BEL	FIN GER SPA	CZE ENG FRA NET ROM SOV SWE SWT	
APIOS Fabr. **Papilionaceae** *(Leguminosae)*					
americana Medic.				USA	
APIUM L. **Apiaceae** *(Umbelliferae)*					
ammi (Jacq.) Urb.			JPN	BRA CHL URU	
australe Thou.				CHL	
leptophyllum (Pers.) F. Muell. *ex* Benth. (Syn. *A. tenuifolium* (Moench) Thell.)		ARG MAU	IND RHO SAF TAI	AUS PER USA	
nodiflorum Reichb. *f.*			LEB POR		
tenuifolium (Moench) Thell. *(See* **A. leptophyllum** (Pers.) F. Muell. *ex* Benth.)	HAW		EGY		
APLUDA L. **Poaceae** *(Gramineae)*					
mutica L.			TAI	PHI	AFG IDO
varia Hack.				IND VIE	

APOCYNUM L. Apocynaceae	S	P	C	X	F
androsaemifolium L.				USA	
cannabinum L.			USA		
sibiricum Jacq.				USA	
APONOGETON L. *f.* Aponogetonaceae					
distachyus L. *f.*				NZ	
monostachyon L. *f.*		IND			
ARABIDOPSIS Heynh. Brassicaceae *(Cruciferae)*					
thaliana (L.) Heynh.			ENG FIN SOV	USA	
ARACHIS L. Papilionaceae *(Leguminosae)*					
hypogaea L.			TAI		IDO ISR
ARALIA L. Araliaceae					
spinosa L.				USA	
ARAUJIA Brot. Asclepiadaceae					
hortorum Fourn.			AUS	NZ	
sericifera Brot.			SAF		ARG
ARBUTUS L. Ericaceae					
menziesii Pursh				USA	
ARCEUTHOBIUM Bieb. Loranthaceae					
oxycedri Bieb.				TUR	
ARCTIUM L. Asteraceae *(Compositae)*					
lappa L.			CAN ENG IRA YUG	ARG CHL NZ URU USA	AFG POL
majus Bernh.			SOV		
minus (Mill.) Bernh.			ARG CAN SOV USA	BRA CHL NZ URU	POL
ARCTOSTAPHYLOS Adans. Ericaceae					
canescens Eastw.				USA	
columbiana Piper				USA	
glandulosa Eastw.				USA	
glauca Lindl.				USA	
hispidula Howell				USA	
manzanita Parry				USA	

ARCTOSTAPHYLOS Adans. Ericaceae *(Cont'd)*	S	P	C	X	F
nevadensis A. Gray				USA	
parryana Lemmon				USA	
patula Greene				USA	
pungens H. B. K.				USA	
uva-ursi (L.) Spreng.				USA	
viscida Parry				USA	
ARCTOTHECA Wendl. **Asteraceae** *(Compositae)*					
calendula (L.) Levyns		AUS	SAF		
repens Wendl.				AUS	
ARCTOTIS L. **Asteraceae** *(Compositae)*					
venusta T. Norl.			SAF		
ARDISIA Sw. **Myrsinaceae**					
humilis Vahl			HAW		
ARENARIA L. **Caryophyllaceae**					
serpyllifolia L.			CHN ENG GER IND JI N	NZ USA	
ARGEMONE L. **Papaveraceae**					
glauca L.			HAW	USA	
intermedia L.				USA	
mexicana L.	IND PK TNZ	AUS KEN MEX RHO	ARG EGY PR SAF USA	BND CHL CUB DR FIJ GHA HON IDO JAM MAD MAU MOZ NIC NZ PER POR URU VEN ZAM	ETH HK
ochroleuca Sweet		AUS		NZ	
platyceras Link & Otto				USA	
subfusiformis G. B. Ownb.			SAF		
ARISTIDA L. **Poaceae** *(Gramineae)*					
adescensionis L.			KEN	GHA	AFG

ARISTIDA L. Poaceae *(Gramineae) (Cont'd)*	S	P	C	X	F
adescensionis L. *(Cont'd)*				IND PER SUD TNZ	HAW
arenaria Gaudich.			AUS		
balansae Henr.				VIE	
congesta Roem. & Schult.			SAF		
depressa Trin. *(See* **A. setacea** Retz.*)*				MAU	
dichotoma Michx.				USA	
hordeacea Kunth				SUD	
latifolia Domin				AUS	
leptopoda Benth.			AUS		
longiseta Steud.				USA	
oligantha Michx.				USA	
pallens Cav.				ARG CHL	
purpurascens Poir.				USA	
setacea Retz. (Syn. *A. depressa* Trin.)				IND	
setifolia H. B. K.				VEN	
ARISTOLOCHIA L. **Aristolochiaceae**					
bracteata Retz. (Syn. *A. bracteolata* Lam.)				IND	
bracteolata Lam. *(See* **A. bracteata** Retz.*)*		SUD			
brasiliensis Mart. & Zucc.				BRA	
clematitis L.		YUG		TUR	
elegans Mast.				AUS	
grandiflora Sw.				HON MEX NIC SAL	
lingua Malme				ARG	
maurorum L.		JOR	IRA LEB	TUR	ISR
rotunda L.				TUR	
ARJONA Comm. *ex* Cav. **Santalaceae**					
tuberosa Cav.				ARG	
ARRHENATHERUM Beauv. **Poaceae** *(Gramineae)*					
elatius Merth. & Koch		CHL	AUS	COL	HAW

ARRHENATHERUM Beauv. Poaceae *(Gramineae) (Cont'd)*	S	P.	C	X	F
elatius Merth. & Koch *(Cont'd)*				IRA	NZ URU USA	IND POR SAF
erianthum Boiss. & Reut.				POR		
ARTEMISIA L. Asteraceae *(Compositae)*						
absinthium L.			·	CAN SOV	JAM NZ TUR USA VEN	
annua L.				ARG IRA	USA	AFG
apiacea Hance					CHN	
arbuscula Nutt.					USA	
austriaca Jacq.				SOV		
biennis Willd.					USA	
californica Less.					USA	
cana Pursh					USA	
capillaris Thunb.				TAI		AFG
dracunculus L.					TUR	
filifolia Torr.					USA	
frigida Willd.					USA	
herba-alba Asso.				IRA		AFG ISR
indica Willd. (See **A. vulgaris** L.)				TAI		
japonica Thunb.				JPN	VIE	AFG
ludoviciana Nutt.					USA	
mexicana Willd. *ex* Spreng.					MEX	URU
nova A. Nels.					USA	
princeps Pampan.			JPN			
siversiana Willd.					SOV	
tridentata Nutt.					USA	
tripartita Rydb.					USA	
verlotorum Lamotte (See **A. vulgaris** L.)			ARG		URU	
vulgaris L. (Syn. *A. indica* Willd.) (Syn. *A. verlotorum* Lamotte)	AFG IDO ITA JPN KOR MAU POL		AST CHN ENG FIN GER HAW TAI	BEL CAN CR IND IRA NEP NOR	CEY HK	

ARTEMISIA L. Asteraceae *(Compositae) (Cont'd)*	S	P	C	X	F
vulgaris L. *(Cont'd)*	SOV		TUN USA YUG	NZ SWE THI TUR VEN	
ARTHRAXON Beauv. **Poaceae** *(Gramineae)*					
hispidus (Thunb.) Makino			IDO JPN TAI	IND	
ARTOCARPUS J. R. & G. Forst. **Moraceae**					
altilis (Park.) Fosb. (Syn. *A. incisa* (Thunb.) L. *f.*)					
incisa (Thunb.) L. *f.* *(See* **A. altilis** (Park.) Fosb.)				JAM	
ARUM L. **Araceae**					
dracunculus L. *(See* **Dracunculus vulgaris** Schott)				TUR	
italicum Mill.		TUN		NZ	
maculatum L.				ITA	
ARUNDINARIA Michx. **Poaceae** *(Gramineae)*					
tecta (Walt.) Muhl.				USA	
ARUNDINELLA Raddi **Poaceae** *(Gramineae)*					
bengalensis (Spreng.) Druce	IND				
hirta (Thunb.) Tanaka			JPN		
leptochloa (Nees) Hook. *f.)*		IND			
setosa Trin.				VIE	
sinensis Rendle				VIE	
ARUNDO L. **Poaceae** *(Gramineae)*					
donax L.			IRA SPA	ARG CHL DR NZ USA	AFG HAW IND ISR PR
formosana Hack.			TAI		
madagascariensis Kunth				VIE	
ASCLEPIAS L. **Asclepiadaceae**					
campestris Decne.			ARG		BRA
curassavica L.	ECU MEL PLW		AUS HAW IND SUR TRI TUR YUG	ARG BOR BRA CAB COL CR FIJ GHA	CHN HK

ASCLEPIAS L. Asclepiadaceae *(Cont'd)*	S	P	C	X	F
curassavica L. *(Cont'd)*				HON IDO IRA JAM JPN MAL MIC NEP NGI NIC PAN PER PR RHO SAL SEG THI USA VEN	
eriocarpa Benth.				USA	
fascicularis Decne.				USA	
fruticosa L. *(See* **Gomphocarpus fruticosus** (L.) R. Br. *ex* W. T. Ait.)			AUS SAF	NZ	
galioides H. B. K. *(See* **A. verticillata** L.)				USA	
incarnata L.				CAN USA	
labriformis Jones				USA	
latifolia Rafin.				MEX USA	
mellodora St. Hil. *(See* **A. nervosa** Decne.)			ARG	URU	
mexicana Cav.				USA	
nervosa Decne. (Syn. *A. mellodora* St. Hil.)					
nivea L.			PR	DR JAM	
ovalifolia Decne.				USA	
physocarpa (E. Mey.) Schlecht.		AUS			
speciosa Torr.				CAN USA	
subverticillata (Gray) Vail				USA	
syriaca L.		CAN	IRQ USA		
tuberosa L.				USA	
verticillata L. (Syn. *A. galioides* H. B. K.)			USA		
ASPARAGUS L. Asparagaceae					
lucidus Lindl.			TAI		

ASPARAGUS L. Asparagaceae (Cont'd)	S	P	C	X	F
officinalis L.			YUG	NZ	AFG ARG AUS
ASPERUGO L. Boraginaceae					
procumbens L.				USA	
ASPERULA L. Rubiaceae					
arvensis L.			LEB SPA	MOR	AFG ISR POR
humifusa (Bieb.) Bess.			IRA		AFG ISR
ASPHODELUS L. Liliaceae					
aestivus Reichb. *(See* **A. fistulosus** L.)			LEB		
albus Willd.			SPA		
fistulosus L. (Syn. *A. aestivus* Reichb.)		AUS		IND NZ SPA	
microcarpus Viv.				TUR	
tenuifolius Cav.	IND PK		ITA NEP	ARB BND EGY IRA IRQ JOR MAL SUD	ISR
ASPILIA Thou. Asteraceae (*Compositae*)					
africana (Pers.) Adams	GHA	NIG			
bupthalmiflora Griseb. (Syn. *A. montevidensis* O. Ktze.)					
helianthoides Benth. & Hook. *f.*			NIG	GHA IVO	
montevidensis O. Ktze. *(See* **A. bupthalmiflora** Griseb.)		BRA			
ASPLENIUM L. Aspleniaceae					
trichomanes L.			TUR		
ASTER L. Asteraceae (*Compositae*)					
ageratoides Turcz. *(See* **A. trinervius** Desf.)			JPN		
ericoides L.				USA	
exillis Ell.				CHL USA	
indicus L. *(See* **Asteromoea indica** (L.) Bl.)				CHN	JPN

ASTER L. Asteraceae *(Compositae) (Cont'd)*	S	P	C	X	F
linariifolius L.				USA	
novae-angliae L.				USA	
novi-belgii L.				USA	
occidentalis (Nutt.) Torr. & Gray				USA	
parryi A. Gray				USA	
pilosus Willd.				USA	
puniceus L.				USA	
scaber Thunb.				JPN	
simplex Willd.				USA	
spectabilis Ait.				USA	
spinosus Benth.				USA	
squamatus (Spreng.) Hieron. *ex* Sod.			ARG MOR SPA	CHL	
subulatus Michx.			AUS IRQ JPN SAF		NZ
trinervius Desf. (Syn. *A. ageratoides* Turcz.)					
tripolium L.				IRQ	
yomena (Kitamura) Honda				JPN	
ASTERACANTHA Nees **Acanthaceae**					
longifolia (L.) Nees *(See* **Hygrophila spinosa** T. Anders.)		CEY		IND NIG	
ASTERISCIUM Cham. & Schlecht. **Apiaceae** *(Umbelliferae)*					
chilense Cham. & Schlecht.				CHL	
ASTEROMOEA Bl. **Asteraceae** *(Compositae)*					
indica (L.) Bl. (Syn. *Aster indicus* L.)					
ASTRAGALUS L. **Papilionaceae** *(Leguminosae)*					
aleppicus Boiss.			LEB		
allochorus A. Gray				USA	
balansae Boiss.				TUR	
bergii Hieron.			ARG	URU	
bisulcatus A. Gray				USA	
cicer L.			SPA		
collinus Boiss.				TUR	
deinacanthus Boiss.			LEB		

ASTRAGALUS L. Papilionaceae *(Leguminosae) (Cont'd)*	S	P	C	X	F
diphysus A. Gray				USA	
earlei Greene				USA	
echinops Aucher *ex* Boiss.			LEB		
garbancillo Cav.		ARG			
hamosus L.			AUS	TUR	
lentiginosus Dougl.				USA	
miser Dougl.				USA	
mollissimus Torr.			USA		
pectinatus Hook.				USA	
tribuloides Delile				PK	AFG
voloratum L.				TUR	
wootonii Sheldon				USA	
ASTRIPOMOEA A. Meeuse **Convolvulaceae**					
hyoscyamoides (Vatke) Verde			KEN		
ASYSTASIA Bl. **Acanthaceae**					
coromandeliana Nees *(See* **A. gangetica** (L.) T. Anders.)	MAL			MAU TRI	
gangetica (L.) T. Anders. (Syn. *A. coromandeliana* Nees)		GHA	HAW	BOR CNK FIJ IND MAU NIG PHI PLW TRI	IDO
schimperi T. Anders.		UGA		KEN	
welwitschii S. Moore				ANG	
ATALAYA Bl. **Sapindaceae**					
hemiglauca F. Muell. *ex* Benth.		AUS			
ATHROISMA Dc. **Asteraceae** *(Compositae)*					
stuhlmannii (O. Hoffm.) Mattf.				KEN	
ATHYRIUM Roth **Athyriaceae**					
thelypteroides (Michx.) Desv.				USA	
ATRACTYLIS L. **Asteraceae** *(Compositae)*					
gummifera L.			MOR		LIB
ATRIPLEX L. **Chenopodiaceae**					
argentea Nutt.				USA	
canescens (Pursh) Nutt.				USA	

ATRIPLEX L. Chenopodiaceae *(Cont'd)*	S	P	C	X	F
chenopodioides Batt.				MOR	
confertifolia (Torr. & Frem.) S. Wats.				USA	
crassifolia C. A. Mey.				PK	AFG
elegans (Moq.) D. Dietr.				USA	
hastata L.			GER	CHL USA	
hortensis L.			GER	IND USA	
laciniata L.			SOV		
patula L.			ENG GER IRA SPA YUG	FIN ICE NZ USA	
rosea L.			ARG	USA	
semibaccata R. Br.			ARG HAW	AUS CHL USA	
tatarica L.			IRQ		
wrightii S. Wats.				USA	
ATROPHA L. Solanaceae					
belladonna L.			AUS	NZ	
ATTALEA H. B. K. Arecaceae *(Palmae)*					
exigua Drude				BRA	
phalerata Mart.				BRA	
ATYLOSIA Wight & Arn. Papilionaceae *(Leguminosae)*					
scarabaeoides (L.) Benth. (Syn. *Cantharospermum scarabaeoides* Baill.)	PLW			FIJ IND	
AUSTRALINA Gaudich. Urticaceae					
acuminata Wedd.		SAF			
AUSTROEUPATORIUM R. M. King & H. Robinson Asteraceae *(Compositae)*					
inulaefolium (H. B. K.) R. M. King & H. Robinson (Syn. *Eupatorium inulaefolium* H. B. K.)					
AVENA L. Poaceae *(Gramineae)*					
alba Vahl			EGY MOR	POR	
barbata Brot.			GRE	LEB	ARG AUS CHL FRA URU USA	HAW IND ISR POR SAF
byzantina C. Koch			ARG		

AVENA L. Poaceae *(Gramineae)* *(Cont'd)*	S	P	C	X	F
byzantina C. Koch *(Cont'd)*		ETH			
cultiformis Malz.			SOV		
fatua L. (Syn. *A. trichophylla* C. Koch)	ARG AUS CAN ENG FRA GRE SAF USA	BEL CHL COL GER IND IRA IRE KEN LEB MEX NZ SPA SWE TUR	EGY HAW IRQ JOR JPN MOR POR SOV YUG	ALG ALK CHN CZE DEN ETH FIN HUN ICE IDO ISR ITA LBY NET NOR PHI PK POL ROM SWT TAI TNZ TUN URU VEN	AFG HK KOR
ludoviciana Dur.	AUS SPA	ETH IND		IRA KEN PK SOV TUN	
sativa L.		ARG	COL YUG		HAW SAF
sterilis L.	AUS ISR POR TUN	MOR	ARG EGY LEB	ALG ARB ENG ETH FRA GRE KEN PER PK SOV	AFG
strigosa Schreb.	ETH		POR SOV		
trichophylla C. Koch (See **A. fatua** L.)			SOV		
AVICENNIA L. Avicenniaceae					
nitida Jacq.			USA		
AXONOPUS Beauv. Poaceae *(Gramineae)*					
affinis A. Chase		BRA MAL	AUS HAW	CEY MAU NZ PLW USA	

AXONOPUS Beauv. Poaceae *(Gramineae) (Cont'd)*	S	P	C	X	F
cimicinus (L.) Beauv. *(See* **Alloteropsis cimicina** (L.) Stapf)				BND	
compressus (Sw.) Beauv. (Syn. *Anastrophus compressus* (Sw.) Schlecht.)	BOR BRA CEY CR GHA IND MAL	IDO NIG THI	ARG AUS HAW TRI	CAM CNK CUB DR GUI LIB MAU NGI PLW SAL USA VEN VIE	BUR FIJ PR SAF
scoparius (Flugge) Kuhlm.				COL	
AXYRIS L. **Chenopodiaceae**					
amaranthoides L.			CAN	SOV USA	
AZADIRACHTA A. Juss. **Meliaceae**					
indica A. Juss.				IND	
AZANZA Moc. & Sesse *ex* Dc. **Malvaceae**					
garckeana (F. Hoffm.) Exell & Hillc. *(See* **Thespesia garckeana** F. Hoffm.)				RHO	
AZOLLA Lam. **Salviniaceae**					
africana Desv. *(See* **A. pinnata** R. Br.)		CNK SAF	KEN	DAH ECU GHA IVO ZAM	
caroliniana Willd.		PR	POR	AFG COL HON ITA SUR USA	CHN HK
filiculoides Lam.		HAW	AUS POR	ARG COL IDO PER USA	
imbricata (Roxb.) Nakai		JPN			
japonica Fr. & Sav. *ex* Nakai			JPN	CHN	
nilotica Decne. *ex* Mett.				CNK MOZ RHO TNZ UGA ZAM	
pinnata R. Br. (Syn. *A. africana* Desv.)		IND THI	PHI	AUS BND BOR BOT	

AZOLLA Lam. **Salviniaceae** *(Cont'd)*	S	P	C	X	F
pinnata R. Br. *(Cont'd)*				CAB CHN CNK IDO MAD NZ PK VIE	
rubra R. Br.				NZ	
AZUKIA Takah. **Papilionaceae** *(Leguminosae)*					
angularis (Willd.) Ohwi (Syn. *Phaseolus angularis* W. F. Wight)					
BACCHARIS L. **Asteraceae** *(Compositae)*					
articulata (Lam.) Pers.				BRA	
coridifolia Dc.		ARG		BRA URU	
dracunculifolia Dc.				BRA	
gillesii A. Gray			ARG		
glutinosa Pers.				PER USA	
halimifolia L.		AUS		USA	
lanceolata H. B. K.				PER	
microphylla H. B. K.				COL	
notosergila Griseb.				ARG	
pilularis Dc.				USA	
pingraea Dc.			ARG		
ramulosa (Dc.) A. Gray				USA	
salicina Torr. & Gray				USA	
sarothroides A. Gray				USA	
trimera Dc.				ARG URU	VEN
ulicina Hook. & Arn.				ARG	
BACOPA Aubl. **Scrophulariaceae** *(Syn.MONIERA P. Br.)*					
aquatica Aubl.				TRI	
calycina (Benth.) Pennell				GHA	
caroliniana (Walt.) Robins. (Syn. *Hydrotrida caroliniana* Small)				USA	
crenata (Beauv.) Hepper	NIG				
cuneifolia (Michx.) Wettst. (Syn. *Moniera cuneifolia* Michx.)					
dianthera (Sw.) Descole & Borsini				ARG	

BACOPA Aubl. Scrophulariaceae *(Cont'd)*	S	P	C	X	F
eisenii (Kell.) Pennell				USA	
erecta Hutch. & Dalz.			GUI SEG		
monnieri (L.) Pennell			IND TAI	CEY	IDO
procumbens (Mill.) Greenm.				IDO	
rotundifolia Wettst.			JPN		
sessiliflora Pulle				TRI	
BAHIA Lag. Asteraceae *(Compositae)*					
oppositifolia Dc.				USA	
BAILEYA Harv. & Gray *ex* Torr. Asteraceae *(Compositae)*					
multiradiata Harv. & Gray				USA	
BALLOTA L. Lamiaceae *(Labiatae)*					
nigra L.			SOV		
BALTIMORA L. Asteraceae *(Compositae)*					
recta L.	MEX	SAL		GUA	IDO
BAMBUSA Schreb. Poaceae *(Gramineae)*					
vulgaris Schrad.				JAM	
BANISTERIA L. Malpighiaceae *(See* **HETEROPTERIS** Kunth*)*					
purpurea L. *(See* **Heteropteris purpurea** (L.) Kunth*)*			PR		
BAPTISIA Vent. Papilionaceae *(Leguminosae)*					
tinctoria (L.) R. Br.				USA	
BARBAREA R. Br. Brassicaceae *(Cruciferae)*					
arcuata Reichb.				SOV	
praecox R. Br.				CHL	
verna (Mill.) Aschers				USA	
vulgaris R. Br.			CAN USA YUG	BEL FIN NZ SOV	
BARLERIA L. Acanthaceae					
mysorensis Heyne				IND	
prionitis L.				MAU	
BARTSIA L. Scrophulariaceae					
latifolia (L.) Sibth. & Sm.				CHL	AUS
odontites (L.) Huds. (Syn. *Euphrasia odontites* L.)				USA	

BASELLA L. Basellaceae	S	P	C	X	F
rubra L.				PHI	
BASSIA L. Sapotaceae					
bicornis (Lindl.) F. Muell.			AUS		
birchii F. Muell.		AUS			
hyssopifolia (Pall.) O. Ktze.			ARG	CAN USA	AFG
quinquecuspis F. Muell.			AUS		
tetracuspis C. T. White				AUS	
BATIS L. Batidaceae					
maritima L.				DR	
BAUHINIA L. Caesalpiniaceae *(Leguminosae)*					
cuyabensis Steud.				BRA	
BECKMANNIA Host Poaceae *(Gramineae)*					
syzigachne (Steud.) Fern.		JPN			
BEGONIA L. Begoniaceae					
decandra Pav. *ex* A. Dc.			PR		
BELLARDIA All. Scrophulariaceae					
trixago (L.) All.				CHL GRE	AUS IDO
BELLIS L. Asteraceae *(Compositae)*					
perennis L.			ENG GER IRQ SPA YUG	CHL NZ USA	
BERBERIS L. Berberidaceae					
aquifolium Pursh				USA	
canadensis Mill.				USA	
fendleri A. Gray				USA	
haematocarpa Woot.				USA	
thunbergii Dc.				CAN USA	
trifoliolata Moric.				USA	
vulgaris L.	NZ		AUS CAN	USA	AFG
BERCHEMIA Neck. *ex* Dc. Rhamnaceae					
scandens (Hill) K. Koch				USA	
BERGIA L. Elatinaceae					

BERGIA L. Elatinaceae *(Cont'd)*	S	P	C	X	F
ammannioides Roxb. (Syn. *B. oryzetorum* Fenzl)			PHI	IND	AFG
aquatica Roxb. *(See **B. capensis** L.)*		SPA			
capensis L. (Syn. *B. aquatica* Roxb.) (Syn. *B. verticillata* Willd.)			IND		IDO
oryzetorum Fenzl *(See **B. ammannioides** Roxb.)*					IDO
verticillata Willd. *(See **B. capensis** L.)*				BND	
BERKHEYA Ehrh. **Asteraceae** *(Compositae)*					
rigida (Thunb.) Bol. & Woll.				AUS	
BERTEROA Dc. **Brassicaceae** *(Cruciferae)*					
incana (L.) Dc.			CAN SOV	USA	
BETA L. **Chenopodiaceae**					
vulgaris L.	EGY		IRQ ISR POR	MOR USA	AFG AUS MEX
BETULA L. **Betulaceae**					
alba L. (Syn. *B. pubescens* Ehrh.)					
alleghaniensis Britt.				USA	
lenta L.				USA	
nigra L.				USA	
occidentalis Hook.				USA	
papyrifera Marsh.				USA	
populifolia Marsh.				USA	
pubescens Ehrh. *(See **B. alba** L.)*			FIN		
BIDENS L. **Asteraceae** *(Compositae)*					
aurea (Ait.) Sherff				CHL	
bipinnata L.			IRA SAF TAI USA		AUS
biternata (Lour.) Merr. & Sherff *ex* Sherff			JPN KEN RHO SAF	ANG	
cernua L.				USA	
chinensis Willd. *(See **B. pilosa** L.)*				IDO	

BIDENS L. Asteraceae *(Compositae) (Cont'd)*	S	P	C	X	F
chrysanthemoides Michx.				CHL	
comosa (Gray) Wieg.				USA	
connata Muhl.				USA	
cynapifolia H. B. K.			HAW TRI		
frondosa L.			IRA JPN POR USA	MEX NZ	
humilis H. B. K.				ECU IND	
leucantha Willd. *(See* **B. pilosa** L.)				CHL	
megapotamica Spreng.				URU	
pilosa L. (Syn. *B. chinensis* Willd.) (Syn. *B. leucantha* Willd.) (Syn. *B. subalternans* Dc.)	BRA GHA HAW MEX PLW SAF TNZ ZAM	BOL COL IND KEN MAU MEL MOZ NGI PER PR RHO SWZ TAI TRI UGA VEN	ANT ARG AUS CHN HK IDO JPN PHI PLE	ANG CAM CHL CNK CUB DR ETH FIJ GUI HON IVO JAM LIB MAL MLI NGR NIG NZ PAN SAL SEG THI USA	AFG URU
polylepis Blake			HAW TRI		
radiata Thuill.				SAL	
reptans (L.) G. Don				JAM	
riparia H. B. K.				VEN	
schimperi Sch.-Bip. *ex* Walp.			KEN		
steppia (Steetz) Sherff			KEN		
subalternans Dc. *(See* **B. pilosa** L.)				ARG	URU
tripartita L.			FIN JPN POR	BEL SPA TUR	AFG
vulgata Greene				USA	

BIFORA Hoffm. Apiaceae *(Umbelliferae)*	S	P	C	X	F
radians Bieb.			IRA SPA	GRE SOV	
testiculata Roth			LEB POR	MOR TUR	
BIGNONIA L. Bignoniaceae					
capreolata L.				USA	
exoleta Vell.				BRA	
unguis-cati L.		MAU		IND	
BIOPHYTUM Dc. Oxalidaceae					
petersianum Klotz. *(See* **B. sensitivum** (L.) Dc.)				GHA	
reinwardtii (Zucc.) Klotz.				IDO	
sensitivum (L.) Dc. (Syn. *B. petersianum* Klotz.)				IDO IND PHI	
BISCUTELLA L. Brassicaceae *(Cruciferae)*					
auriculata L. *(See* **Iondraba auriculata** (L.) Schulz)	SPA			MOR	
columnae (Ten.) Halacs. *(See* **B. didyma** L.)				LEB	
didyma L. (Syn. *B. columnae* (Ten.) Halacs.)		TUN			
laevigata L.				SPA	
BLACKSTONIA Huds. Gentianaceae					
serotina G. Beck				TUR	
BLAINVILLEA Cass. Asteraceae *(Compositae)*					
rhomboidea Cass.				BRA IND	
tampicana Hemsl. *(See* **Calyptocarpus vialis** Less.)					
BLECHNUM L. Blechnaceae					
capense (L.) Schlecht.				NZ	
orientale L.				MAL	
BLECHUM P. Br. Acanthaceae					
brownei Juss. *(See* **B. pyramidatum** (Lam.) Urb.)			CUB	ANT JAM TRI	
pyramidatum (Lam.) Urb. (Syn. *B. brownei* Juss.)			PR	DR MIC PAN PLW	

BLEPHARIS Juss. Acanthaceae	S	P	C	X	F
maderaspatensis (L.) Roth		GHA			
molluginifolia Pers.				IND	
BLIGHIA Kon. Sapindaceae					
sapida Kon.				JAM	
BLUMEA Dc. Asteraceae *(Compositae)*					
aurita Dc. *(See* **Laggera aurita** Sch.-Bip. *ex* C. B. Clarke)				GHA IVO THI	
balsamifera (L.) Dc.				BOR IDO PHI	
lacera (Burm. *f.*) Dc.			MAU TAI	IND PK	
laciniata (Roxb.) Dc.			HAW	PHI	IDO
sinuata (Lour.) Merr.			PHI		
wightiana Dc.				IND	
BLYXA Noronh. *ex* Thou. **Hydrocharitaceae**					
auberti Rich.			JPN		
ceratosperma Maxim.			JPN		
echinosperma (Clarke) Hook. *f.*			JPN TAI THI		
japonica Maxim. *ex* Archers & Gurcke	THI		JPN		
shimadai Hay.			TAI		
BOEHMERIA Jacq. Urticaceae					
frutescens Thunb. *(See* **Villebrunea frutescens** Bl.)			TAI		
nivea (L.) Gaudich.			HAW JPN		CAB
platyphylla D. Don (Syn. *B. spicata* Thunb.)					
spicata Thunb. *(See* **B. platyphylla** D. Don)			JPN		
tricuspis Makino			JPN		
BOERHAVIA L. Nyctaginaceae					
adscendens Willd. *(See* **B. diffusa** L.)			KEN		
caribaea Jacq. *(See* **B. diffusa** L.)				PER	
coccinea Mill. *(See* **B. diffusa** L.)			KEN	BRA COL DR GHA	PR

BOERHAVIA L. Nyctaginaceae *(Cont'd)*	S	P	C	X	F
coccinea Mill. *(Cont'd)*				USA	
decumbens Vahl *(See* **B. paniculata** Rich.)			COL		
diffusa L. (Syn. *B. adscendens* Willd.) (Syn. *B. caribaea* Jacq.) (Syn. *B. coccinea* Mill.) (Syn. *B. repens* L.)	COL ECU GHA PHI PK TRI	IDO IND PER THI TNZ UGA	HAW HON KEN MAU NIG PR RHO SAF SEG ZAM	AFG ANT AUS BOL BOT BRA CAB CHN DAH DR EGY ETH FIJ GUA IVO JAM MAL MEL MEX MIC MOZ NEP NGI PLW SAL SUD VEN VIE	
erecta L.	HON KEN MEX	COL GUA IDO PER VEN	MAU PR SAF SAL UGA	ANT ECU GHA JAM MEL NGI NIC NIG PLW RHO THI TNZ USA ZAM	
paniculata Rich. (Syn. *B. decumbens* Vahl)					
repanda Willd.				IND	
repens L. *(See* **B. diffusa** L.)				IND PK SUD USA	IDO ISR
scandens L.				JAM	
BOLBOSCHOENUS Palla **Cyperaceae**					
maritimus Asch. & Godr. *(See* **Scirpus maritimus** L.)	HUN				
BONGARDIA C. A. Mey. **Berberidaceae**					
chrysogonum (L.) Boiss.		JOR	IRA		AFG

BONGARDIA C. A. Mey. Berberidaceae *(Cont'd)*	S	P	C	X	F
chrysogonum (L.) Boiss. *(Cont'd)*			LEB		ISR
BONNAYA Link & Otto **Scrophulariaceae**					
brachiata Link & Otto *(See* **Lindernia ciliata** (Colsm.) Pennell)		IND	TAI	BND	
multiflora Bonati				CAB	
oppositifolia Spreng.				IND	
veronicaefolia Spreng.				VIE	
BORAGO L. **Boraginaceae**					
officinalis L.				SPA	
BOREAVA Jaub. & Spach **Brassicaceae** *(Cruciferae)*					
orientalis Jaub. & Spach				TUR	
BORRERIA G. F. W. Mey. **Rubiaceae**					
alata (Aubl.) Dc. (Syn. *B. latifolia* (Aubl.) Schum.) (Syn. *Spermacoce latifolia* Aubl.)		BOR IDO	SEG	ANT BRA CAB CEY CR GHA GUI IVO LIB MAL MEX NIG PLW SUR THI VIE	
articularis (L. *f.*) F. N. Williams (Syn. *B. hispida* (L.) Schum.) (Syn. *B. scabra* Schum.) (Syn. *Spermacoce hispida* L. *f.*)				PHI	
bartlingiana Dc.				TRI	
centranthoides Cham. & Schl.				BRA	
hispida (L.) Schum. *(See* **B. articularis** (L.f.) F. N. Williams)	IND		TAI	PHI	IDO
laevis (Lam.) Griseb. (Syn. *Spermacoce laevis* Lam.)	HAW		IDO JAM PLW SAL TRI	FIJ GUA NGI PER PHI	PR
latifolia (Aubl.) Schum. *(See* **B. alata** (Aubl.) Dc.)	MAL	IDO	GHA TRI	IVO	
ocymoides (Burm.f.) Dc.		MEX	IDO PHI PR TRI	CEY DR IND PER PLW	
poaya Dc.			BRA		

	S	P	C	X	F
BORRERIA G. F. W. Mey. **Rubiaceae** *(Cont'd)*					
princeae K. Schum.		KEN			
radiata Dc.				GHA IVO	
ramisparsa Dc.				CNK	
repens Dc.			IDO		PK
ruelliae Schum.				GHA	
scabra Schum. (*See* **B. articularis** (L. *f.*) F. N. Williams)	GHA		RHO SAF	IVO	
stachydea (Dc.) Hutch. & Dalz.				IVO	
stricta (L.f.) G. F. W. Mey.			KEN		
verticillata (L.) G. F. W. Mey.	TRI		JAM PR	BRA IVO MEX VEN	FIJ
BOTHRIOCHLOA O. Ktze. **Poaceae** *(Gramineae)*					
ambigua S. T. Blake			AUS		HAW
decipiens (Hack.) C. E. Hubb.			AUS		
intermedia (R. Br.) A. Camus (Syn. *Andropogon intermedius* R. Br.)				PLW	
ischaemum (L.) Keng (Syn. *Andropogon ischaemum* L.)					
pertusa (L.) A. Camus (Syn. *Amphilophis pertusa* (L.) Nash *ex* Stapf) (Syn. *Andropogon pertusus* (L.) Willd.)			HAW		
BOTHRIOSPERMUM Bunge **Boraginaceae**					
tenellum (Hornem.) Fisch. & Mey.		MAU	JPN TAI		
BOUCHEA Cham. **Verbenaceae**					
ehrenbergii Cham. (Syn. *B. prismatica* (L.) O. Ktze.)					
prismatica (L.) O. Ktze. (*See* **B.** enbergii Cham.)			COL DR		
BOURRERIA P. Br. **Boraginaceae**					
acimoides (Burm.) Dc.	MEX				
andreuxii Hemsl.				MEX	
BOUSSINGAULTIA H. B. K. **Basellaceae**					
baselloides H. B. K.	SAF				AUS HAW KEN
cordifolia Ten.				NZ	
gracilis Miers			HAW	USA	AUS

BOUTELOUA Lagasca **Poaceae** *(Gramineae)*	S	P	C	X	F
aristidoides (H. B. K.) Griseb.				USA	
barbata Lag.				USA	
pilosa Benth. *ex* S. Wats.				NIC	
BOWLESIA Ruiz & Pav. **Apiaceae** *(Umbelliferae)*					
incana Ruiz & Pav.				USA	
tenera Spreng.			ARG	BRA	
BRACHIARIA Griseb. **Poaceae** *(Gramineae)*					
brizantha (Hochst.) Stapf				TRI	
ciliatissima (Buckl.) Chase				USA	
deflexa (Schum.) C. E. Hubb.	MOZ	GHA		IVO NIG TNZ	IND
distachya (L.) Stapf			IDO PHI TAI TRI	IND VIE	BUR CEY MAL
eruciformis (J. E. Sm.) Griseb. (Syn. *B. isachne* (Roth) Stapf) (Syn. *Panicum eruciforme* J. E. Sm.) (Syn. *P. isachne* Roth)	MOZ SUD	SAF TNZ	ANT EGY LEB MAU MOR	FIJ IND	AFG ISR
extensa A. Chase			CUB		
isachne (Roth) Stapf *(See.* **B. eruciformis** (J. E. Sm.) Griseb.)	MOZ				
lata (Schumach.) C. E. Hubb.		NIG		GHA IVO SUD	IND
milliiformis (Presl) A. Chase	AUS		MAL	BRA FIJ	CEY IND
mutica (Forsk.) Stapf (Syn. *B. purpurascens* (Raddi) Henr.) (Syn. *Panicum barbinode* Trin.) (Syn. *P. muticum* Forsk.) (Syn. *P. purpurascens* Raddi)	AUS FIJ THI	CEY COL HAW JAM MAL PER PHI PR TRI	BOR MAU	ANG ANT ARG BRA CAB CR CUB DR ECU MEX NZ PK PLW SUR TAI USA VEN VIE	IND ISR
paspaloides (Presl) C. E. Hubb.	BOR MAL	GRE	MAU	FIJ IDO PLW THI TRI	BUR CEY HAW IND

	S	P	C	X	F
BRACHIARIA Griseb. **Poaceae** *(Gramineae) (Cont'd)*					
paspaloides (Presl) C. E. Hubb. *(Cont'd)*				USA	
piligera D. K. Hughes		AUS		USA	
plantaginea (Link) Hitchc.	BRA HON	GUA MEX USA	HAW	CR FRA MEL NIC PR	
platyphylla (Griseb.) Nash			ARG TRI USA		
purpurascens (Raddi) Henr. *(See* **B. mutica** (Forsk.) Stapf)		THI TRI		SUR	CEY COL
ramosa (L.) Stapf (Syn. *Panicum ramosum* L.)				AUS MAU	
reptans (L.) Gard. & C. E. Hubb. (Syn. *Panicum reptans* L.)	NGI	MAD THI	MAU TAI	PHI PK PLW	AFG BOR HAW IND MAL
subquadripara (Trin.) Hitchc.		CUB	MAU	AUS PLW	
BRACHYACHNE Stapf **Poaceae** *(Gramineae)*					
convergens (F. Muell.) Stapf		AUS			
BRACHYAPIUM (Baill.) Maire **Apiaceae** *(Umbelliferae)*					
involucratum Maire				MOR	
BRACHYPODIUM Beauv. **Poaceae** *(Gramineae)*					
pinnatum (L.) Beauv.			SPA		
BRACHYSTEGIA Benth. **Caesalpiniaceae** *(Leguminosae)*					
spicaeformis Benth.		RHO			
BRASENIA Schreb. **Nymphaeaceae**					
schreberi Gmel.				USA	
BRASSAIA Endl. **Araliaceae**					
actinophylla F. Muell.			HAW		
BRASSICA L. **Brassicaceae** *(Cruciferae)*					
adpressa Boiss.				ARG	
alba Boiss.			IRA	URU	
armoracioides Czern. *ex* Turcz.			SOV		
arvensis (L) O. Ktze. *(See* **B. kaber** (Dc.) L. C. Wheeler *var.* pinnatifida (Stokes) L. C. Wheeler	ALK	COL	IRA JPN LEB	IND	
campestris L.	ARG CAN GUA HON	BRA CHL COL DEN	EGY FIN GER HAW	CAB CHN CR ENG	AUS KOR

BRASSICA L. Brassicaceae *(Cruciferae) (Cont'd)*	S	P	C	X	F
campestris L. *(Cont'd)*	MEX VEN	FRA ITA KEN NZ PER PR	IRA JPN POL POR SOV TNZ UGA	ICE IND IRQ NEP NOR PK SAL SWE THI TUR URU USA	
hirta Moench				PER USA	
incana (L.) F. W. Schultz				USA	
juncea (L.) Coss.		CAN	ARG AUS	FIJ MEX USA	CAB
kaber (Dc.) L. C. Wheeler *(See* **B. kaber** (Dc.) L. C. Wheeler *var.* pinnatifida (Stokes) L. C. Wheeler		CAN	LEB USA		AUS
kaber (Dc.) L. C. Wheeler *var.* pinnatifida (Stokes) L. C. Wheeler (Syn. *B. arvensis* (L.) O. Ktze.) (Syn. *B. kaber* (Dc.) L. C. Wheeler) (Syn. *Sinapis arvensis* L.)					
napus L.			FIN KEN	ARG BRA ECU NZ TUR	AUS
nigra (L.) Koch			ARG CAN EGY ENG HAW IRA ISR POR SOV SPA USA	BRA CHL GRE ITA MOR NZ VEN	AFG
rapa L.		BRA	ENG FIN KEN	SOV USA	
rugosa Prain				ARG	
sinapistrum Boiss.				SOV	
tournefortii Gouan		AUS	EGY	NZ	ISR
BRAYULINEA Small **Amaranthaceae**					
densa (Humb. & Bonpl.) Small			SAF	USA	
BREEA Less. **Asteraceae** *(Compositae)*					
setosum (Bieb.) Kitam.			JPN		

BRILLANTAISIA Beauv. Acanthaceae	S	P	C	X	F
lamium Benth.			GHA		
nitens Lindau			GHA		
BRIZA L. Poaceae *(Gramineae)*					
maxima L.			POR SPA	ARG AUS CHL NZ	BRA HAW IND SAF
media L.			SPA	ECU USA	IND
minor L.			JPN MAU POR	ARG AUS BRA CHL NZ URU USA	CEY HAW IND ISR SAF
BROMELIA L. Bromeliaceae					
antiacantha Bertol. (*See* **B. fastuosa** Lindl.)			BRA		
bicolor Ruiz & Pav. (*See* **Rhodostachys bicolor** Benth. & Hook.)			PER		
fastuosa Lindl. (Syn. *B. antiacantha* Bertol.)					
pinguin L.			PR	DR JAM	
BROMUS L. Poaceae *(Gramineae)*					
adoensis Hochst. *ex* Steud.			KEN		
alopecuros Poir.			MOR		
arenarius Labill.			USA		
arvensis L.			SOV	USA	
brevis Steud.			ARG		
brizaeformis Fisch. & Mey.			USA		
catharticus Vahl (Syn. *B. unioloides* H. B. K.)		COL	EGY JPN	ARG AUS PER URU	HAW SAF
commutatus Schrad. (*See* **B. racemosus** L.)			AFG USA		ARG HAW SAF
danthoniae (Desf.) Trin.			IRA		AFG ISR
diandrus Roth. (Syn. *B. gussonii* Parl.)			AUS SAF	NZ	HAW
gussonii Parl. (*See* **B. diandrus** Roth.)			POR		AUS SAF

BROMUS L. Poaceae *(Gramineae) (Cont'd)*	S	P	C	X	F
hordeaceus L. *(See* **B. racemosus** L.)			MAL POR	CHL MOR	AUS HAW POL
inermis Leyss.				CAN USA	
japonicus Thunb.			JPN USA	CAN IRA PK	AFG IND ISR SAF
madritensis L.	AUS		MOR POR	CHL IRA	AFG ISR
mollis L.	AUS		ARG MOR SOV SPA	CHL NZ USA	HAW IND ISR POL
racemosus L. (Syn. *B. commutatus* Schrad.) (Syn. *B. hordeaceus* L.)				USA	
rigidus Roth (Syn. *B. villosus* Forsk.)			ARG HAW MOR POR	AUS USA	
rubens L.			AUS	GRE JOR MOR USA	HAW ISR
secalinus L.			ARG AUS CAN FIN GER SPA USA	CHL COL SOV	POL
squarrosus L.			SOV		
sterilis L.			SPA	GRE NZ USA	AUS HAW ISR POL
tectorum L.		TUR	CAN LEB MOR SOV USA	IRQ JOR	AFG AUS HAW ISR POL
unioloides H. B. K. *(See* **B. catharticus** Vahl)			ARG JPN	AUS CHL NZ PER	BRA HAW IND
villosus Forsk. *(See* **B. rigidus** Roth)				GRE	AUS HAW
willdenowii Kunth			SAF	USA	
BROUSSONETIA L'herit. *ex*. Ventenat **Moraceae**					
papyrifera (L.) Vent.				USA	

	S	P	C	X	F
BROWALLIA L. **Solanaceae**					
americana L. (Syn. *B. demissa* L.)				TRI	
demissa L. *(See* **B. americana** L.)				TRI	
BRUNNICHIA Banks & Gaertn. **Polygonaceae**					
cirrhosa Gaertn.			USA		
BRYONOPSIS Arn. **Cucurbitaceae**					
laciniosa (L.) Naud.				AUS	
BRYOPHYLLUM Salisb. **Crassulaceae** *(See* **KALANCHOE** Adans.)					
pinnatum (Lam.) Kurz *(See* **Kalanchoe pinnata** (Lam.) Pers.)			HAW PR	DR FIJ JAM MAU	
tubiflorum Harvey *(See* **Kalanchoe verticillata** Elliot)			HAW	AUS	
BUCHLOE Engelm. **Poaceae** *(Gramineae)*					
dactyloides (Nutt.) Engelm.			USA		
BUCHNERA L. **Scrophulariaceae**					
ternifolia H. B. K.				AUS	
urticifolia R. Br.				AUS	
BUDDLEJA L. **Buddlejaceae**					
asiatica Lour.			HAW	PHI	
BULBOSTYLIS Kunth **Cyperaceae** *(Syn.STENOPHYLLUS* Rafin.)					
barbata (Rottb.) C. B. Clarke (Syn. *Fimbristylis barbata* (Rottb.) Benth.) (Syn. *Stenophyllus barbatus* (Rottb.) Cooke)	NIG TAI	JPN	GHA IND PHI VIE		
filamentosa C. B. Clarke			GHA IVO		
puberula (Poir.) C. B. Clarke	CEY				
BUMELIA Sw. **Sapotaceae**					
lanuginosa (Michx.) Pers.			USA		
lycioides (L.) Gaertn.			USA		
BUNIAS L. **Brassicaceae** *(Cruciferae)*					
erucago L.			POR SPA		POL
orientalis L.			SOV		
BUPLEURUM L. **Apiaceae** *(Umbelliferae)*					
falcatum L.				TUR	

BUPLEURUM L. Apiaceae *(Umbelliferae) (Cont'd)*	S	P	C	X	F
gerardi Jacq.			IRA	TUR	ISR
lancifolium Hornem. *(See* **B. protractum** Hoffm. & Link)			ISR MOR POR		TUN
protractum Hoffm. & Link (Syn. *B. lancifolium* Hornem.) (Syn. *B. subovatum* Link *ex* Spreng.)					
rotundifolium L.			IRA SPA	GRE NZ TUR	AUS
semicompositum L.				MOR	AUS ISR
subovatum Link *ex* Spreng. *(See* **B. protractum** Hoffm. & Link)			ISR		LEB
tenuissimum L.				NZ	
BURSARIA Cav. **Pittosporaceae**					
spinosa Cav.			AUS		
BUTOMUS L. **Butomaceae**					
umbellatus L.			IRA IRQ POR	BEL FRA GER TUR	AFG ISR
CABOMBA Aubl. **Nymphaeaceae**					
caroliniana A. Gray				USA	
CACABUS Bernh. **Solanaceae**					
prostratus Bernh.				PER	
CACALIA L. **Asteraceae** *(Compositae)*					
tuberosa Nutt.				USA	
CACHRYS L. **Apiaceae** *(Umbelliferae)*					
eriantha Dc.			IRA		
CAESALPINIA L. **Caesalpiniaceae** *(Leguminosae)*					
bonduc (L.) Roxb. (Syn. *C. bonducella* Flem.)			HAW	JAM USA	
bonducella Flem. *(See* **C. bonduc** (L.) Roxb.)				JAM	
coriaria Willd.				JAM	
crista L.			HAW		
decapetala (Roth) Alston (Syn. *C. sepiaria* Roxb.)			KEN RHO SAF	AUS	
gilliesii (Hook.) Wall.				ARG USA	ISR

	S	P	C	X	F
CAESALPINIA L. **Caesalpiniaceae** *(Leguminosae) (Cont'd)*					
sepiaria Roxb. *(See* **C. decapetala** (Roth) Alston)		HAW		USA	KEN SAF
CAJANUS Dc. **Papilionaceae** *(Leguminosae)*					
cajan (L.) Huth			IDO	COL JAM PHI VEN	AUS
CAKILE L. **Brassicaceae** *(Cruciferae)*					
maritima Scop.			SPA		
CALADIUM Vent. **Araceae**					
arboreum H. B. K.				COL	
bicolor (Ait.) Vent. (Syn. *Cyrtospadix bicolor* Britton & Wils.)			TRI	ANT SUR	HAW
esculenta Vent. *(See* **Colocasia esculenta** (L.) Schott)				PR	
CALAMAGROSTIS Adans. **Poaceae** *(Gramineae)*					
arundinacea (L.) Roth			FIN JPN		
canadensis (Michx.) Nutt.				USA	HAW
epigeios Roth			FIN		
montevidensis Nees			ARG		
neglecta Gaertn.			FIN		
CALAMINTHA Lam. **Lamiaceae** *(Labiatae)*					
acinos Clairv.			SOV		
CALEA L. **Asteraceae** *(Compositae)*					
jamaicensis L.			JAM		
CALENDULA L. **Asteraceae** *(Compositae)*					
aegyptiaca Desf.			JOR		
algeriensis Boiss. & Reut.			MOR		
arvensis L. (Syn. *C. micrantha* Tin. & Guss.)	TUN		SPA	AUS GRE JOR MOR POR TUR	ISR
micrantha Tin. & Guss. *(See* **C. arvensis** L.)			EGY		
CALEPINA Adans. **Brassicaceae** *(Cruciferae)*					
corvini Desv. (Syn. *C. irregularis* Thell.)			ITA		
irregularis Thell. *(See* **C. corvini** Desv.)			POR		

CALLICARPA L. Verbenaceae	S	P	C	X	F
americana L.				USA	
CALLITRICHE L. **Callitrichaceae**					
fallax Petrov.		JPN			
japonica Engelm. *ex* Hegelm.			JPN		
palustris L.				BEL	
platycarpa Kuetz.				BEL NET	
stagnalis Scop.				AUS BEL ENG	
verna L.		JPN		USA	
CALLUNA Salisb. **Ericaceae**					
vulgaris Salisb.			FIN	NZ	
CALONYCTION Choisy **Convolvulaceae** *(See* **IPOMOEA** L.*)*					
aculeatum (L.) House *(See* **Ipomoea alba** L.*)*		PR		HON	AUS USA
muricatum G. Don *(See* **Ipomoea muricata** (L.) Jacq.*)*				USA	
CALOPOGONIUM Desv. **Papilionaceae** *(Leguminosae)*					
mucunoides Desv.	PHI	MAL	IDO	COL IND NGI NIG PLW	AUS
CALOTIS R. Br. **Asteraceae** *(Compositae)*					
lappulacea Benth.				AUS	
CALOTROPIS R. Br. **Asclepiadaceae**					
gigantea (Willd.) Dryand. *ex* W. T. Ait.				IND	CAB
procera (Willd.) Dryand. *ex* W. T. Ait.		PR		AUS DR IND PK SUD VEN	AFG ISR USA
CALTHA L. **Ranunculaceae**					
palustris L.				AUS FIN GER	BEL USA	AFG
CALYCERA Cav. **Calyceraceae**					
leucanthema (Poepp. *ex* Less.) O. Ktze.				CHL	
CALYCOCARPUM Nutt. *ex* Torr. & Gray **Menispermaceae**					
lyonii (Pursh) Gray				USA	

CALYPTOCARPUS Less. Asteraceae *(Compositae)*	S	P	C	X	F
vialis Less. (Syn. *Blainvillea tampicana* Hemsl.)		MEX			
CALYSTEGIA R. Br. **Convolvulaceae**					
hederacea Wall.			CHN JPN		
japonica Choisy		JPN			
sepium (L.) R. Br. (Syn. *Convolvulus sepium* L.)			JPN POR SPA	BEL CHL NZ	AFG
CAMELINA Crantz **Brassicaceae** *(Cruciferae)*					
alyssum Thell.				ARG CHL	
dentata Pers.			CAN	USA	ARG
glabrata Fritsch			SOV		
linicola Sch. & Sp.			SOV		
microcarpa Andrz.			CAN USA	SOV TUR	
parodii Ibar. & La Porte			ARG CAN		
pilosa (Dc.) Zinger			SOV		
rumelica Vel.			IRA		AFG
sativa (L.) Crantz			ARG CAN JPN POR	SOV TUR USA	AFG AUS POL
CAMPANULA L. **Campanulaceae**					
glomerata L.			FIN	GER	
patula L.			FIN GER SOV		
rapunculoides L.			GER	NZ USA	POL POR SOV
rotundifolia L.			FIN GER		
strigosa Soland.			LEB		
CAMPSIS Lour. **Bignoniaceae**					
radicans (L.) Seem.			USA		
CAMPTOSEMA Hook. & Arn. **Papilionaceae** *(Leguminosae)*					
grandiflorum Benth.			BRA		
CAMPYLONEURUM Presl **Polypodiaceae**					
phyllitidis (L.) Presl (Syn. *Polypodium phyllitidis* L.)					

CANAVALIA Dc. Papilionaceae (Leguminosae)	S	P	C	X	F
cathartica Thou. (Syn. C. microcarpa (Dc.) Piper)				HAW	PLW	
ensiformis (L.) Dc.			HAW PR	HON PER	AUS IDO RHO
maritima Thou.				PR TRI	DR PLW	
microcarpa (Dc.) Piper (See C. cathartica Thou.)				HAW	
sericea A. Gray				PLW	
CANELLA P. Br. Canellaceae					
alba Murr. (Syn. C. winterana Gaertn.)					
winterana Gaertn. (See C. alba Murr.)				JAM	
CANNA L. Cannaceae					
coccinea Mill.			PR	BRA COL DR MEX TRI	IDO
edulis Ker-Gawl.				ANT TRI	
glauca L.			PR	SUR	IDO
hybrida Hort. ex Back.				IDO	
indica L.	PLW	ITA	GHA HAW PHI	ARG AUS BOR CHN CNK COL HON JOR JPN MEL NEP NZ PR RHO SAL SEG THI UGA USA	CAB HK
CANNABIS L. Cannabaceae					
ruderalis Janisch.				SOV	
sativa L.	AFG	AUS CHN HK PK PR	HUN USA	BOT CAB CAN CNK IND IRA IRQ	

	S	P	C	X	F
CANNABIS L. **Cannabaceae** (*Cont'd*)					
sativa L. (*Cont'd*)				JAM	
				JOR	
				JPN	
				KOR	
				NEP	
				NZ	
				RHO	
				SAF	
				SEG	
				SOV	
				THI	
				TUR	
CANOTIA Torr. **Koeberliniaceae**					
holacantha Torr.				USA	
CANSCORA Lam. **Gentianaceae**					
decussata Schult.			GHA		
CANTHAROSPERMUM Wight & Arn. **Papilionaceae** (*Leguminosae*)					
scarabaeoides Baill. (*See* **Atylosia scarabaeoides** (L.) Benth.)					IDO
CANTHIUM Lam. **Rubiaceae**					
hispidium Benth.			TNZ		
CAPERONIA St. Hil. **Euphorbiaceae**					
castanaefolia (L.) St. Hil.				USA	
palustris (L.) St. Hil.			TRI	ANT	
				COL	
				SAL	
				SUR	
serrata Presl				NIG	
CAPILLIPEDIUM Stapf **Poaceae** (*Gramineae*)					
filiculme (Hook. *f.*) Stapf (Syn. *Andropogon filiculmis* Hook. *f.*)					
parviflorum (R. Br.) Stapf (Syn. *Andropogon serratus* Thunb. *non* Miq.)					
CAPNOPHYLLUM Gaertn. **Apiaceae** (*Umbelliferae*)					
peregrinum (L.) Lange			MOR		ISR
			POR		
CAPPARIS L. **Capparidaceae**					
aphylla Roth				IND	
				PK	
erythrocarpos Isert				GHA	
sandwichiana Dc.				USA	
spinosa L.			IRA	PK	AFG
			IRQ	TUR	ISR
tomentosa Lam.				GHA	

CAPRARIA L. Scrophulariaceae	S	P	C	X	F
biflora L.			PR TRI	GHA PAN SAL	
CAPSELLA Medic. **Brassicaceae** *(Cruciferae)*					
bursa-pastoris (L.) Medic.	ALK IRE NEP SPA	BEL CAN COL FIN GRE JPN NOR NZ TAI TUN YUG	ARG AUS CHN EGY ENG GER HAW HK IRA IRQ KEN LEB MOR POR SAF SOV USA	BRA BUL CHL CR FRA HUN ICE ITA JOR KOR PK TUR URU VEN	AFG ETH ISR POL TNZ UGA
rubella Reut.				POR	
CAPSICUM L. **Solanaceae**					
annuum L.				AUS IND	IDO
baccatum L.				TRI	
frutescens L.				IND JAM PLW TRI	
CARAGANA Lam. **Papilionaceae** *(Leguminosae)*					
arborescens Lam.				USA	
CARBENIA Adans. **Asteraceae** *(Compositae)*					
benedicta (L.) Adans. *(See* **Cnicus benedictus** L.)				TUR	
CARDAMINE L. **Brassicaceae** *(Cruciferae)*					
debilis Banks *ex* Dc.				NZ	
flexuosa With.		JPN	TAI	NZ	KOR
hirsuta L.			CHN ENG POR	CHL ITA NZ USA	ISR
lyrata Bunge			JPN		
parviflora L.			TAI		
pensylvanica Muhl.				USA	
pratensis L.			GER SPA	CHN NZ USA	
sarmentosa Forst.f.				FIJ	

CARDAMINE L. Brassicaceae *(Cruciferae) (Cont'd)*	S	P	C	X	F
scutata Thunb.			JPN		
CARDANTHERA Buch.-Ham. *ex* Nees **Acanthaceae**					
difformis (L. *f.*) Druce	IND				
uliginosa Buch.-Ham.				IND	
CARDARIA Desv. **Brassicaceae** *(Cruciferae)*					
chalepensis (L.) Hand.-Mazz.			CAN		
draba (L.) Desv. (Syn. *Lepidium draba* L.)	AFG AUS HUN ITA SOV USA	GRE IRA JOR YUG	CAN ENG GER IRQ LEB POR SAF TUN TUR	ARG BEL CHL CZE GUA ISR NET NZ PK RHO TAS	
pubescens (C. A. Mey.) Rollins			CAN	USA	
CARDIOSPERMUM L. **Sapindaceae**					
grandiflorum Sw.				GHA NIG	
halicacabum L. (Syn. *C. microcarpum* H. B. K.)		AUS	HAW IND TAI	GHA MAU MOZ USA VIE	IDO
microcarpum H. B. K. *(See* **C. halicacabum** L.)				TRI	
CARDOPATIUM Juss. **Asteraceae** *(Compositae)*					
corymbosum Pers.				TUR	
CARDUUS L. **Asteraceae** *(Compositae)*					
acanthoides L.		ARG	CAN ENG SOV	NZ USA	
argentatus L.			LEB		ISR
arvensis Robs.			MOR		
crispus L.		TUN	FIN JPN SOV	NZ USA	AUS POL
hamulosus Ehrh.			SOV		
kikuyorum R. E. Fries			KEN		
lanatus Brot.		AUS			
macrocephalus Desf.				SOV	
marianus L. *(See* **Silybum marianum** (L.) Gaertn.)					AUS

CARDUUS L. Asteraceae *(Compositae) (Cont'd)*	S	P	C	X	F
meonanthus Hoffmgg. & Link				MOR	
myriacanthus Salzm.				MOR	
nutans L. (Syn. *C. thoermeri* Weinm.)	USA	ARG AUS HUN NZ	AST CAN ENG GER ITA POL SOV TUR	FRA IRA PK TAS URU YUG	
pteracanthus Dur.				MOR	
pycnocephalus L.	AFG	AUS BRA GRE IRA NZ	IRQ ITA JOR LEB MOR SAF	ARG CHL EGY TUR URU USA	ISR TAS
tenuiflorus Curt.		AUS	ARG	NZ USA	POR TAS
theodori R. E. Fries			KEN SOV		
thoermeri Weinm. *(See* **C. nutans** L.*)*				AUS SOV	
uncinatus Bieb.			SOV		
CAREX L. Cyperaceae					
acuta L. (Syn. *C. gracilis* Curt.)				BEL	
aquatilis Wahlenb.				TUR USA	
atherodes Spreng.				USA	
baccans Nees				IDO	
bonariensis Desf.			ARG		BRA
brongniartii Kunth				CHL	
canescens L.			FIN		
dietrichiae Boeck.				FIJ	
disticha Huds.				BEL GER	
divisa Huds.				TUR	
divulsa Good.			ARG	NZ	
eurycarpa Holm.				USA	
glauca Scop.				TUR	
gracilis Curt. *(See* **C. acuta** L.*)*			GER		
graeffeana Boeck.				FIJ	

CAREX L. Cyperaceae *(Cont'd)*	S	P	C	X	F
hudsonii A. Benn.				BEL	
lacustris Willd.				USA	
lasiocarpa Ehrh.				USA	
leporina L.			FIN		
longebrachiata Steud. (*See* **C. myosurus** Nees)	NZ		AUS		
lupulina Muhl.				USA	
maximowiczii Miq. (*See* **C. picta** Boott)			JPN		
myosurus Nees (Syn. *C. longebrachiata* Steud.)					
nebraskensis Dewey			USA		
nigra Siev.			FIN	BEL ENG	
nubigena D. Don				NEP	
paniculata L.			POR		
picta Boott (Syn. *C. maximowiczii* Miq.)					
pumila Thunb.			TAI	NZ	
remota L.				TUR	
riparia Curt.			POR	BEL ENG	
senta Boott				USA	
vulpina L.			GER		
CAREYA Roxb. **Barringtoniaceae**					
arborea Roxb.				CEY	
CARISSA L. **Apocynaceae**					
edulis Vahl				KEN RHO	
CARLINA L. **Asteraceae** *(Compositae)*					
involucrata Poir.			LEB		ISR
lanata L.				MOR	
racemosa L.			MOR		ISR POR
CARPESIUM L. **Asteraceae** *(Compositae)*					
divaricatum Sieb. & Zucc.				JPN	
CARPINUS L. **Betulaceae**					
americana Michx. (Syn. *C. caroliniana* Walt.)					

CARPINUS L. Betulaceae *(Cont'd)*	S	P	C	X	F
caroliniana Walt. *(See* **C. americana** Michx.*)*				USA	
CARTHAMUS L. Asteraceae *(Compositae)*					
caeruleus L.				MOR	POR
calvus Batt				MOR	
flavescens Willd.		LEB			
glaucus Bieb.			IRA		AUS ISR
lanatus L.	AUS		ARG MOR	CHL GRE NZ TUR URU	POR
oxyacantha Bieb.	IND PK		IRA IRQ		
syriacus (Boiss.) Dinsm.			LEB		ISR
tenuis (Boiss. & Blanche) Bornm.			ISR		
tinctorius L.				TUR	AUS POR
CARUM L. Apiaceae *(Umbelliferae)*					
carvi L.			CAN FIN	NZ USA	POL SOV
CARYA Nutt. Juglandaceae					
aquatica (Michx. *f.*) Nutt.				USA	
cordiformis (Wangenh.) C. Koch				USA	
floridana Sarg.				USA	
glabra (Mill.) Sweet				USA	
illinoensis (Wangenh.) C. Koch				USA	
laciniosa (Michx. *f.*) Loud.				USA	
leiodermis Sarg.				USA	
ovata (Mill.) C. Koch				USA	
pallida (Ashe) Engl. & Graebn.				USA	
texana C. Dc.				USA	
tomentosa Nutt.				USA	
x lecontei Little				USA	
CARYOCAR L. Caryocaraceae					
brasiliense St. Hil.				BRA	
CASSIA L. Caesalpiniaceae *(Leguminosae)* *(Syn.HERPETICA Rafin.)*					
absus L.			RHO	GHA	

CASSIA L. Caesalpiniaceae *(Leguminosae) (Cont'd)*	S	P	C	X	F
alata L. (Syn. *Herpetica alata* Rafin.)				GHA NGI SAL	CAB IDO PR
auriculata L.		IND			
barclayana Sweet *(See* **C. sophera** L.*)*				AUS	
bauhinioides A. Gray				USA	
bicapsularis L.			HAW		
brewsterii F. Muell.		AUS			
eremophila A. Cunn. *ex* Vog.		AUS			
fasciculata Michx.				USA	
fistula L.				JAM	
floribunda Cav. (Syn. *C. laevigata* Willd.)			AUS		
glauca Lam. *(See* **C. surattensis** Burm. *f.)*			HAW		
hirsuta L.				IND	
laevigata Willd. *(See* **C. floribunda** Cav.*)*			HAW		
leiophylla Vogel				PER	
leptocarpa Benth.				USA	
lechenaultiana Dc.			HAW	PLW	
ligustrina L.				JAM	
marylandica L.				USA	
mimosoides L.		GHA	RHO	IVO MAU VIE	HAW
newtonii Mend. & Torre				ANG	
nictitans L.				USA	
obovata Collad. *(See* **C. obtusa** Roxb.*)*				JAM PER PK VIE	
obtusa Roxb. (Syn. *C. obovata* Collad.)					
obtusifolia L. *(See* **C. tora** L.*)*				IDO IND MAL USA	
occidentalis L. (Syn. *Ditremexa occidentalis* (L.) Britton & Rose)	COL PLW	IDO KOR	ARG AUS CUB HAW IND NIG	ANT BOL BOR BOT BRA CAB	CHN HK

CASSIA L. Caesalpiniaceae *(Leguminosae)* *(Cont'd)*	S	P	C	X	F
occidentalis L. *(Cont'd)*				CNK CR DAH ECU GHA GUA HON IRA JAM JPN KEN MAL MAU MEL MEX MIC NGI NIC PER PHI PK PR RHO SAL SEG SUD SUR THI TNZ TRI UGA USA VEN VIE	
pleurocarpa F. Muell.				AUS	
senna L.				SUD	
sophera L. (Syn. *C. barclayana* Sweet)			AUS		
surattensis Burm. *f.* (Syn. *C. glauca* Lam.)					
tomentosa L. *f.*				PER	
tora L. (Syn. *Emelista tora* (L.) Britton & Rose) (Syn. *C. obtusifolia* L.)	COL ECU FIJ MEL PLW	CUB IND USA	AUS CAB HON IDO MIC PR TAI	BND BOL BRA CHN CNK DAH GHA GUA GUI JPN LIB MAL MAU MEX MLI NGI NGR NIG PER PHI PK RHO	CEY HK KOR

CASSIA L. Caesalpiniaceae *(Leguminosae) (Cont'd)*	S	P	C	X	F
tora L. *(Cont'd)*				SAL SEG SUD SUR THI TRI VIE	
CASSINIA R. Br. **Asteraceae** *(Compositae)*					
arcuata R. Br.			AUS	NZ	
fulvida Hook. *f.)*				NZ	
laevis R. Br.				AUS	
leptophylla (Forst. *f.)* R. Br.			NZ		
CASSYTHA L. **Cassythaceae**					
filiformis L.		PR		DR FIJ IND PLW	
CASTALIA Salisb. **Nymphaeaceae** *(See* **NYMPHAEA** L.)					
alba (L.) Wood *(See* **Nymphaea alba** L.)			IRA		
CASTANEA L. **Fagaceae**					
alnifolia Nutt.				USA	
dentata (Marsh.) Borkh.				USA	
pumila (L.) Mill.				USA	
CASTANOPSIS Spach. **Fagaceae**					
chrysophylla (Doug.) A. Dc.				USA	
sempervirens (Kell.) Dudl.				USA	
CASTILLEJA Mutis. **Scrophulariaceae**					
coccinea (L.) Spreng.				USA	
sessiliflora Pursh				USA	
CATALPA Scop. **Bignoniaceae**					
bignonioides Walt.				USA	
speciosa Warder				USA	
CATHARANTHUS G. Don **Apocynaceae**					
pusillus G. Don				IND	
roseus (L.) G. Don (Syn. *Vinca rosea* L.)			HAW	DR FIJ	AUS PR
CAUCALIS L. **Apiaceae** *(Umbelliferae)*					
bifrons Coss. & Dur. *ex* Ball				MOR	
daucoides L.			IRA	SOV	POL

CAUCALIS L. Apiaceae *(Umbelliferae) (Cont'd)*	S	P	C	X	F
daucoides L. *(Cont'd)*			SPA	TUR	
lappula (Weber) Grande			POR		
latifolia L.			MOR POR SOV		AFG
leptophylla L.			LEB MOR		IRA ISR
platycarpos L.				MOR	
scabra (Thunb.) Makino *(See* **Torilis scabra** (Thunb.) Dc.)			JPN		
CAYRATIA Juss. **Vitaceae**					
japonica (Thunb.) Gagnep. (Syn. *Vitis japonica* Thunb.)			JPN TAI		
trifolia (L.) Domin				PHI	
CEANOTHUS L. **Rhamnaceae**					
americanus L.				CHL	
cordulatus Kellogg				USA	
cuneatus (Hook.) Nutt.				USA	
cyaneus Eastw.				USA	
integerrimus Hook. & Arn.				USA	
leucodermis Greene				USA	
lemmonii Parry				USA	
megacarpus Nutt.				USA	
prostratus Benth.				USA	
sanguineus Pursh				USA	
sorediatus Hook. & Arn.				USA	
spinosus Nutt.				USA	
thyrsiflorus Esch.				USA	
velutinus Dougl.				USA	
CECROPIA Loefl. **Moraceae**					
adenopus Mart. *ex* Miq.				BRA	
ferreyrae Cuatr.				PER	
juranyiana Alád. Richter				PER	
lyratiloba Miq.				PER	
peltata L.				HAW PR	DR HON JAM TRI
tessmannii Mildbr.				PER	

CELOSIA L. Amaranthaceae	S	P	C	X	F
argentea L.	AFG	GHA IND KOR PHI SUD	IDO MIC TAI	BND CAB CHN CNK COL DAH ETH HK IVO JPN MEL NEP NGI PK PR SEG SUR THI UGA VIE	
cristata L.		IND			
laxa Schur. & Thonn.				GHA	
polygonoides Retz.				IND	
trigyna L.		ZAM	RHO	ANG GHA IVO NIG	
CELTIS L. Ulmaceae					
laevigata Willd.				USA	
occidentalis L.				CHL USA	
pallida Torr.				USA	
reticulata Torr.				USA	
spinosa Spreng.			MEX		
CENCHRUS L. Poaceae *(Gramineae)*					
australis R. Br.				AUS	
biflorus Roxb.			GHA		
brownii Roem. & Schult.			COL	SAF	VEN	PR
calyculatus Cav.			TAI		
ciliaris L.			IND		GHA PK TNZ VEN	HAW SAF
echinatus L. (Syn. *C. viridis* Spreng.)	BRA CEY COL FIJ HAW PER PHI VEN	CUB GUA JAM THI	ARG AUS PAR PLE PR USA	BOL CHL DR HON HUN MAU MEL MEX MIC	CR NGI

CENCHRUS L. Poaceae *(Gramineae) (Cont'd)*	S	P	C	X	F
echinatus L. *(Cont'd)*				NIG PAN PLW SAL TRI URU	
incertus Curt.			SAF USA		
longispinus (Hack.) Fern.			USA		
macrocephalus Scribn.				URU	
myosuroides H. B. K.			ARG	CHL URU	
pauciflorus Benth.		AUS	ARG	CHL MEX PR URU USA	AFG IND LEB POR SAF
pilosus H. B. K.				PER VEN	
tribuloides L.			AUS	MOR	
viridis Spreng. *(See* **C. echinatus** L.*)*	MEX			PHI	
CENTAUREA L. Asteraceae *(Compositae)* *(Syn. ACROPTILON* Cass.*)*					
acaulis L.			TUN		
americana Nutt.				MEX	
aspera L.			SPA		POR
calcitrapa L.			AUS EGY LEB SPA	ARG CHL MOR NZ TUR URU USA	AFG POR
cyanus L.		FIN	ARG GER MOR SAF SOV SPA	ICE ITA NZ PK TUR USA	POL
depressa Bieb.		IRA TUR			AFG BRA
diffusa Lam.			CAN USA		ARG
diluta Dryand.			MOR		
dubia Suter *(See* **C. nigra** L.*)*			ARG		
iberica Trev. *ex* Spreng.			ARG IRA IRQ	USA	AFG ISR

CENTAUREA L. Asteraceae *(Compositae) (Cont'd)*	S	P	C	X	F
iberica Trev. *ex* Spreng. *(Cont'd)*			LEB		
jacea L.			CAN FIN GER SOV SPA	AUS USA	POL
lippii L.			KEN		
maculosa Lam.			USA	NZ	CAN
maroccana Ball			MOR		
melitensis L.			ARG AUS HAW KEN	CHL MOR NZ URU USA	LEB POR SAF
montana L.			SPA	TUR	
muricata L.			KEN		
napifolia L.			MOR		
nigra L. (Syn. *C. dubia* Suter)			CAN	NZ USA	
pallescens Delile			LEB	IRQ	
phrygia L.			SOV		
phyllocephala Boiss.			IRA		AFG
pieris Pall.		SOV		CAN	AFG AUS SAF
praecox Oliv. & Hiern.			GHA		
repens L. (Syn. *Acroptilon picris* (L.) Dc.) (Syn. *Acroptilon repens* (L.) Dc.)	AUS CAN	IRA TUR	ARG SAF USA	IND SOV	AFG TRI
salmantica L.			SAF		
scabiosa L.			FIN GER SOV		POL
solstitialis L.		ARG JOR POL TUR	AUS IRA ITA LEB SAF SPA USA	CAN CHL EGY GRE IRQ NZ RHO SOV SWZ URU	TRI
squarrosa Roth				USA	
sulphurea Willd.				MOR	
triumfetti All.				TUR	
tweedieii Hook. & Arn.			ARG		

CENTAUREA L. Asteraceae *(Compositae) (Cont'd)*	S	P	C	X	F
variegata Lam.			IRA		AFG
verutum L.			ISR LEB		
vochinensis Bernh.			USA		
CENTAURIUM Gilib. **Gentianaceae**					
minus Moench			HAW		
CENTELLA L. **Apiaceae** *(Umbelliferae)*					
asiatica (L.) Urb. (Syn. *Hydrocotyle asiatica* L.)		HAW IND MAU PR	AUS CNK GHA IDO JPN PHI TAI	BRA CAB CEY CHL CHN DAH HK IRA JAM KOR MEL MIC NEP PLW RHO SEG SUD THI UGA USA VIE	
triflora Nannf.				CHL	
CENTIPEDA Lour. **Asteraceae** *(Compositae)*					
minima (L.) A. Br. & Aschers.		JPN TAI		KOR THI	AUS CEY IDO
orbicularis Lour.				IND NZ VIE	
CENTOTHECA Desv. **Poaceae** *(Gramineae)*					
lappacea (L.) Desv.		MAL	BOR	GHA IDO NGI PHI PLW	AUS BUR CHL IND
CENTRANTHERA R. Br. **Scrophulariaceae**					
muticum (H. B. K.) Less.			TRI		
CENTRANTHUS Dc. **Valerianaceae**					
ruber Dc.			SPA		
CENTROPOGON Presl **Campanulaceae**					
surinamensis Presl			TRI		

	S	P	C	X	F
CENTROSEMA Benth. **Papilioaceae** *(Leguminosae)*					
plumieri (Turp. *ex* Pers.) Benth.			PR	PHI SAL	IDO
pubescens Benth.		AUS	PR TRI	BOR COL DR NGI NIG PHI	IND
virginianum (L.) Benth.				GUA HON	PR
CEPHAELIS Sw. **Rubiaceae** *(See* **PSYCHOTRIA** L.)					
ruelliaefolia Cham. & Schlecht. *(See* **Phychotria ruelliaefolia** Muell.-Arg.)		BRA			
CEPHALANDRA Schrad. *ex* Eckl. & Zeyh. **Cucurbitaceae**					
indica Naud. *(See* **Coccinia indica** Wight & Arn.)				IND	
CEPHALANTHUS L. **Rubiaceae**					
occidentalis L.				USA	
CEPHALARIA Schrad. *ex* Roem. & Schult. **Dipsacaceae**					
alpina Schrad.				TUR	
syriaca (L.) Schrad.		JOR TUR	IRA IRQ ISR LEB	SOV	AFG
CEPHALOSTIGMA A. Dc. **Campanulaceae**					
perrottetii A. Dc.				IVO	
CERASTIUM L. **Caryophyllaceae**					
arvense L.		NZ	CAN POL SOV SPA USA	CHL COL ITA	
caespitosum Gilib. *(See* **C. glomeratum** Thuill.)			FIN JPN SOV	ICE URU	
capense Sond.			SAF		
glomeratum Thuill. (Syn. *C. caespitosum* Gilib.)		BRA NZ	AUS JPN SPA	TUR	AFG POR URU
glutinosum H. B. K.				TUR	
holosteoides Fries			ENG		
indicum Wight & Arn.		TNZ			
perfoliatum L.				TUR	
triviale Link.				CHN	

CERASTIUM L. Caryophyllaceae *(Cont'd)*	S	P	C	X	F
viscosum L.		AUS	ARG CHN JPN	CHL USA	AFG ISR
vulgatum L.			CAN ENG HAW IND USA	CHL	
CERATOCARPUS L. **Chenopodiaceae**					
arenarius L.			IRA SOV		
CERATOCEPHALUS Moench **Ranunculaceae** *(See* **RANUNCULUS** L.*)*					
falcatus Pers. *(See* **Ranunculus falcatus** L.*)*				TUR	
CERATOPHYLLUM L. **Ceratophyllaceae**					
demersum L.	GHA NZ	AUS IND PAN THI USA	JPN POR TAI	ARG BEL BND BUR CAN CHN COL EGY ENG ETH FIJ FRA GER HUN IDO KOR MAL NET PHI PR RHO SOV SUD SWE VIE YUG	AFG BER ISR
echinatum Gray				USA	
submersum L.				TAI	
CERATOPTERIS A. Brogn. **Parkeriaceae**					
cornuta Le Prieur *(See* **C. thalictroides** (L.) Brongn.*)*				ARG CNK DAH GHA NIG RHO SEG	
pteridoides (Hook.) Hieron.				ARG AUS USA	

CERATOPTERIS A. Brogn. Parkeriaceae *(Cont'd)*	S	P	C	X	F
siliquosa (L.) Copel. *(See* **C. thalictroides** (L.) Brongn.)					
thalictroides (L.) Brongn. (Syn. *C. cornuta* Le Prieur) (Syn. *C. siliquosa* (L.) Copel.)		IND JPN	TAI THI	AUS BOT CAB CEY CHN IDO IRQ MIC PHI RHO VIE	KOR
CERATOTHECA Endl. **Pedaliaceae**					
integribracteata Engl.				ANG	
sesamoides Endl.				GHA	
CERCIDIUM Tul. **Papilionaceae** *(Leguminosae)*					
floridum Benth.				USA	
macrum Johnst.				USA	
microphyllum (Torr.) Rose & Johnston				USA	
CERCIS L. **Caesalpiniaceae** *(Leguminosae)*					
canadensis L.				CHL USA	
occidentalis Torr.				USA	
CERCOCARPUS H. B. K. **Rosaceae**					
betuloides Nutt.				USA	
ledifolius Nutt.				USA	
montanus Rafin.				USA	
CERINTHE L. **Boraginaceae**					
aspera Roth *(See* **C. major** L.)			SPA		
major L. (Syn. *C. aspera* Roth)			SPA		POR
minor L.			TUR		
CEROPTERIS Link **Gymnogrammaceae**					
tartarea (Cav.) Link				IDO	
CERUANA Forsk. **Asteraceae** *(Compositae)*					
pratensis Forsk.			EGY		
CESTRUM L. **Solanaceae**					
aurantiacum Lindl.			AUS	NZ	RHO
diurnum L.				USA	
parqui L'herit.		ARG	AUS	CHL	

CESTRUM L. Solanaceae *(Cont'd)*	S	P	C	X	F
parqui L'herit. *(Cont'd)*				NZ URU USA	
CHAENARRHINUM Reichb. **Scrophulariaceae**					
minus (L.) Lange				USA	
CHAEROPHYLLUM L. **Apiaceae** *(Umbelliferae)*					
bulbosum L.				SOV	
CHAMAEBATIA Benth. **Rosaceae**					
foliolosa Benth.				USA	
CHAMAECRISTA Moench **Caesalpiniaceae** *(Leguminosae)*					
aeschynomene (Dc.) Greene		PR		DR	
CHAMAECYPARIS Spach **Cupressaceae**					
lawsoniana (A. Murr.) Parl.				USA	
CHAMAEDAPHNE Moench **Ericaceae**					
calyculata (L.) Moench				USA	
CHAMAEMELUM Adans. **Asteraceae** *(Compositae)*					
mixtum (L.) All.		POR			
CHAMAENERION Adans. **Onagraceae** *(See* **EPILOBIUM** L.)					
angustifolium Scop. *(See* **Epilobium angustifolium** L.)			FIN SOV		
CHAMAERAPHIS R. Br. **Poaceae** *(Gramineae)*					
brunoniana (Hook. *f.*) A. Camus				VIE	
gracilis Hack.				BND	
spinescens Poir.				IND	
CHAMAESARACHA A. Gray *ex* Franch. & Sav. **Solanaceae**					
coronopus (Dunal) Gray				USA	
CHAMAESYCE S. F. Gray **Euphorbiaceae**					
hirta (L.) Millsp. *(See* **Euphorbia hirta** L.)		PR		ARG DR MEX	
CHAMISSOA Kunth **Amaranthaceae**					
altissima H. B. K.				JAM	
CHAPTALIA Vent. **Asteraceae** *(Compositae)*					
nutans Hemsl.		PR		HON JAM SAL	

CHARA L. Characeae	S	P	C	X	F
fragilis Desv. *(See* **C. globularis** Thuill.)				SUD	
globularis Thuill. (Syn. *C. fragilis* Desv.)				SUD	
globularis Thuill *var.* kraussi *forma* kraussiana (J. Gr. & Steph.) R. D. W. (Syn. *C. kraussiana* J. Gr. & Steph.)					
kraussiana J. Gr. & Steph. *(See* **C. globularis** Thuill *var.* kraussi *forma* kraussiana (J. Gr. & Steph.) R. D. W.		TNZ			
vulgaris L.			USA		
zeylanica Willd.	IND				
CHASMANTHE N. E. Br. **Iridaceae**					
aethiopica (L.) N. E. Brown				AUS	
CHEILANTHES Sw. **Sinopteridaceae**					
sieberi Kunze				AUS	
CHELIDONIUM L. **Papaveraceae**					
majus L.			JPN POL SOV SPA	USA	
CHENOPODIUM L. **Chenopodiaceae**					
acuminatum Willd.			TAI		
album L. (Syn. *C. viride* L.)	ALK ARG FIN FRA GER IND IRA IRE ITA NZ PK ROM SAF SOV USA	ALG AUS BEL BRA BUL CAN CHL COL CZE ENG ETH HAW HUN JPN KOR LEB MEX NOR POR SPA SWE TNZ TUN TUR YUG	BOR CHN EGY HK IRQ ISR KEN RHO TAI	BND DEN DR ICE MOR MOZ NEP NET SWT ZAM	AFG POL
ambrosioides L.	AFG ITA MOZ	ARG BRA ECU	CHN EGY HAW HK HON IND JPN	AUS BND BOR BOT CAB CAN CHL	ISR

CHENOPODIUM L. Chenopodiaceae *(Cont'd)*	S	P	C	X	F
ambrosioides L. *(Cont'd)*			KEN POL PR RHO SAF TAI ZAM	CNK COL CR DR ETH FIJ GHA GUA HUN JAM JOR MAU MEX NEP NZ PER PK SAL SEG SUR UGA URU USA VEN	
anthelminticum L.			JPN		
aristatum L.				SOV	
berlandieri Moq.				CAN	
bontei Aellen		SAF			
bonus-henricus L.			ENG SPA	USA	
botrys L.				USA	
capitatum (L.) Aschers.				CAN USA	
carinatum R. Br.			AUS HAW KEN SAF		
cordobense Aellen			ARG		
cristatum F. Muell.			AUS		
fasciculosum Aellen		KEN			
ficifolium Sm.	ITA TAI	KOR	HUN JPN NET POL TUR	EGY ENG IRA	CHN
foetidum Schrad. (Syn. *C. schraderianum* Roem. & Schult.)					
foliosum (Moen.) Aschers.				SOV	AFG
gigantospermum Aellen				CAN	
glaucum L.		CAN	SOV	MEX TUR USA	AFG AUS POL

CHENOPODIUM L. Chenopodiaceae *(Cont'd)*	S	P	C	X	F
hircinum Schrad.		ARG		PER URU	
hybridum L.			ENG SOV	USA	POL
leptophyllum Nutt. *ex* Moq.				ARG USA	
macrospermum Hook. *f.*			ARG		
multifidum L.			ARG SAF	URU	AUS
murale L.	AFG ISR ITA MEX PK	ARB EGY GRE IND JOR PER	ARG HAW HUN IRQ KEN MOR POL RHO SAF	AUS CAN CHL CNK DR HON IRA LEB MEL NEP NZ SAL SUD TNZ TUR URU USA VEN	
oahuense (Meyen) Aellen			HAW		
opulifolium Schrad.	LEB	TNZ	ISR KEN POR RHO	TUR	AUS POL
paganum Reichb.				CAN USA	
paniculatum Hook.		COL			
polyspermum L.			ENG FIN GER SOV SPA	USA	ISR POL
pratericola Rydb.				ARG USA	
procerum Hochst. *ex* Moq.			ETH KEN		
rubrum L.			ENG SOV	TUR USA	AFG ISR POL
schraderianum Roem. & Schult. (*See* **C. foetidum** Schrad.)			KEN RHO SAF		
strictum Roth				CAN	
triangulare R. Br. (Syn. *C. trigonon* Roem. & Schult.)					

CHENOPODIUM L. Chenopodiaceae *(Cont'd)*	S	P	C	X	F
trigonon Roem. & Schult. *(See* C. triangulare R. Br.)			AUS		
urbicum L.			SOV	USA	
viride L. *(See* C. album L.)			SOV		
vulvaria L.		LEB	IRA	ISR	
zobelii Ludw.			ARG		
CHILIOTRICHUM Cass. **Asteraceae** *(Compositae)*					
diffusum O. Ktze.				CHL	
CHIOCOCCA L. **Rubiaceae**					
alba (L.) Hitchc.			PR		
CHIONACHNE R. Br. **Poaceae** *(Gramineae)*					
hubbardiana Henr.			AUS		
CHLORIS Sw. **Poaceae** *(Gramineae)*					
barbata (L.) Sw. (Syn. *C. inflata* Link.)	AUS KOR THI	CAB IND	HAW MAL SUD	ANT DAH EGY FIJ GHA IDO IRA JAM KEN MAU MEL NGI PHI PLW PR TNZ USA VIE	CHN HK
chloridea (Presl) Hitchc.	MEX				
ciliata Sw.		VEN			HAW TRI
distichophylla Lag.			AUS		HAW
divaricata R. Br.			HAW	USA	AUS
gayana Kunth		PER	HAW	BRA NZ USA	AUS IND ISR SAF
halophila Parodi		PER			
inflata Link. *(See* C. barbata (L.) Sw.)		IND	HAW PR TRI	DR MEX USA VEN	
polydactyla Sw.			COL PHI	VEN	HAW

CHLORIS Sw. Poaceae *(Gramineae) (Cont'd)*	S	P	C	X	F
pycnothrix Trin. *(See* **C. radiata** (L.) Sw.)			KEN SAF	MAU TNZ	
radiata (L.) Sw. (Syn. *C. pycnothrix* Trin.)			HAW TRI	MAU NIC PER USA VEN	
truncata R. Br.			HAW		
verticillata Nutt.			CHL USA		
virgata Sw.		AUS PER	HAW SAF	SUD TNZ USA	IND
CHLOROGALUM (Lindl.) Kunth **Liliaceae**					
pomeridianum (Ker.) Kunth				USA	
CHONDRILLA L. **Asteraceae** *(Compositae)*					
graminea Bieb.			SOV		
juncea L.	AUS	ITA USA YUG	IRA LEB POL SOV SPA	GRE HUN IRQ JOR NGI TUR	AFG ISR POR
CHORISPORA R. Br. *ex* Dc. **Brassicaceae** *(Cruciferae)*					
syriaca Boiss.			LEB		
tenella (Willd.) Dc.			SOV	USA	
CHROMOLAENA Dc. **Asteraceae** *(Compositae)*					
odorata (L.) R. M. King & H. Robinson (Syn. *Eupatorium odoratum* L.)	IND	AUS THI		GHA MAL NIG TRI	
CHROZOPHORA A. Juss. **Euphorbiaceae**					
plicata (Vahl) A. Juss.		SUD	EGY	IND	
rottleri (Geisel) A. Juss. *ex* Spreng.				IND	
tinctoria (L.) A. Juss.	IRA		ISR	SOV TUR	
verbascifolia (Willd.) A. Juss.			IRQ LEB		AFG
CHRYSANTHELLUM Rich. **Asteraceae** *(Compositae)*					
americanum (L.) Vatke				GHA IVO JAM	
CHRYSANTHEMUM L. **Asteraceae** *(Compositae)*					
coronarium L.	GRE TUN	JOR	MOR SAF SPA	AUS CHL CHN	AFG KOR

CHRYSANTHEMUM L. Asteraceae *(Compositae) (Cont'd)*	S	P	C	X	F
coronarium L. *(Cont'd)*				EGY ETH IRA ISR JPN LEB POR THI	
indicum L.			CHN TAI		
inodorum L. *(See* **Matricaria inodora** L.*)*				SOV	
leucanthemum L.	CAN HUN	CHL ITA SOV SWE	ARG AST AUS COL ENG FIN GER HAW IRA NOR POL SPA TUN USA	BEL CHN NZ PK TUR	
myconis L.			BRA	CHL	POR
parthenium (L.) Bernh.				CHL NZ USA	AFG ARG AUS
segetum L.	ENG GRE	ITA SWE TUN	GER ISR NET NOR POL SAF SPA	AST BEL CAN CHL ICE IRE JOR LEB MOR NZ POR TNZ TUR	AUS
suaveolens Aschers.				SOV	
tanacetum Vis.				TUR	
vulgare (Lam.) Parsa			FIN	CHL	ARG
CHRYSOPOGON Trin. **Poaceae** *(Gramineae)*					
aciculatus (Retz.) Trin. (Syn. *Andropogon aciculatus* Retz.)	AUS BOR MEL	IDO IND	HAW MAL PHI TAI THI	BND CAB CEY FIJ MIC PLW VIE	BUR CHN HK
montanus Trin.				IND	
orientalis (Desv.) A. Camus				VIE	

	S	P	C	X	F
CHRYSOTHAMNUS Nutt. **Asteraceae** *(Compositae)*					
graveolens (Nutt.) Greene				USA	
greenei (A. Gray) Greene				USA	
nauseosus (Pallas) Britt.				USA	
parryi (A. Gray) Greene				USA	
pulchellus (A. Gray) Greene				USA	
viscidiflorus (Hook.) Nutt.				USA	
CICHORIUM L. **Asteraceae** *(Compositae)*					
endivia L.	EGY				
intybus L.	AFG ARG HUN ITA PK SOV	EGY GRE IND JOR POL	AUS CAN IRA IRQ LEB MOR SAF SPA USA	BEL CHL CHN ENG IDO JPN NZ RHO TUR URU VEN YUG	POR
pumilum Jacq.		ISR	EGY LEB		POR
spinosum L.				GRE	
CICUTA L. **Apiaceae** *(Umbelliferae)*					
bulbifera L.			CAN		
douglasii (Dc.) Coult. & Rose				USA	
mackenzieana Raup	ALK				
maculata L.			CAN USA	ARG	
virosa L.			JPN SPA	BEL	
CINERARIA L. **Asteraceae** *(Compositae)*					
lyrata Dc.	SAF				
CINNAMOMUM Schaef. **Lauraceae**					
zeylanicum Nees				JAM	
CIRSIUM Adans. **Asteraceae** *(Compositae)*					
acarna (L.) Moench		LEB		AUS MOR TUR	AFG ISR
acaule Web. *ex* Wigg.			GER		
altissimum (L.) Spreng.				USA	
arvense (L.) Scop. (Syn. *C. incanum* Bieb.) (Syn. *C. lanatum* Spreng.)	FIN LEB POR	BEL BUL CAN	AUS CHN JPN	CHL CZE FRA	AFG ALK RHO

CIRSIUM Adans. **Asteraceae** *(Compositae) (Cont'd)*	S	P	C	X	F
(Syn. *Cnicus arvensis* Hoffm.) (cont'd)	TUR USA	ENG GER GRE IND IRA ITA NZ PK POL ROM SAF SOV TUN YUG	SPA	ICE KOR MEX NET NOR SUD SWZ	
eriophorum Scop.			ENG	TUR	
flodmanii (Rydb.) Arthur			CAN	USA	
heterophyllum Hill			FIN		
horridulum Michx.				USA	
incanum Bieb. *(See* **C. arvense** (L.) Scop.)					SOV
japonicum Dc.			CHN	JPN	
lanatum Spreng. *(See* **C. arvense** (L.) Scop.)				ARG	
lanceolatum Hill *(See* **C. vulgare** (Savi) Tenore)		NZ	AUS CAN GER IRA SOV	ARG BRA CHL SPA TUR	HAW POL SAF
libanoticum Dc.			LEB		
oleraceum (L.) Scop.			GER		
palustre (L.) Scop.			CAN ENG FIN GER	BEL NZ	
pumilum Spreng.				USA	
segetum Bunge				CHN	
serrulatum Bieb.			SOV		
syriacum (L.) Gaertn. *(See* **Notobasis syriaca** (L.) Cass.)			LEB MOR		
undulatum (Nutt.) Spreng.			CAN	USA	
vulgare (Savi) Tenore (Syn. *C. lanceolatum* Hill)	HUN ITA POL	AUS NZ	ARG AST CAN ENG HAW KEN SAF TUN USA	AFG BEL CHL FIN GER GRE GUA IRA IRQ MEL NOR SOV	

CIRSIUM Adans. Asteraceae *(Compositae) (Cont'd)*	S	P	C	X	F
vulgare (Savi) Tenore *(Cont'd)*				SWE TUR URU	
CISSAMPELOS L. **Menispermaceae**					
glaberrima St. Hil.				BRA	VEN
mucronata A. Rich. *(See* C. **pareira** L.)				ANG GHA	
pareira L. *(See* C. **mucronata** A. Rich.)	ARG	PR		DR TRI VEN	
CISSUS L. **Vitaceae**					
cornifolia (Baker) Planch.				RHO	
incisa (Nutt.) Des Moul.				USA	
sicyoides L.		PR		DR HON JAM SAL	COL
trifoliata L.				JAM	
CISTANCHE Hoffm. & Link **Orobanchaceae**					
tubulosa Wight				PK	
CITHAREXYLUM Mill. **Verbenaceae**					
spinosum L.			HAW		
CITRULLUS Schrad. **Cucurbitaceae**					
colocynthis Schrad.			AUS	IND PK	AFG ISR
lanatus (Thunb.) Mansf. (Syn. *C. vulgaris* Schrad.) (Syn. *Colocynthis citrullus* (L.) O. Ktze.)			AUS		
vulgaris Schrad. *(See* C. **lanatus** (Thunb.) Mansf.)		TAI	AUS		CAB
CLADIUM P. Br. **Cyperaceae**					
jamaicense Crantz				USA	
mariscoides (Muhl.) Torr.				USA	
CLADRASTIS Rafin. **Papilionaceae** *(Leguminosae)*					
lutea (Michx. *f.*) K. Koch				USA	
CLEMATIS L. **Ranunculaceae**					
apiifolia Dc.			JPN		
dioica L.				BRA	
drummondii Torr. & Gray				MEX	
ligusticifolia Nutt.				USA	

CLEMATIS L. Ranunculaceae *(Cont'd)*	S	P	C	X	F
maximowicziana Fr. & Sav. (Syn. *C. terniflora* Dc.)					
taiwaniana Hay.			TAI		
terniflora Dc. *(See* **C. maximowicziana** Fr. & Sav.)			JPN		
virginiana L.				CHL USA	
CLEOME L. Capparidaceae					
aculeata L.			PR TRI	AUS	
aspera Koen. *ex* Dc.			IDO		
brachycarpa Vahl				PK	
burmanni Wight & Arn.	CEY	MAL			
chelidonii L. *f.* (Syn. *Polanisia chelidonii* Dc.)				IND	
ciliata Schum. & Thonn. *(See* **C. rutidosperma** Dc.)	BOR TRI	ANG MAL		CNK GHA JAM NIG SUR	
diffusa Banks *ex* Dc.				BRA	
gynandra L. (Syn. *Gynandropsis gynandra* (L.) Briq.) (Syn. *G. pentaphylla* (L.) Dc.) (Syn. *Pedicellaria pentaphylla* Schrank)		COL	CNK KEN MAU PHI PR SEG SUD THI TNZ UGA	BOR CAB CHN EGY ETH FIJ GHA HAW IND JPN MIC NEP NIG PK PLW	
hirta (Klotz.) Oliv.			KEN		
icosandra L.				IND	
integrifolia Torr. & Gray (Syn. *C. serrulata* Pursh)					
lutea Hook.				USA	
monophylla L.		RHO SAF SWZ UGA	IND KEN	TNZ	
moritziana Klotz. *ex* Eichl.				VEN	
rutidosperma Dc. (Syn. *C. ciliata* Schum. & Thonn.)	JAM	BOR	CAB SUD VIE	CNK DAH GHA GUI	

CLEOME L. Capparidaceae *(Cont'd)*	S	P	C	X	F
rutidosperma Dc. *(Cont'd)*				IVO LIB MAL NIG PHI PLW SEG THI TNZ UGA ZAM	
serrata Jacq.				TRI	
serrulata Pursh *(See* C. integrifolia Torr. & Gray)				MEX USA	
spinosa Jacq.			PR TRI	BRA DR USA VEN	AUS
stenophylla Klotz.			COL		
viscosa L. (Syn. *Polanisia icosandra* (L.) Wight & Arn.) (Syn. *P. viscosa* Dc.)	NGI	MAD		AUS FIJ IND MAU PK PLW	CAB MAL
welwitschii Exell				ANG	
CLERODENDRUM L. Verbenaceae					
buchanani Roxb. *ex* Wall.				PLW	
fragrans (Vent.) Willd. *(See* C. philippinum Schauer)	PLW		HAW PR	FIJ USA	
indicum (L.) O. Ktze. (Syn. *C. siphonanthus* R. Br.)					
infortunatum Gaertn.				IND	
philippinum Schauer (Syn. *C. fragrans* (Vent.) Willd.)				PLW	
phlomoidis L.f.				IND	
siphonanthus R. Br. *(See* C. indicum (L.) O. Ktze.)				VIE	
splendens (Thunb.) G. Don				CNK	
CLIBADIUM L. Asteraceae *(Compositae)*					
surinamense L.				JAM	
CLIDEMIA D. Don Melastomataceae					
dentata D. Don			TRI		
dependens D. Don			TRI		
hirta (L.) D. Don	MEL	FIJ HAW MAD MAL	IDO PR TRI	BOR COL JAM MAU	

CLIDEMIA D. Don **Melastomataceae** (*Cont'd*)	S	P	C	X	F
hirta (L.) D. Don (*Cont'd*)				MIC PER PLW USA	
rubra (Aubl.) Mart.			TRI		
CLIFTONIA Banks & Gaertn. *f.* **Cyrillaceae**					
monophylla Sarg.				USA	
CLINOPODIUM L. **Lamiaceae** (*Labiatae*)					
chinense (Benth.) O. Ktze.			JPN TAI		
CLITORIA L. **Papilionaceae** (*Leguminosae*)					
ternatea L.	AUS	PR		DR GHA IND JAM PAN SAL	IDO
CNICUS L. **Asteraceae** (*Compositae*)					
arvensis Hoffm. (*See* **Cirsium arvense** (L.) Scop.)	PK	IND			AFG AUS
benedictus L. (Syn. *Carbenia benedicta* (L.) Adans.)			ARG LEB SAF	GRE USA	ISR POR
CNIDIUM Cusson **Apiaceae** (*Umbelliferae*)					
formosanum Yabe			TAI		
CNIDOSCOLUS Pohl **Euphorbiaceae**					
stimulosus (Michx.) Gray				USA	
texanus (Muell.-Arg.) Small				USA	
COCCINIA Wight & Arn. **Cucurbitaceae**					
cordifolia Cogn. (*See* **C. indica** Wight & Arn.)				THI	IDO
indica Wight & Arn. (Syn. *C. cordifolia* Cogn.) (Syn. *Cephalandra indica* Naud.)				IND	
COCCOCYPSELUM P. Br. **Rubiaceae**					
herbaceum Aubl. (*See* **C. repens** Sw.)			PR		
repens Sw. (Syn. *C. herbaceum* Aubl.)					
COCCOLOBA P. Br. **Polygonaceae**					
acapulcensis Standl.		MEX			
acuminata H. B. K.				HON NIC	

COCCOLOBA P. Br. Polygonaceae *(Cont'd)*	S	P	C	X	F
mollis Casar. *(See* **C. polystachya** Wedd.)				BRA	
polystachya Wedd. (Syn. *C. mollis* Casar.)					
schiedeana Lindau		MEX		SAL	
uvifera (L.) L.				USA	
COCCULUS Dc. **Menispermaceae**					
carolinus (L.) Dc.				USA	
COCHLOSPERMUM Kunth *ex* Dc. **Cochlospermaceae**					
insigne St. Hil.				BRA	
COELORACHIS Brongn. **Poaceae** *(Gramineae)*					
glandulosa (Trin.) Stapf			BOR		IDO MAL PHI
COGNIAUXIA Baill. **Cucurbitaceae**					
trilobata Cogn.				CNK	
COIX L. **Poaceae** *(Gramineae)*					
aquatica Roxb.				THI	
gigantea Koenig	CEY				
lachryma L. *(See* **C. lacryma-jobi** L.)				IND	
lacrima L. *(See* **C. lacryma-jobi** L.)					IRA
lacryma-jobi L. (Syn. *C. lachryma* L.) (Syn. *C. lacrima* L.)	PLW	ITA KOR	HAW IRA JPN MIC PR	BOR CAB CNK COL CR DR FIJ GHA GUA HON IND IRQ MEL NEP PER PHI PK RHO SEG SUD THI USA VEN	AUS BUR CHN HK SAF
COLA Schott & Endl. **Sterculiaceae**					
acuminata Schott & Endl.				JAM	

COLCHICUM L. Liliaceae	S	P	C	X	F
autumnale L.			AUS GER SPA	ITA	
kotschyi Boiss. *(See* **C. laetum** Stev.*)*			IRA		
laetum Stev. (Syn. *C. kotschyi* Boiss.)					
COLDENIA L. **Boraginaceae**					
procumbens L.				IND	
COLEOGYNE Torr. **Rosaceae**					
ramosissima Torr.				USA	
COLEUS Lour. **Lamiaceae** *(Labiatae)*					
blumei Benth.			PR	PLW	
COLLINSIA Nutt. **Scrophulariaceae**					
parviflora Dougl.				USA	
COLLOMIA Nutt. **Polemoniaceae**					
biflora (R. & Pav.) A. Brand				CHL	
linearis Nutt.				USA	
COLOCASIA Schott **Araceae**					
antiquorum Schott *(See* **C. esculenta** (L.) Schott*)*				IND MAU	
esculenta (L.) Schott (Syn. *C. antiquorum* Schott) (Syn. *Caladium esculenta* Vent.)			PR	BND JAM	IDO
gigantea Hook.				IDO	
COLOCYNTHIS L. **Cucurbitaceae**					
citrullus (L.) O. Ktze. *(See* **Citrullus lanatus** (Thunb.) Mansf.*)*				AUS	
vulgaris Schrad.				AUS	
COLOPHOSPERMUM Kirk *ex* J. Leonard **Caesalpiniaceae** *(Leguminosae)*					
mopane (Kirk *ex* Benth) J. Leonard				RHO	
COLUBRINA Rich. *ex* Brongn. **Rhamnaceae**					
asiatica (L.) Brongn.				PLW	
texensis (Torr. & Gray) Gray				USA	
COMARUM L. **Rosaceae**					
palustre L.			FIN		
COMBRETUM Loefl. **Combretaceae**					
apiculatum Sond.		RHO			
hereroense Schinz		RHO			

COMBRETUM Loefl. Combretaceae *(Cont'd)*	S	P	C	X	F
imberbe Wawra				RHO	
mechowianum O. Hoffm.				ANG	
paniculatum Vent.				ANG	
COMMELINA L. Commelinaceae					
africana L.		UGA			KEN
agraria Kunth				BRA	
benghalensis L.	BOR IND MOZ PHI TNZ ZAM	ANG IDO KEN RHO SAF SWZ THI UGA	CEY HAW JPN TAI	AUS BND CHN GHA GUI IVO MAD MAL MAU NIG PLW SEG VIE	HK
coelestis Willd.				ARG MEX	
communis L.	JPN	BOL TAI	CHN	ARG USA VIE	
condensata C. B. Clarke				IVO	
cyanea R. Br.				AUS	
diffusa Burm. *f.*	MEX PLW	HAW JAM PR THI	PHI	AUS BND BOL COL CR FIJ GHA GUA GUY IDO IVO MAL NGI NIC RHO TAI TRI	
elegans H. B. K.		CUB TRI	BOR PLW	ANT BRA JAM	PR
erecta L.			SAL	GUA NIG	
forskalaei Vahl			IND		
gerrardii C. B. Clarke		GHA			
jacobi Fischer				IND	
kotschyi Hassk.		SUD			

COMMELINA L. Commelinaceae *(Cont'd)*	S	P	C	X	F
lagosensis C. B. Clarke				GHA NIG	
latifolia Hochst. *ex* A. Rich.		KEN			
longicaulis Jacq.		JAM	COL		PR
nudiflora L. *(See* **Murdannia nudiflora** (L.) Brenan)	BOR	CEY GUI IDO IND MAL PHI SAF TAI UGA	NIG	ANG BND CNK FIJ GHA MAU SUR THI TRI VEN VIE	CHN COL HAW HK MEX NGI PR
salicifolia Roxb.		IND		CAB	IDO
subulata Roth		KEN			
sulcata Hoffm. *(See* **C. virginica** L.)				URU	
virginica L. (Syn. *C. sulcata* Hoffm.)				ARG DR	
COMMICARPUS Standl. **Nyctaginaceae**					
scandens (L.) Standl.			PR	DR	
COMMIPHORA Jacq. **Burseraceae**					
mollis Engl.				RHO	
COMOCLADIA P. Br. **Anacardiaceae**					
glabra (Schult.) Spreng.			PR	DR	
COMPTONIA L'herit. **Myricaceae**					
peregrina (L.) Coult.				USA	
CONDALIA Cav. **Rhamnaceae**					
obovata Hook.				USA	
obtusifolia (Hook.) Weberb.				USA	
CONIUM L. **Apiaceae** *(Umbelliferae)*					
maculatum L.	AFG	ARG AUS ITA NZ SWE	CAN COL ENG HUN IRA LEB POL SOV SPA USA	CHL IRQ ISR JOR NOR PER RHO TUR URU VEN YUG	
CONRINGIA L. **Brassicaceae** *(Cruciferae)*					
orientalis (L.) Dum. (Syn. *Erysimum orientale* Mill.)		IRA	CAN SOV	AUS GER	AFG ISR

CONRINGIA L. Brassicaceae *(Cruciferae)* *(Cont'd)*	S	P	C	X	F
orientalis (L.) Dum. *(Cont'd)*			USA	TUR	
CONSOLIDA S. F. Gray **Ranunculaceae**					
ajacis (L.) Schur (Syn. *Delphinium ajacis* L.)					
orientalis (J. Gray) Schroed.			IRA		AFG
CONVALLARIA L. **Liliaceae**					
majalis L.				SPA USA	
CONVOLVULUS L. **Convolvulaceae**					
aegyptius L.				USA	
althaeoides L.			LEB MOR SPA	POR	ISR
arvensis L.	ARG AUS BOR CEY FRA GER GRE IND IRA LEB NZ PK SAF YUG	ARB BEL BUL CAN CHL CZE ENG HAW PER PHI POR ROM SOV SPA SWZ TAS TUN TUR USA	EGY FIN IRQ ISR JPN MOR	AFG BRA CNK HUN ICE ITA MEX UGA URU	JOR POL
erubescens Sims			AUS		
galaticus Rost. *ex* Choisy			TUR		
hederaceus L. *(See* **Ipomoea hederacea** (L.) Jacq.)			THI		
hirsutus Bieb.			ISR		
humilis Jacq.			MOR		
japonicus Thunb.			USA		
pluricaulis Choisy				IND PK	POR
sepium L. *(See* **Calystegia sepium** (L.) R. Br.)			GER USA	CAN TUR	
tricolor L.		TUN	MOR	POR	ISR
ulosepalus Hallier *f.*			SAF		
CONYZA Less. **Asteraceae** *(Compositae)*					
abyssinica Sch.-Bip. *ex* A. Rich.			ANG		

CONYZA Less. Asteraceae *(Compositae)* *(Cont'd)*	S	P	C	X	F
aegyptiaca Dryand.			EGY KEN	GHA MAU	
ambigua Dc.				IND	
bonariensis (L.) Cronq. (Syn. *C. floribunda* H. B. K.) (Syn. *Erigeron bonariensis* L.)		ARG	HAW KEN	COL PER PLW TNZ URU VEN	USA
canadensis (L.) Cronq. (Syn. *Erigeron canadensis* L.)			ENG HAW	USA	PR
chilensis Spreng.			ARG		
dioscoridis Desf.			EGY		
floribunda H. B. K. *(See* **C. bonariensis** (L.) Cronq.)			KEN	PER	
japonica (Thunb.) Less.				IND	AFG IDO
lineariloba Dc.				MAU	
podocephala Dc.			SAF		
pyrrhopappa Sch.-Bip. *ex* A. Rich.				ANG	
schimperi Sch.-Bip. *ex* A. Rich. *(See* **C. stricta** Willd.)			KEN		
steudelii Sch.-Bip. *ex* A. Rich.			KEN		
stricta Willd. (Syn. *C. schimperi* Sch.-Bip. *ex* A. Rich.)			KEN	ANG PK	
COPERNICIA Mart. **Arecaceae** *(Palmae)*					
australis Becc.		PAR			
CORBICHONIA Scop. **Aizoaceae**					
decumbens (Forsk.) Exell			RHO		
CORCHOROPSIS Sieb. & Zucc. **Tiliaceae**					
tomentosa Makino			JPN		
CORCHORUS L. **Tiliaceae**					
acutangulus Lam. *(See* **C. aestuans** L.)	PHI	IND	TAI		IDO
aestuans L. (Syn. *C. acutangulus* Lam.)			BOR	NIG THI	
angolensis Exell & Mendonca				ANG	
antichorus Raeusch.		IND		PK	AFG
argutus H. B. K.				BRA	
capsularis L.	PHI	THI	IND	CAB	
depressus Stocks				PK	

CORCHORUS L. Tiliaceae *(Cont'd)*	S	P	C	X	F
fascicularis Lam.		IND SUD			
hirsutus L.				DR	
hirtus L.				MEX	
olitorius L.	AUS EGY MOZ PHI SEG THI	SUD	AFG IND KEN NEP TUR ZAM	BOT CAM CNK GHA IDO IRA IRQ ISR IVO JAM JOR LIB NIG PK RHO UGA	CHN
orinocensis H. B. K.		COL		MEX PER	
siliquosus L.				JAM PAN SAL	
tridens L.		PK	IND RHO	IVO	
trilocularis L.	MOZ	SUD	IND	FIJ PK TNZ	IDO ISR
CORDIA L. Boraginaceae					
boissieri A. Dc.		MEX			
corymbosa G. Don			PR	MEX	
curassavica Roem. & Schult.		TRI			
cylindristachya Roem. & Schult.			MAL	JAM	
globosa H. B. K.				JAM	
macrostachya Roem. & Schult.				MAU	
oxyphylla Dc.				NIC	
polycephala (Lam.) I. M. Johnston				PER	
COREOPSIS L. Asteraceae *(Compositae)*					
tinctoria Nutt.				USA	
CORIANDRUM L. Apiaceae *(Umbelliferae)*					
sativum L.			ISR LEB MOR TAI		
tordylioides Boiss.			LEB		

CORISPERMUM L. Chenopodiaceae	S	P	C	X	F
declinatum Steph. *(See* **C. hyssopifolium** L.*)*				SOV	
hyssopifolium L. (Syn. *C. declinatum* Steph.)				USA	
CORNUS L. **Cornaceae**					
drummondii C. A. Meyer				USA	
florida L.				USA	
nuttallii Aud.				CHL USA	
rugosa Lam.				USA	
stolonifera Michx.				USA	
torreyi S. Wats.				USA	
CORONILLA L. **Papilionaceae** *(Leguminosae)*					
repanda Boiss.				MOR	
scorpioides (L.) Koch			ISR LEB MOR POR SPA		
varia L.			IRA	SOV USA	POL
CORONOPUS Zinn. **Brassicaceae** *(Cruciferae)*					
didymus (L.) Sm. (Syn. *C. pinnatifidus* Dulac)	HAW NZ	AUS BRA IND ITA TNZ	ARG CHN EGY ENG POL	AFG CAN CHL ETH JPN MEL NEP PER PK PLW RHO URU USA	
niloticus (Del.) Spreng.			EGY		
pinnatifidus Dulac *(See* **C. didymus** (L.) Sm.*)*				CHL	
procumbens Gilib.				USA	
squamatus (Forsk.) Asch. (Syn. *Senebiera coronupus* Poir.)			EGY ENG	NZ	
CORRIGIOLA L. **Caryophyllaceae**					
littoralis L.		ANG	KEN SPA		
CORYDALIS Vent. **Papaveraceae**					
aurea Willd.				USA	

	S	P	C	X	F
CORYDALIS Vent. **Papaveraceae** *(Cont'd)*					
edulis Maxim.				CHN	
CORYLUS L. **Betulacea**					
americana Walt.				USA	CHL
cornuta Marsh.				USA	
COSMOS Cav. **Asteraceae** *(Compositae)*					
bipinnatus Cav.			SAF	AUS	
caudatus H. B. K.				AUS FIJ PHI	
COSTUS L. **Zingiberaceae**					
cylindricus Rosc.				PER	
COTONEASTER Rupp. **Rosaceae**					
vulgaris Lindl.				TUR	
COTULA L. **Asteraceae** *(Compositae)*					
abyssinica (Sch.-Bip.) A. Rich.		ETH			
anthemoides L.		KEN	EGY		
australis (Sieber) Hook. *f.*			ARG AUS	CHL NZ USA	
coronopifolia L.			AUS POR	CHL COL NZ	
scariosa Franch.				CHL	
COURTOISIA Nees **Cyperaceae**					
cyperoides Nees	IND	MAD		BOT ETH RHO	
COUSINIA Cass. **Asteraceae** *(Compositae)*					
minuta Boiss.				PK	
COUTOUBEA Aubl. **Gentianaceae**					
spicata Aubl.			TRI		
CRAMBE L. **Brassicaceae** *(Cruciferae)*					
orientalis L.				TUR	AFG ISR LEB
CRANTZIA Nutt. **Apiaceae** *(Umbelliferae)*					
ambigua (Urb.) Britton			PR		
CRASSOCEPHALUM Moench **Asteraceae** *(Compositae)*					
crepidioides (Benth.) S. Moore (Syn. *Gynura crepidioides* Benth.)			CEY IND	GHA ZAM	

CRASSOCEPHALUM Moench Asteraceae *(Compositae) (Cont'd)*	S	P	C	X	F
rubens (Juss.) S. Moore			JPN RHO	GHA	
CRASSULA L. **Crassulaceae**					
helmsii A. Berger				AUS	
CRATAEGUS L. **Rosaceae**					
crus-galli L.				USA	
douglasii Lindl.				USA	
marshallii Eggelst.				USA	
oxyacantha L.				TUR	
rivularis Nutt.				USA	
saligna Greene				USA	
succulenta Link				USA	
CREPIS L. **Asteraceae** *(Compositae)*					
biennis L.				USA	
bullosa Tausch.		TUN			
capillaris (L.) Wallr.		NZ	CAN ENG	ARG CHL USA	AUS
fraasii Sch.-Bip.				GRE	
japonica (L.) Benth. *(See **Youngia japonica** (L.) Dc.)*		MAU	CHN IDO	FIJ PK VIE	AUS HAW JPN TAI
occidentalis Nutt.				USA	
parviflora Desf.			IRA		
setosa Hall. *f.*			ARG	NZ	
taraxacifolia Thuill.		NZ			POR
tectorum L.			CAN FIN SOV	USA	AUS
vesicaria L.			MOR		
virens L.				CHL POR	
CRESCENTIA L. **Bignoniaceae**					
cujete L.				JAM	
CRESSA L. **Convolvulaceae**					
cretica L.			IRQ		
CRINUM L. **Amaryllidaceae**					
defixum Ker-Gawl.				IND	

	S	P	C	X	F
CROCUS L. **Iridaceae**					
sativus L.			SPA		
CROSSOPHORA Link **Euphorbiaceae**					
tinctoria A. Juss.		ISR	IRA		
CROTALARIA L. **Papilionaceae** (*Leguminosae*)					
aculeata De Wild.		ANG			
anagyroides H. B. K.			IDO	NGI PLW TRI	
berteriana Dc.			HAW	USA	
bracteata Roxb.			PHI		
breviflora Dc.			BRA		
chrysochlora Bak. *f. ex* Harms			KEN		
falcata Vahl *ex* Dc.				ANG	
fulva Roxb.				USA	
goreensis Guill. & Perr.		GHA		AUS	
incana L.			HAW KEN PHI TRI	COL GUA HON PER SAL USA	AUS IDO
intermedia Kotschy			KEN		AUS
juncea L.			PHI	SUR	
laburnifolia L.			KEN		
lanceolata E. Mey.				COL	
linifolia L. *f.*				PHI	IDO
longirostrata Hook. & Arn.			HAW	USA	
medicaginea Lam.				IND	AFG IDO
mucronata Desv. (Syn. *C. striata* Dc.)			AUS HAW	COL PLW USA	FIJ
mysorensis Roth				IND	
polysperma Kotschy			KEN		
pycnostachya Benth.				SUD	
quinquefolia L.			PHI	IND	
retusa L.			AUS PR	DR GHA HON IDO IND NIC PLW	

CROTALARIA L. Papilionaceae *(Leguminosae) (Cont'd)*	S	P	C	X	F
retusa L. *(Cont'd)*				TRI	
sagittalis L.				MEX SAL USA	
saltiana Andr.			SUD	PHI TAI	HAW
senegalensis Bacle			SUD		
spectabilis Roth				HAW	AUS HON USA
stipularia Desv.				PR	
striata Dc. *(See* C. mucronata Desv.)				IDO PLW PR	IND SUR USA	AUS FIJ HAW MEL NGI
usaramoensis Baker *(See* C. zanzibarica Benth.)				IDO	AUS
verrucosa L.				IND	PLW TRI	IDO NGI
vitellina Ker-Gawl.				COL GUA SAL	
zanzibarica Benth. (Syn. *C. usaramoensis* Baker)					
CROTON L. Euphorbiaceae					
bonplandianus Baill.				IND	MAU
campestris St. Hil.				BRA	
capitatus Michx.				USA	
ciliato-glanduliferus Ort.			MEX		
dioicus Cav.				MEX	
glandulosus L.				IDO	BRA CEY USA
hirtus L'herit.			MAL	TRI	IDO SUR VIE	ARG
humilis L.				JAM	
leptostachyus H. B. K.				COL	
lindheimerianus Scheele				USA	
linearis Jacq.				JAM	
lobatus L.	TRI	CUB GHA	ARG PR	ANT COL DR IVO NIG	

CROTON L. Euphorbiaceae *(Cont'd)*	S	P	C	X	F
lobatus L. *(Cont'd)*				SUR	
monanthogynus Michx.				USA	
sparsiflorus Morong				BND IND	
texensis (Klotzsch) Muell.-Arg.				USA	
tinctorius L.			LEB	SOV SPA	
trinitatis Millsp.				COL PER	
verbascifolia Willd.			LEB		
wilsonii Griseb.				JAM	
CRUPINA L. Asteraceae *(Compositae)*					
crupinastrum (Moris.) Vis.			LEB		ISR
CRYPSIS Ait. Poaceae *(Gramineae)* (Syn.*HELEOCHLOA* Host *ex* Roem.)					
aculeata (L.) Ait.				IRQ	ISR
schoenoides (L.) Lam. (Syn. *Heleochloa schoenoides* (L.) Host)					
CRYPTOCORYNE Fisch. Ex Wydl. Araceae					
ciliata (Roxb.) Schott				VIE	
CRYPTOSTEGIA R. Br. Asclepiadaceae					
grandiflora R. Br.			AUS	USA	
CRYPTOSTEMMA R. Br. Asteraceae *(Compositae)*					
calendula (L.) Druce	AUS			NZ	
CRYPTOTAENIA Dc. Apiaceae *(Umbelliferae)*					
canadensis (L.) Dc.			HAW		
japonica Hassk.			JPN		
CUCUMIS L. Cucurbitaceae					
anguria L.				AUS PER	
callosus Cogn.				IND	
dipsaceus Ehrenb. *ex* Spach			HAW	PER SUD USA	
hirsutus Sond.				ANG	
melo L.			COL	GHA PLW SUD	AFG CAB IDO
metuliferus E. Mey. *ex* Schrad.	AUS				
myriocarpus Naud.			AUS	SAF	

CUCUMIS L. Cucurbitaceae *(Cont'd)*	S	P	C	X	F
sativus L.			AUS	CAB IND	AFG IDO
trigonus Roxb.				AUS IND	AFG
CUCURBITA L. Cucurbitaceae					
andreana Naud.			ARG		
digitata A. Gray			USA		
foetidissima H. B. K.			USA		
pepo L.			JAM PLW		
CUDRANIA Trec. Moraceae					
javanensis Trec.			AUS		
CUPHEA P. Br. Lythraceae					
balsamona Cham. & Schlecht.			IDO	BRA COL	
carthagenensis (Jacq.) Macbr.	HAW	FIJ	BRA	PLW USA	
densiflora Koehne				BRA	
glutinosa Cham. & Schlecht.				URU	
parsonsia R. Br. *ex* Steud.				JAM	
petiolata Pohl *ex* Koehne				USA	
racemosa Spreng.				COL	
wrightii A. Gray		MEX			
CURATELLA L. Dilleniaceae					
americana L.				BRA	NIC
CURCAS Adans. Euphorbiaceae					
curcas (L.) Britton *(See* **Jatropha curcas** L.)					PR
CUSCUTA L. Convolvulaceae					
americana L.		MEX		BRA DR GER VEN	
approximata Bab.	IRA		ITA MOR USA	AFG IRQ ISR JOR PK SOV TUR YUG	
arvensis Beyrich *ex* Hook.				AUS CAN DEN	

CUSCUTA L. Convolvulaceae *(Cont'd)*	S	P	C	X	F
arvensis Beyrich *ex* Hook. *(Cont'd)*				FRA GER ITA JAM MAL PR SOV TUR USA	
australis R. Br.		ITA	AUS JPN MOR	BND CHN DEN FIJ FRA IDO MAL MEL NGI SOV USA	GER KOR PK
babylonica Aucher & Choisy				ISR	
breviflora Vis.				SOV	
brevistyla A. Braun *ex* A. Rich.				ISR	
burrellii Yuncker				USA	
californica Choisy				GER	
campestris Yuncker	AFG BOT HUN	ARB PK SOV	CAN MOR RHO SAF	AST CHL EGY ENG GER IDO IND ISR MEX NET ROM UGA USA YUG	
chilensis Ker-Gawl.				CHL MAU	AFG JPN
chinensis Lam.				CHN IND MAU SOV	
coryli Engelm.				USA	
cupulata Engelm.				USA	
epilinum Weihe			SOV	MOR	AST BEL CAN FRA GER ISR ITA SAF TUR USA

CUSCUTA L. Convolvulaceae *(Cont'd)*	S	P	C	X	F
epithymum (L.) Murr.	AFG HUN POL TUR	ITA POR SOV	CAN IRA SAF SPA VEN	ARG AST AUS BEL CHL EGY ENG FRA GER GRE ISR JOR JPN MOR NOR NZ ROM SWE USA YUG	
europaea L.	AFG POL	CZE SOV		AUS BEL BUL DEN FIN FRA GER GRE ITA JOR PK ROM SWE TUR USA YUG	
gigantea Griff.				ISR PK	
glandulosa (Engelm.) Small				CUB MEX PR USA	
glomerata Choisy				USA	
gronovii Willd. *ex* Roem. & Schult.			CAN	AST DEN FRA GER LEB SOV USA	
hyalina Roth				IND PK SUD	
indecora Choisy	ARG		MOR USA	CUB GRE IND JAM MEX PR VEN	
japonica Choisy				CHN	

CUSCUTA L. Convolvulaceae *(Cont'd)*	S	P	C	X	F
japonica Choisy *(Cont'd)*				FRA JPN	
kotschyi Des Moul.				ISR	
lehmanniana Bunge				SOV	
lupuliformis Krock.			SOV	GER IND NET TUR	BRA
maroccana Trabut		PK		MOR	
monogyna Vahl.		IRA PK SOV	ITA MOR SPA TUR	AFG EGY IRQ ISR JOR ROM YUG	
obtusiflora H. B. K.				BRA SOV	IND
palaestina Boiss.			LEB	IRQ ISR	
pedicellata Ledeb.			EGY	ISR	
pentagona Engelm.			ARG USA	CAN GER ITA JAM PR SOV YUG	
planiflora Tenore		PK	ITA	AFG IRA IRQ ISR JOR LEB RHO SOV USA	
polygonorum Engelm.				USA	
pulchella Engelm.				PK	
racemosa Mart.				ARG CHL SOV USA	
reflexa Roxb.	AFG NEP	IND PK	TUN	BND ENG IDO MAU USA	
rhodesiana Yuncker				SAF	
sandwichiana Choisy			HAW	USA	
suaveolens Ser.			ARG RHO	SOV	

CUSCUTA L. Convolvulaceae *(Cont'd)*	S	P	C	X	F
suaveolens Ser. *(Cont'd)*			SAF SPA		
subinclusa Dur. & Hilg.				GER	
trifolii Bab.		CZE	GER	HUN YUG	
triumvirati Lange				MOR	
umbellata H. B. K.				JAM MEX PR USA	
umbrosa Beyrich *ex* Hook.				CAN USA	
CYANOTIS D. Don Commelinaceae					
axillaris (L.) D. Don	IND SUD THI	PHI		BND VIE	IDO
cristata D. Don				PHI VIE	IDO
cucullata Kunth				IND	
lanata Benth.				IVO	
papilionacea Schult. *f.*		CAB			
villosa Schult. *f.*		CEY		IND	
CYATHULA Bl. Amaranthaceae					
achyranthoides Moq.				CNK PER	
cylindrica Moq. (Syn. *C. schperiana* Moq.)			KEN		
polycephala Bak.			KEN		
prostrata (L.) Bl.	GHA		KEN	CNK FIJ NIG PHI PLW VIE	IDO
schperiana Moq. *(See* **C. cylindrica** Moq.)			KEN		
uncinulata (Schrad.) Schinz				KEN SAF	
CYCLEA Arn. *ex* Wight Menispermaceae					
burmanni Arn. *ex* Wight				CEY	
CYCLOLOMA Moq. Chenopodiaceae					
atriplicifolium (Spreng.) Coult.				USA	
platyphyllum Moq.				ARG	

CYCLOSORUS Link Thelypteridaceae	S	P	C	X	F
afrus (Christ) Ching				GHA	
aridus (Don) Ching (Syn. *Dryopteris arida* O. Ktze.)					
dentatus (Forsk.) Ching				PHI	
gongylodes (Schkuhr) Link				MAL	
parasiticus (L.) Farwell			HAW		
CYCLOSPERMUM Caruel **Apiaceae** *(Umbelliferae)*					
leptophyllum (Pers.) Sprague			HAW		
CYMBALARIA Hill **Scrophulariaceae**					
muralis Gaertn., Mey.& Scherb.				USA	
CYMBONOTUS Cass. **Asteraceae** *(Compositae)*					
lawsonianus Gaudich.				AUS	
CYMBOPOGON Spreng. **Poaceae** *(Gramineae)*					
afronardus Stapf			KEN		
caesius (Nees) Stapf				IND	
citratus (Dc.) Stapf				JAM	
giganteus (Hochst.) Chiov.				GHA	
proximus (Hochst.) Stapf		SUD		GHA	
refractus (R. Br.) A. Camus			HAW		AUS USA
schoenanthus Spreng.				IND	
validus (Stapf) Stapf *ex* Burtt-Davy			KEN		
CYMOPTERUS Rafin. **Apiaceae** *(Umbelliferae)*					
watsoni (Coult. & Rose) Jones				USA	
CYNANCHUM L. **Asclepiadaceae**					
acutum L.			EGY IRQ	SOV TUR	AFG ISR
nigrum (L.) Pers.				USA	
vincetoxicum (L.) Pers.				USA	
CYNARA L. **Asteraceae** *(Compositae)*					
cardunculus L.		ARG AUS	SPA	CHL MOR NZ URU	
humilis L.			MOR		
syriaca Boiss.			ISR		

CYNODON L. Poaceae *(Gramineae)*		S	P	C	X	F
dactylon (L. C. Rich) Pers.		ANG	ARB	CHN	AFG	BUR
(Syn. *Panicum dactylon* L.)		ARG	BOL	IRQ	ANT	
		AUS	BUL	MOR	BOR	
		BND	CUB	MOZ	CAB	
		BRA	ETH	PLW	CHL	
		CEY	FRA	SUR	CR	
		COL	GUA		DAH	
		EGY	HAW		DR	
		GRE	ITA		ECU	
		GUI	JAM		FIJ	
		IDO	JOR		GHA	
		IND	JPN		HK	
		IRA	MAL		HON	
		ISR	MAU		HUN	
		KEN	MEX		IVO	
		LEB	NZ		KOR	
		NGI	PAR		LIB	
		PHI	PER		MEL	
		PK	PR		MIC	
		POR	RHO		MLI	
		SPA	SAF		NGR	
		SUD	SAL		NIC	
		TAI	SOV		NIG	
		TRI	SWZ		SEG	
		TUN	THI		URU	
		UGA	TNZ		VIE	
		YUG	TUR		VOL	
		ZAM	USA			
			VEN			
hirsutus Stent				ARG		
plectostachyum (Schum.) Pilger				AUS		HAW
				KEN		IND
CYNOGLOSSUM L. **Boraginaceae**						
coeruleum Hochst. *ex* Dc.		TNZ		KEN		POR
creticum Mill.				ARG	CHL	ISR
enerve Turcz.				SAF		
geometricum Bak. & C. H. Wright				KEN		
lanceolatum Forsk.				KEN		
				SAF		
officinale L.				CAN	USA	
				POL		
				SOV		
				SPA		
pictum Soland.				CHL		
				TUR		
CYNOSURUS L. **Poaceae** *(Gramineae)*						
cristatus L.				CHL		AUS
				NZ		HAW
				USA		IND
echinatus L.				POR	CHL	AUS
					NZ	IND
						ISR
						SAF

CYPERUS L. Cyperaceae	S	P	C	X	F
albo-marginatus Mart. & Schrad. *ex* Nees *(See* **Pycreus albo-marginatus** Nees)				MEX	
alopecuroides Rottb. (Syn. *Juncellus alopecuroides* (Rottb.) C. B. Clarke)			EGY		ISR SUD THI
alternifolius L.				FIJ USA	
amabilis Vahl		IND			
amuricus Maxim.			JPN	KOR	
aristatus Rottb.		TNZ			IDO
articulatus L.	PER	SUR	EGY	IND JAM TRI USA VEN	GHA ISR TUR
babakan Steud.			MAL	THI	
bancanus Miq. *(See* **C. trialatus** (Boeck.) Kern)				VIE	
brevifolius (Rottb.) Hassk. (Syn. *Kyllinga brevifolia* Rottb.)	MAL	ECU HAW PER SOV SUR	AUS CAB IDO JPN THI	ANT ARG BOR CHN FIJ IRA MEL MIC NGI NZ PHI PLW TAI USA	CR TRI
bulbosus Vahl	TNZ	IND	KEN	PK	AUS
castaneus Willd.				VIE	
cayennensis Link *(See* **C. haspan** L.)			ARG BRA		
compactus Retz.			PHI	THI	
compressus L.	HON IND	TAI VIE	CAB CHN IDO JPN MAL PHI	BOR BOT DAH EGY FIJ GHA IVO MAU MIC MOZ NGI PK PLW RHO SEG SUD SUR THI	CR

CYPERUS L. Cyperaceae *(Cont'd)*	S	P	C	X	F
corymbosus Rottb.		PER		ARG BND CEY IND	
cyperinus (Retz.) Valck. Sur. (Syn. *Mariscus cyperinus* (Retz.) Vahl)			TAI	PHI PLW	
cyperoides (L.) O. Ktze. (Syn. *Mariscus sieberianus* Nees)			IDO	BOR FIJ NEP PLW	
dehiscens Nees *(See* **C. tegetum** Roxb.*)*			CEY	BND	
diandrus Torr.				USA	
difformis L.	AUS CEY CHN EGY FIJ GHA HAW ITA JPN MAD MEX NGR NIG PHI PK ROM SWZ TAI USA	KOR POR SPA THI	HK IDO IND	BUR FRA IVO MAU NGI RHO SAF VIE	AST GER GRE MAL TUR
diffusus Vahl	TRI		COL	PHI	CR
digitatus Roxb.		SEG	BOR		IDO
distachyos All. *(See* **C. laevigatus** L.*)*					AFG ISR
distans L. *f.*	TRI	GHA SWZ	PHI	CAB FIJ MAU NIG VIE	IDO
eragrostis Vahl			ARG CHN POR	CHL USA	AUS IDO NZ
erythrorhizos Muhl.			IDO	USA	
esculentus L. (Syn. *C. repens* Ell.)	ANG CAN KEN MAD MOZ PER RHO SAF TNZ USA	AUS HAW IND MEX SWZ	ARG IRA POR	ALK CAM CHL COL CUB DAH ETH FRA GHA GUI IDO IVO	CAB

CYPERUS L. Cyperaceae *(Cont'd)*	S	P	C	X	F
esculentus L. *(Cont'd)*				MLI MRE NIC NIG PR SEG TAI VEN	
exaltatus Retz.		IND		AUS	
fasciculatus Ell. *(See* **Pycreus propinquus** Nees)				SUR	
ferax L. C. Rich. *(See* **C. odoratus** L.)		COL PER SAL	TAI	MEX PAN	CR IDO
flavescens L. *(See* **Pycreus flavescens** Beauv. *ex* Reichb.)		POR			AFG
flavidus Retz.	IND			BND CEY JPN	
fulvus (Vahl) Nees			COL		
fuscus L.		POR			AFG ISR
giganteus Vahl				SUR	
globosus All.		JPN TAI	IDO		AFG ISR
gracilinux C. B. Clarke				GHA	
gracilis R. Br.			HAW	AUS	
hakonensis Fr. & Sav.		JPN		KOR	
haspan L. *(Syn. C. cayennensis* Link)	CEY IND	BRA DAH PER VIE	BOR CAB HON IDO JPN MAL PK SUD SUR TAI ZAM	ARG BOT CNK FIJ GHA IVO NIG PHI RHO SEG THI USA	CHN HK KOR
hermaphroditus (Jacq.) Standl.		BRA		MEX	
imbricatus Retz. *(Syn. C. radiatus* Vahl)		PHI		THI	AFG
iria L.	CEY IND PHI	AUS FIJ JPN KOR MAL SEG SWZ TAI THI USA	IDO	BND CAB CHN CUB DR PK SOV TRI VIE	AFG HK

CYPERUS L. Cyperaceae *(Cont'd)*	S	P	C	X	F
javanicus Houtt.		HAW		FIJ PLW	
kyllingia Endl. (Syn. *Kyllinga monocephala* Rottb.)	PHI	IND	HAW	IDO NIG PLW THI USA	
laetus J. & C. Presl		BRA		CHL	
laevigatus L. (Syn. *C. distachyos* All.)			EGY	PK	AFG ISR
ligularis L. *(See* **Mariscus rufus** H. B. K.)	BRA			COL PAN TRI	
longus L.			EGY	MOR	TUR	AFG ISR POR
luzulae Rottb. *ex* Willd. (Syn. *C. surinamensis* Rottb.)		BRA PER	ARG TRI	PAN	CR
malaccensis Lam.				PHI THI	IDO
maranguensis K. Schum.			KEN	TNZ	
maritimus Poir.		SEG			
michelianus Delile *(See* **C. pygmaeus** Rottb.)			IRQ	BND	
microiria Steud.	JPN	KOR		CHN	TAI
monti L. (Syn. *C. serotinus* Rottb.)					
mutisii (H. B. K.) Griseb. *(See* **Mariscus mutisii** H. B. K.)			SAL	COL MEX	
nipponicus Fr. & Sav. *var.* nipponicus (Syn. *Juncellus nipponicus* (Fr. & Sav.) C. B. Clarke)					
novae-hollandiae Boeck.		AUS			`
odoratus L. (Syn. *C. ferax* L. C. Rich.) (Syn. *Mariscus ferax* (L. C. Rich.) C. B. Clarke) (Syn. *Torulinium ferax* (L. C. Rich.) Ham.)			IRQ MAL	PLW	IDO
orthostachyus Fr. & Sav. *(See* **C. truncatus** C. A. Mey. *ex* Turcz.)			JPN		
pangorei Rottb.				IND	
papyrus L.				DAH	EGY
pilosus Vahl	JPN	FIJ MAL	SUD TAI	ANT BOR CAB CEY IDO IND NGI PHI PLW	CHN HK

CYPERUS L. Cyperaceae *(Cont'd)*	S	P	C	X	F
pilosus Vahl *(Cont'd)*				THI VIE	
polystachyos Rottb. (Syn. *C. tenuis* Muhl.) (Syn. *Pycreus odoratus* Urb.) (Syn. *P. polystachyos* (Rottb.) Beauv.)		BRA FIJ IND	HAW	PLW VIE	AUS IDO ISR JPN
procerus Rottb.		THI		CEY	IDO
pulcherrimus Willd. *ex* Kunth				THI	IDO
pumilus L.	IND				IDO
pygmaeus Rottb. (Syn. *C. michelianus* Delile)			IDO IND	IRQ THI	ISR
radians Nees & Mey.				VIE	
radiatus Vahl *(See* **C. imbricatus** Retz.)		PHI			AFG IDO
reduncus Hochst. *ex* Boeck.		NIG			
repens Ell. *(See* **C. esculentus** L.)		IDO			
retzii Nees		AUS			
rigidifolius Steud.	KEN		ETH		
rotundus L. (Syn. *C. tuberosus* Rottb.)	ANT ARG AUS BOR BRA CEY CNK COL CR FIJ GHA GRE GUI HAW HON IDO IND IRA ISR ITA JAM JPN KEN MAD MAL MEL MEX MOZ NIG PAN PER PHI PK PLW POR RHO SAF SOV	AFG ARB CAB CHL CUB EGY ETH FRA IVO MAU MOR NGI NIC PAR PR SAL SUR VIE	CHA CHN IRQ JOR	ANG BER BND BOL CAM DR GAB GUA KOR LEB MLI MRE NGR NZ PLE SEG URU	

CYPERUS L. Cyperaceae *(Cont'd)*	S	P	C	X	F
rotundus L. *(Cont'd)*	SPA SUD SWZ TAI THI TNZ TRI TUN TUR UGA USA VEN YUG ZAM				
sanguinolentus Vahl (Syn. *Pycreus eragrostis* (Vahl) Palla) (Syn. *Pycreus sanguinolentus* (Vahl) Nees)			JPN		AFG
schweinfurthianus Boeck.			NIG		
seemannianus Boeck.				PLW	
serotinus Rottb. *(See* **C. monti** L.)	AFG JPN	ITA KOR	POR	IDO LEB PK	
seslerioides H. B. K.				MEX	
sesquiflorus (Torr.) Mattf. & Kuk.		BRA		AUS	
sesquiflorus (Torr.) Mattf. & Kuk. *var.* subtriceps (Nees) T. Koyama (Syn. *Kyllinga cylindrica* Nees)					
sphacelatus Rottb.	MOZ NIG	MAL		CNK GHA IVO TRI	
strigosus L.		MEX		BND FIJ USA	
surinamensis Rottb. *(See* **C. luzulae** Rottl. *ex* Willd.)	TRI		ARG		
tegetiformis Roxb.				IND THI VIE	ISR
tegetum Roxb. (Syn. *C. dehiscens* Nees)				IND	
tenuiculmis Boeck.			NIG	CEY	
tenuis Muhl. *(See* **C. polystachyos** Rottb.)				SAL	
tenuispica Steud.			NIG	CEY JPN	
trialatus (Boeck.) Kern (Syn. *C. bancanus* Miq.)					
truncatus C. A. Mey. *ex* Turcz. (Syn. *C. orthostachyus* Fr. & Sav.)					

CYPERUS L. Cyperaceae *(Cont'd)*	S	P	C	X	F
tuberosus Rottb. *(See C. rotundus L.)*				BND MAU	
uncinatus Poir.				PHI	IDO
zollingeri Steud.		MAL		PHI	
CYRTOCOCCUM Stapf **Poaceae** *(Gramineae)*					
accrescens (Trin.) Stapf				MAL	
oxyphyllum (Hochst. *ex* Steud.) Stapf				MAL PLW	
patens (L.) A. Camus		IND	TAI	PLW	BUR
trigonum (Retz.) A. Camus (Syn. *Panicum trigonum* Retz.)			BOR	CEY PLW	IND
CYRTOSPADIX C. Koch **Araceae**					
bicolor Britton & Wils. *(See Caladium bicolor (Ait.) Vent.)*					PR
CYSTOPTERIS Bernh. **Aspidiaceae**					
fragilis (L.) Bernh.				TUR	
CYTISUS L. **Papilionaceae** *(Leguminosae)*					
monspessulanus L.			HAW	CHL NZ USA	
scoparius (L.) Link		NZ	HAW IRA	IND USA	AUS
DACTYLIS L. **Poaceae** *(Gramineae)*					
glomerata L. (Syn. *D. hispanica* Roth)	AUS	NZ	COL FIN IRA JPN LEB SPA	ARG GRE IRQ ITA PER USA	HAW IND ISR SAF
hispanica Roth *(See D. glomerata L.)*					LEB
DACTYLOCTENIUM Willd. **Poaceae** *(Gramineae)*					
aegyptium (L.) Richt.	AUS GHA MOZ NIG PHI THI UGA	IND KEN SAL SEG TAI TNZ USA	COL EGY ETH HAW PR ZAM	ANG ANT ARB BND BRA CEY CHN CUB DAH DR GUI IDO ISR IVO JAM MAL MAU MLI	ARG BUR FIJ PK SAF

DACTYLOCTENIUM Willd. Poaceae *(Gramineae) (Cont'd)*	S	P	C	X	F
aegyptium (L.) Richt. *(Cont'd)*				MRE NGI NGR NIC PER PLW SUD TRI VIE VOL	
giganteum Fish. & Schweick.				MOZ	
DALEA L. **Papilionaceae** *(Leguminosae)*					
alopecuroides Willd.				MEX	AUS
DAMASONIUM Mill. **Alismataceae**					
alisma Mill.			POR		
australe Salisb. (Syn. *D. minus* Buchen.)					
minus Buchen. *(See* **D. australe** Salisb.)	AUS				
DANTHONIA Dc. **Poaceae** *(Gramineae)*					
spicata (L.) Beauv.			USA		
DASYLIRION Zucc. **Agavaceae**					
texanum Scheele			USA		
wheeleri S. Wats.			USA		
DATURA L. **Solanaceae**					
arborea L.			BRA		
discolor Bernh.			USA		
fastuosa L. *(See* **D. metel** L.)		GHA		IND PK	CAB IDO
ferox L.	SAF	ARG	AUS RHO	CHL URU	
innoxia Mill.			EGY		
leichhardtii F. Muell.			AUS		
metel L. (Syn. *D. fastuosa* L.)		GHA IND	AUS KEN	BOR FIJ MAU SUD THI TUR USA	AFG ISR SAF
meteloides Dc. *ex* Dun.		.		AUS USA	
stramonium L. *var.* chalybaea Koch (Syn. *D. tatula* L.)	AFG BOT CHL ETH GHA	HON ISR POR SWZ USA	ARG AST AUS EGY ENG	ANG ANT ARB BEL BND	KOR

DATURA L. Solanaceae *(Cont'd)*	S	P	C	X	F
stramonium L. *var.* chalybaea Koch *(Cont'd)*	GUA KEN MOZ PER POL SAF TNZ UGA		GER HAW HUN IDO IRA LEB MOR PR RHO SOV TUR	BRA BUL CAN CNK COL CR CZE DR FIJ GRE IND IRQ ITA JAM JOR MEX NEP NOR NZ PK PLW SAL SEG SUD THI VEN YUG	
tatula L. *(See* **D. stramonium** L. *var.* chalybaea Koch)			AUS RHO		SAF
DAUBENTONIA Dc. Papilionaceae *(Leguminosae)*					
punicea (Cav.) Dc.			USA		
texana Pierce				USA	
DAUCUS L. Apiaceae *(Umbelliferae)*					
aureus Desf.			ISR LEB		ITA
carota L.	AFG GRE HUN POL	JOR MAU PR SWE TUN	AST CAN EGY ENG GER IRA IRQ PLW SOV USA	BEL BRA CAB CHL CR DR FRA ICE ISR ITA JPN MOR NEP NET NOR NZ PER SUD THI TUR VEN YUG	AUS KOR POR
glochidiatus (Lab.) Fisch., Mey. & Ave-Lall.				AUS NZ	
guttatus Sibth. & Sm.				GRE	

DAUCUS L. Apiaceae *(Umbelliferae)* *(Cont'd)*	S	P	C	X	F
maximus Desf.			ISR LEB		
montanus Humb. & Bonpl. *ex* Schult.			PER		
montevidensis Link *ex* Spreng.			ARG		
muricatus L.			MOR		POR
pusillus Michx.			HAW		
setulosus Guss. *ex* Dc.			LEB		
DECODEN J. F. Gmel. **Lythraceae**					
verticillatum (L.) Ell.			USA		
DEINOSTEMA Yamasaki **Scrophulariaceae**					
violaceum Yamasaki			JPN		
DELPHINIUM L. **Ranunculaceae**					
ajacis L. *(See* **Consolida ajacis** (L.) Schur)			POR	MOR	
andersonii A. Gray				USA	
axilliflorum Dc.			LEB		
barbeyi Huth			USA		
bicolor Nutt.				USA	
consolida L.			GER IRA SOV SPA		
cossonianum Battand.			MOR		
geyeri Greene				USA	
glaucum S. Wats.		ALK			
halteratum Sibth. & Sm.			POR		
megacarpum Nels. & Macbr.				USA	
nelsonii Greene				USA	
orientale Gay				MOR TUR USA	
scaposum Greene				USA	
scopulorum Gray				USA	
tricorne Michx.				USA	
virescens Nutt.				USA	
DENDROPHTHOE Mart. **Loranthaceae**					
falcata Bl.			IND		

	S	P	C	X	F
DENNSTAEDTIA Bernh. **Dennstaedtiaceae**					
punctilobula (Michx.) Moore				USA	
DENTELLA J. R. & G. Forst. **Rubiaceae**					
repens (L.) J. R. & G. Forst.		IDO		IND	
DERRIS Lour. **Papilionaceae** (*Leguminosae*)					
elliptica (Roxb.) Benth.	PLW				FIJ
DESCHAMPSIA Beauv. **Poaceae** (*Gramineae*)					
caespitosa (L.) Beauv. (Syn. *Aira caespitosa* L.)	FIN	NOR	GER IRA SOV	BEL NZ	AUS HAW IND SAF
elongata (Hook.) Munro *ex* Benth.				USA	
flexuosa (L.) Trin.			FIN		
DESCURAINIA Webb & Berth. **Brassicaceae** (*Cruciferae*)					
argentina O. E. Schulz			ARG		
longipedicellata O. E. Schulz				USA	
pinnata (Walt.) Brit.			CAN USA		
richardsonii (Sweet) O. E. Schulz				CAN USA	
sophia (L.) Webb *ex* Prantl		CAN IRA	SOV	TUR USA	AFG ARG
DESMANTHUS Willd. **Mimosaceae** (*Leguminosae*)					
virgatus Willd.			HAW TRI	BRA GUA MAU	PER
DESMODIUM Desv. **Papilionaceae** (*Leguminosae*)					
adscendens (Sw.) Dc.			TRI	ANG BRA CNK GHA PER	
adscendens (Swartz) Dc. *var.* trifoliastrum (Miq.) Schindl. (Syn. *D. trifoliastrum* Miq.)					
axillare Dc.				DR HON JAM PER	
baccatum (Miq.) Schindl.				PHI	
barbatum (L.) Benth. & Oerst.				MEX	
cajanifolium (H. B. K.) Dc.				PER	
canadense (L.) Dc.				USA	
canum (Gmel.) Schinz & Thel.			HAW	BRA PLW VEN	

DESMODIUM Desv. **Papilionaceae** *(Leguminosae) (Cont'd)*	S	P	C	X	F
capitatum (Burm. *f.*) Dc. *(See* **D. styracifolium** (Osb.) Merr.)				PHI	
dichotomum (Willd.) Dc.				SUD	
diffusum (Willd.) Dc.		IND			
frustescens (Jacq.) Schindl.			TRI	ANT MAU	
gangeticum (L.) Dc.				GHA NIG PHI	IDO
heterocarpum (L.) Dc.			IDO	PLW	CEY
heterophyllum (Willd.) Dc.			IDO TAI	FIJ PLW VIE	CEY
lasiocarpum (Beauv.) Dc.			TAI		
polycarpum (Poir.) Dc.				PER	
polygaloides Chod. & Hassl.				BRA	
procumbens Hitchc.		PHI		JAM TRI	
ramosissimum G. Don				GHA	
sandwicense E. Mey.			HAW	USA	
scorpiurus Desv.				GUA PLW SAL	
sequax Wall.			TAI		
sericophyllum Schlecht.				COL	
styracifolium (Osb.) Merr. (Syn. *D. capitatum* (Burm. *f.*) Dc.)					
supinum Dc.				JAM	PR
tortuosum (Sw.) Dc.			COL HAW PR	HON MEX PER PLW USA	AUS
triflorum (L.) Dc.	GHA IND		HAW PHI TAI TRI	MAU NGI NIG PER PLW SAL THI USA	AUS CEY IDO
trifoliastrum Miq. *(See* **D. adscendens** (Swartz) Dc. *var.* trifoliastrum (Miq.) Schindl.				BND	
uncinatum (Jacq.) Dc.			HAW	USA	
vargasianum B. G. Schubert				PER	
velutinum (Willd.) Dc.				GHA	

	S	P	C	X	F
DESMODIUM Desv. **Papilionaceae** *(Leguminosae) (Cont'd)*					
velutinum (Willd.) Dc. *(Cont'd)*				NIG	
DESMOSTACHYA Stapf **Poaceae** *(Gramineae)*					
bipinnata (L.) Stapf (Syn. *Eragrostis cynosuroides* (Retz.) Beauv.)	PK		EGY IRQ	IRA	IND
DIANTHERA Gronov. **Acanthaceae**					
pectoralis J. F. Gmel.				JAM	
DIANTHUS L. **Caryophyllaceae**					
anatolicus Boiss.				TUR	
armeria L.				HAW	USA
prolifer L. (Syn. *Tunica prolifera* Scop.)				CHL	
DICEROCARYUM Boj. **Pedaliaceae**					
zanguebaricum (Lour.) Merr.				RHO SAF	
DICHANTHELIUM Hitchc. **Poaceae** *(Gramineae)*					
clandestinum (L.) Gould (Syn. *Pennisetum clandestinum* Hochst.)					
DICHANTHIUM Willemet **Poaceae** *(Gramineae)*					
annulatum (Forsk.) Stapf (Syn. *Andropogon annulatus* Forsk.)		IND SUD	EGY IRQ	MAU PK	AFG AUS BUR HAW
aristatum (Poir.) C.E. Hubb. (Syn. *Andropogon nodosus* (Willem.) Nash)				MAU	AUS HAW IND
caricosum (L.) A. Camus (Syn. *Andropogon caricosus* L.)				IND TRI	
DICHAPETALUM Thou. **Dichapetalaceae**					
cymosum (Hook.) Engl.		RHO SAF			
guineense (Dc.) Keay				GHA	
DICHONDRA J. R. & G. Forst. **Convolvulaceae**					
repens G. Forst.				ARG AUS KEN RHO TAI	CHL NZ URU USA VEN
DICHROCEPHALA Dc. **Asteraceae** *(Compositae)*					
bicolor (Roth) Schlecht. (Syn. *D. latifolia* (Lam.) Dc.)		TAI			
latifolia (Lam.) Dc. *(See **D. bicolor** (Roth) Schlecht.)*				IDO TAI	MAU

DICHROMENA Michx. **Cyperaceae**	S	P	C	X	F
ciliata Vahl			PR TRI	COL DR NIC PER	CR
DICHROSTACHYS Wight & Arn. **Mimosaceae** *(Leguminosae)*					
cinerea (L.) Wight & Arn.		GHA	KEN		IDO
glomerata (Forsk.) Chiov.	SAF	CUB	KEN RHO	ANG	
nutans Benth.			USA		
DICLIPTERA Juss. **Acanthaceae**					
chinensis Juss.			HAW		
DICLIS Benth. **Scrophulariaceae**					
ovata Benth.			TNZ		
DICRANOPTERIS Bernh. **Gleicheniaceae**					
linearis (Burm.) Und. *(See* **Gleichenia linearis** (Burm.) Clarke)		CEY MAL	HAW TAI		
DIEFFENBACHIA Schott **Araceae**					
seguine (Jacq.) Schott			PR	DR FIJ	
DIERVILLA Adans. **Caprifoliaceae**					
lonicera Mill.			USA		
DIGERA Forsk. **Amaranthaceae**					
alternifolia (L.) Aschers. *(See* **D. muricata** (L.) Mart.)		SUD	IND		AFG IDO ISR
arvensis Forsk.		IND		PK	AFG
muricata (L.) Mart. (Syn. *D. alternifolia* (L.) Aschers.)					IND
DIGITALIS L. **Scrophulariaceae**					
orientalis Lam.			TUR		
purpurea L.			AUS SPA	BRA CHL NZ TUR USA	
DIGITARIA Heist. *ex* Fabr. **Poaceae** *(Gramineae)*					
abyssinica (Hochst.) Stapf			KEN		
adscendens (H. B. K.) Henr. (Syn. *D. chinensis* Horn.) (Syn. *D. henryi* Rend.) (Syn. *D. marginata* Link) (Syn. *Panicum fimbriatum* (Link) Kunth)	AUS JPN MAL MOZ SEG TAI THI	CEY GUI JAM SAF	HAW IDO IND	BOR CHN ITA KOR MAU MEL PHI PLW	BUR

DIGITARIA Heist. Ex Fabr. **Poaceae** *(Gramineae) (Cont'd)*	S	P	C	X	F
adscendens (H. B. K.) Henr. *(Cont'd)*				SOV USA VIE	
biformis Willd.				MAU	
chinensis Horn. *(See* **D. adscendens** (H. B. K.) Henr.)			TAI		JPN
ciliaris (Retz.) Koch	JPN			AUS	
debilis (Desf.) Willd.	NIG	ANG		GHA	HAW SAF
decumbens Stent			HAW	DR TRI USA VEN	SAF
didactyla Willd.				AUS MAU	IND
fauriei Ohwi			TAI		
filiformis (L.) Koel.				USA	HAW
fuscescens Presl		MAL		BRA PLW	BUR HAW IND
hayatae Honda			TAI		
henryi Rend. *(See* **D. adscendens** (H. B. K.) Henr.)	HAW		TAI	USA	CEY
horizontalis Willd. *non* Ohwi *(See* **D. microbachne** (Presl) Henr.)	GHA MEX SUR	SAF	TRI	ANG ANT BRA CNK COL HON IVO MAU NIG	HAW
insularis (L.) Mez *ex* Ekman				ARG	
ischaemum (Schreb.) Schreb.	MEX		CAN GER USA	BND KOR NZ	AUS HAW IND POL
longiflora (Retz.) Pers.	IND MOZ NGI SEG TRI	CEY IDO MAL	BOR TAI TUN ZAM	AST BUR CNK FIJ GHA IRA PHI PLW RHO SUD THI	CHN HAW HK IRQ ISR SAF
macractenia (Benth.) Hughes				AUS	
magna (Honda) Honda			TAI		

DIGITARIA Heist. *ex* Fabr. **Poaceae** *(Gramineae)* *(Cont'd)*	S	P	C	X	F
marginata Link *(See* **D. adscendens** (H. B. K.) Henr.)	MAL	CEY	BOR	VIE	JPN SAF THI
microbachne (Presl) Henr. (Syn. *D. horizontalis* Willd. *non* Ohwi)	PHI	PLW SUR	TAI	CAB IDO NGI THI VIE	HAW IND
pentzii Stent		MOZ			
pruriens (Trin.) Buese			HAW	BND PLW	USA
pseudo-ischaemum Buese			HAW		
sanguinalis (L.) Scop. (Syn. *Panicum sanguinale* L.) (Syn. *Paspalum sanguinale* (L.) Lam.)	ARG AUS BND BRA COL GUY HAW IDO IND ISR MEX NZ PHI POR ROM SAF SOV SPA SWZ TAI THI TUR USA	CAN CUB GER GUA ITA PAR URU VEN	CHN EGY IRQ LEB MAL MOR PR TRI	ANG ANT BOR CHL DR FRA HON IRA JAM JPN KEN KOR NGI NIC PER VIE	HK MIC POL YUG
scalarum (Schweinf.) Chiov.	KEN TNZ UGA	ETH IND		SAF	
sericea (Honda) Honda	TAI				
ternata (A. Rich.) Stapf	KEN TNZ UGA		BOR SUD	CHN CNK GHA RHO THI ZAM	SAF
timorensis (Kunth) Bal.		MAU	JPN TAI	PHI PLW	CEY HAW IND
velutina (Forsk.) Beauv.	IVO KEN TNZ UGA	SEG	SUD ZAM	BOT CNK DAH RHO	SAF
violascens Link	HAW	TAI	JPN	ANT MAU PLW TRI	IND USA

DINEBRA Jacq. Poaceae *(Gramineae)*	S	P	C	X	F
arabica Jacq.				IND	
retroflexa (Vahl) Panz.	EGY	SUD			AFG ISR
DIODIA L. **Rubiaceae**					
scandens Sw.				IVO	
teres Walt.			USA		
virginiana L.				USA	
DIOSCOREA L. **Dioscoreaceae**					
bulbifera L.	PLW				
pentaphylla L.				PLW	
DIOSPYROS L. **Ebenaceae**					
texana Scheele				USA	
virginiana L.				CHL USA	PER
DIPLACHNE Beauv. **Poaceae** *(Gramineae)*					
fusca (L.) Beauv.		AUS		EGY	AFG BUR IND SAF
malabarica (L.) Merr.			EGY		
DIPLAZIUM Sw. **Athyriaceae**					
esculentum (Retz.) Sw.			IDO TAI		
japonicum (Thunb.) Bedd. (Syn. *D. petersenii* Christ)			TAI		
petersenii Christ (*See* **D. japonicum** (Thunb.) Bedd.)		IDO			
DIPLORHYNCHUS Welw. *ex* Ficalho & Hiern **Apocynaceae**					
angustifolia Stapf				ANG	
DIPLOTAXIS Dc **Brassicaceae** *(Cruciferae)*					
catholica (L.) Dc.			POR		
erucoides (L.) Dc.	SPA	ISR ITA JOR TUN	IRQ LEB TUR	EGY IRA SUD	
muralis (L.) Dc.	SPA		ARG IRA SAF	CHL NZ USA	AUS
tenuifolia (L.) Dc.		ARG AUS		NZ USA	
tenuisiliqua Delile				MOR	
virgata (Cav.) Dc.			MOR		

DIPLOTAXIS Dc Brassicaceae *(Cruciferae) (Cont'd)*	S	P	C	X	F
virgata (Cav.) Dc. *(Cont'd)*			SAF		
DIPSACUS L. **Dipsacaceae**					
fullonum L.				AUS CHL NZ	SAF
laciniatus L.				TUR USA	
sylvestris Huds.			SPA USA	ARG	AUS
DIPTERACANTHUS Nees **Acanthaceae**					
prostratus (Poir.) Nees				IND	
DIRCA L. **Thymelaeaceae**					
palustris L.				USA	
DISCARIA Hook. **Rhamnaceae**					
toumatou Raoul				NZ	
DISSOTIS Benth. **Melastomataceae**					
debilis Triana				ANG	
irvingiana Hook.				GHA	
rotundifolia (Sm.) Triana				GHA NIG	
DISTICHLIS Rafin. **Poaceae** *(Gramineae)*					
spicata (L.) Greene		PER	ARG	CHL USA	AUS
stricta (Torr.) Rydb.				USA	
thalassica Desv.				CHL	
DITREMEXA Rafin. **Caesalpiniaceae** *(Leguminosae)*					
occidentalis (L.) Britton & Rose *(See* **Cassia occidentalis** L.)			PR	DR	
DODARTIA L. **Scrophulariaceae**					
orientalis L.				SOV	
DODONAEA Mill. **Sapindaceae**					
eriocarpa Sm.				HAW	
viscosa Jacq.		AUS		HAW	AFG USA
DOLICHOLUS Medic. **Papilionaceae** *(Leguminosae)* *(See* **RHYNCHOSIA** Lour.)					
minimus (L.) Medic. *(See* **Rhynchosia minima** (L.) Dc.)			PR		
DOLICHOS L. **Papilionaceae** *(Leguminosae)*					
lablab L.				HAW	IDO

	S	P	C	X	F
DOLICHOS L. **Papilionaceae** *(Leguminosae) (Cont'd)*					
lablab L. *(Cont'd)*					ISR
minimus L.				DR	
DOMBEYA Cav. **Sterculiaceae**					
rotundifolia Planch.			RHO		
DOPATRIUM Buch.-Ham. *ex* Benth. **Scrophulariaceae**					
junceum (Roxb.) Buch.-Ham. *ex* Benth.	AFG JPN	CHN HK IND KOR TAI	GHA	CAB HAW IDO RHO SEG THI VIE	CEY
DORYCNIUM L. **Papilionaceae** *(Leguminosae)*					
pentaphyllum Scop.			SPA		
DRABA L. **Brassicaceae** *(Cruciferae)*					
nemorosa L.				CAN JPN	SOV
DRACOCEPHALUM L. **Lamiaceae** *(Labiatae)*					
parviflorum Nutt.			CAN	USA	
thymiflorum L.			CAN SOV		
DRACUNCULUS Mill. **Araceae**					
vulgaris Schott (Syn. *Arum dracunculus* L.)					
DRYMARIA Willd. *ex* Roem. & Schult. **Caryophyllaceae**					
arenarioides H. B. K.				MEX	
cordata (L.) Willd. *ex* Roem. & Schult. (Syn. *D. diandra* Bl.)	CEY ECU HON JAM MEX	IND TAI	CNK HAW IDO NEP PR TNZ TRI VIE	ANT BRA COL CR GHA GUA JPN KEN MEL NGI PHI RHO SAL SUD SUR THI UGA USA VEN	AUS CHN HK
diandra Bl. *(See **D. cordata** (L.) Willd. ex Roem. & Schult.)*			TAI	AUS	
hirsuta Bart. *ex* Presl			IDO		
pachyphylla Woot. & Standl.				USA	

	S	P	C	X	F
DRYMOGLOSSUM Presl **Polypodiaceae**					
heterophyllum (L.) C. Chr.			IDO		
DRYOPTERIS Adans. **Aspidiaceae**					
arida O. Ktze. *(See* **Cyclosorus aridus** (Don) Ching)				IDO	
filix-mas (L.) Schott				BND USA	
spinulosa (Mull.) O. Ktze.				FIN	
splendens (Hook.) O. Ktze. (Syn. *D. sprengeli* O. Ktze.)					
sprengeli O. Ktze. *(See* **D. splendens** (Hook.) O. Ktze.)			PR		
vivipara (Raddi) C. Chr. (Syn. *Goniopteris prolifera* Fee)					
DUBOISIA R. Br. **Solanaceae**					
myoporoides R. Br.			AUS		
DUCHESNEA Sm. **Rosaceae**					
chrysantha Miq. *(See* **Fragaria indica** Andr.)			JPN		
indica (Andr.) Focke *(See* **Fragaria indica** Andr.)			CHN TAI	USA	NZ
DUGGENA Vahl **Rubiaceae**					
hirsuta Brit.			TRI		IDO PR
DYSCHORISTE Nees **Acanthaceae**					
depressa Nees			IND		
DYSODIA Cav. **Asteraceae** *(Compositae)*					
papposa (Vent.) Hitchc.			USA		
ECBALLIUM A. Rich. **Cucurbitaceae** *(Syn.ELATERIUM* Mill.)					
ciliatum (Cogn.) A. Rich. (Syn. *Elaterium ciliatum* Cogn.)					
elaterium (L.) A. Rich.	ISR JOR TUN	AUS LEB MOR	GRE		
ECDYSANTHERA Hook. & Arn. **Apocynaceae**					
rosea Hook. & Arn.			TAI		
ECHEANDIA Ort. **Liliaceae**					
reflexa Rose				MEX	
ECHINARIA Desf. **Poaceae** *(Gramineae)*					
capitata (L.) Desf.			LEB		

ECHINOCHLOA Beauv. Poaceae *(Gramineae)*	S	P	C	X	F
colona (L.) Link (Syn. *Panicum colonum* L.)	ARG AUS CEY COL ECU EGY FIJ GHA GUY HAW IND ISR MEX MOZ PER PHI PK PR SUR TAI THI TRI	CUB IDO JAM KEN MAD MAL NIG SAF SAL SEG SUD TNZ USA VEN	BOR IRQ MOR ZAM	ANG ANT ARB BND BOL BRA CAB CHL CHN CNK DR ETH FRA GUI IVO JPN MAU MLI MRE NGI NGR NIC PLW POR URU VIE VOL	BUR CR HK
crus-galli (L.) Beauv. (Syn. *Panicum crus-galli* L.)	ARG AUS BRA CEY COL EGY FIJ FRA GER GRE HUN IDO IND IRA ITA JPN KOR MEX MOZ NEP NGI NZ PER PHI POL POR ROM SOV SPA TAI USA YUG	BUL CAN CHL CZE HAW ISR MAD SWZ TNZ TUR	CHN IRQ LEB MOR	AST BEL BND CUB DR GUI MAL MAU MEL NET PK PLW SAF SAL SUD THI TUN URU VIE	AFG BUR CR JAM
crus-pavonis (H. B. K.) Schult.	AUS BRA PER PHI USA	SUR	ARG TRI	CHL FIJ GUA	HAW SAF
helodes (Hack.) Parodi			ARG		

ECHINOCHLOA Beauv. Poaceae *(Gramineae) (Cont'd)*	S	P	C	X	F
holubii (Stapf) Stapf	SWZ	SAF			
macrocarpa Vas.				SOV	
macrocorvi Nakai				KOR	
orizicola Vas. (Syn. *Panicum orizicola* Vas.)				SOV	
oryzoides (Ard.) Fritsch	POR			ARG JPN	
phyllopogon (Stapf) Koss	POR				
polystachya (H. B. K.) Hitchc. *(See* **Pseudechinolaena polystachya** (H. B. K.) Stapf)	ARG	IND MEX			HAW
pungens (Poir.) Rydb.			CAN	USA	
pyramidalis (Lam.) Hitchc. & A. Chase	NIG	MAD SUD SWZ	BOT CNK SEG TUN	DAH IVO PLW RHO TNZ ZAM	AUS GHA SAF
stagnina (Retz.) Beauv. (Syn. *Panicum stagninum* Retz.)	CEY		IDO	BND CAM EGY PHI SUD VIE	BUR GHA HAW IND MAD SAF
walteri (Pursh) Heller				AUS	
ECHINOCYSTIS Torr. & Gray **Cucurbitaceae**					
lobata (A. Mich.) Torr. & Gray		MEX		USA	
oregana Cogn.				USA	
ECHINODORUS Rich. **Alismataceae**					
cordifolius (L.) Griseb.				USA	
ECHINOPHORA L. **Apiaceae** *(Umbelliferae)*					
sibthorpiana Guss.				TUR	
ECHINOPS L. **Asteraceae** *(Compositae)*					
echinatus Roxb.				IND PK	
ritro L.			SOV		
sphaerocephalus L.			SOV		
spinosus L.				MOR	
strigosus L.				MOR	POR
viscosus Dc.				LEB	
ECHINOSPERMUM Sw. **Boraginaceae**					
lappula Rgl.				SOV	

ECHITES P. Br. Apocynaceae	S	P	C	X	F
repens Jacq.				DR	
umbellata Jacq.				JAM	
ECHIUM L. Boraginaceae					
amoenum Fisch. & Mey.			IRA		
australe Lam.				MOR	
glomeratum Poir.			LEB		
italicum L.			AUS IRA LEB	TUR	ISR
plantagineum L.		AUS TUN	ARG MOR SAF SPA	URU	POR
rauwolfii Delile			EGY		
vulgare L.		AUS	CAN SAF SOV SPA	CHL NZ TUR USA	BRA POL
ECLIPTA L. Asteraceae *(Compositae)*					
alba (L.) Hassk. *(See* **E. prostrata** (L.) L.*)*	TAI THI	HAW IDO IND JPN	CHN EGY IRQ PHI POR PR RHO TRI	ANG ARB ARG AUS BND BRA CEY COL CR CUB FIJ GHA IVO KOR MAL MEX PER PK SOV SUD SUR USA VIE	AFG HK SAF
erecta L.				MAU	
prostrata (L.) L. (Syn. *E. alba* (L.) Hassk.)		IND JPN SUD TAI	CHN HK IDO PHI SAF	ARB BND POR	HAW KOR
prostrata (L.) L. *var.* zippeliana (Bl.) Koster (Syn. *E. zippeliana* Bl.)					
zippeliana Bl. *(See* **E. prostrata** (L.) L. *var.* zippeliana (Bl.) Koster			PHI		

EGERIA Planch. **Hydrocharitaceae**	S	P	C	X	F
densa Planch. (Syn. *Elodea densa* (Planch.) Casp.)		HAW NZ USA		ARG AUS BOR SAF	
EGLETES Cass. **Asteraceae** *(Compositae)*					
prostrata O. Ktze.		TRI			
EICHHORNIA Kunth **Pontederiaceae**					
azurea Kunth		IND	ARG	VEN	
crassipes (Mart.) Solms	ANG BOR BRA CNK ETH FIJ IDO IND MAD MEL NGI PK THI USA	AUS CAB EGY HAW MEX NZ PAN SUD	ARG JPN KEN MAL PLW PR RHO SAF	BND BUR CEY CHN COL CR CUB DR HK HON MOZ NIC PHI POR SEG TAI TNZ TRI VEN VIE	GUY KOR NIG SAL SUR UGA
natans Solms		NIG			
ELAEAGNUS L. **Elaeagnaceae**					
argentea Pursh (Syn. *E. commutata* Bernh.)					SOV
commutata Bernh. *(See* **E. argentea** Pursh)				CHL USA	
ELATERIUM Mill. **Cucurbitaceae** *(See* **ECBALLIUM** A. Rich.)					
ciliatum Cogn. *(See* **Ecballium ciliatum** (Cogn.) A. Rich.)			SAL		
ELATINE L. **Elatinaceae**					
americana (Pursh) Arn.				USA	AUS
minima (Nutt.) Fisch. & Mey.				USA	
orientalis Makino *(See* **E. triandra** Schk.)		KOR	JPN		
triandra Schk. (Syn. *E. orientalis* Makino)		JPN			TAI
ELATINOIDES Wettst. **Scrophulariaceae**					
elatine Well. *(See* **Linaria elatine** Mill.)				CHL	

ELEOCHARIS R. Br. Cyperaceae	S	P	C	X	F
acicularis (L.) Roem. & Schult.	JPN TAI	AUS KOR	USA	BOR CAN CHN GER IND ITA MEX NET NOR SOV SWE THI	NIG
acuta R. Br.				AUS	
acutangula (Roxb.) Schult. (Syn. *E. fistulosa* Link)			NIG		THI
afflata Steud.	IND				
articulata Kunth (*See* **E. interstincta** (Roem. & Schult.) R. Br.)				FIJ	
atropurpurea (Retz.) Presl				IND	
attenuata (Fr. & Sav.) Palla *var.* laeviseta (Nakai) Hara (Syn. *E. laeviseta* Nakai)					
caribaea (Rottb.) S. F. Blake (*See* **E. geniculata** (L.) Roem. & Schult.)			SUR		AFG
cellulosa Torr.				USA	
chaetaria Roem. & Schult.	IND MAL			VIE	
congesta D. Don			JPN		
congesta D. Don *var.* japonica (Miq.) T. Koyama (Syn. *E. japonica* Miq.)					
dulcis (Burm. *f.*) Henschel (Syn. *E. plantaginea* (Retz.) Roem. & Schult.)		CAB MAD PR VIE	TAI THI	CEY CNK DAH GHA GUI IDO JPN MAL PHI PK PLW RHO TNZ UGA USA ZAM	ANG AUS IVO MAU MLI NGR SEG TRI VOL
equisetina Presl			PHI	AUS CAB VIE	
erecta Schumac.		NIG			
filiculmis Schur (*See* **E. palustris** (L.) R. Br.)				TRI	
fistulosa Link (*See* **E. acutangula** (Roxb.) Schult.)	THI		TAI		

ELEOCHARIS R. Br. Cyperaceae *(Cont'd)*	S	P	C	X	F
flaccida Urb. *(See* E. ochreata Nees)		SUR			
geniculata (L.) Roem. & Schult. (Syn. *E. caribaea* (Rottb.) S. F. Blake)		NIG		COL CR FIJ SAL	
interstincta (Roem. & Schult.) R. Br. (Syn. *E. articulata* Kunth)		SUR	PR	DR	
japonica Miq. *(See* E. congesta D. Don *var.* japonica (Miq.) T. Koyama)			JPN TAI		
kuroguwai Ohwi		JPN KOR			
laeviseta Nakai *(See* E. attenuata (Fr. & Sav.) Palla *var.* laeviseta (Nakai) Hara)				KOR	
mamillata Lindb. *f.*			JPN		
multicaulis Sm.			POR		
mutata R. Br.	GUY SUR			DAH GHA HON JAM	
nodulosa Schult.			BRA		
obtusa (Willd.) Schult. *(See* E. ovata (Roth) Roem. & Schult.)				USA	
ochreata Nees (Syn. *E. flaccida* Urb.)			BRA		
ovata (Roth) Roem. & Schult. (Syn. *E. obtusa* (Willd.) Schult.)			JPN		
palustris (L.) R. Br. (Syn. *E. filiculmis* Schur)		HUN SWE	CHN POR	BEL CAN CHL ECU EGY GER IRA IRQ JOR JPN MEX NOR PK ROM TUR USA	AFG ISR
parvula (Roem. & Schult.) Link				USA	
plantaginea (Retz.) Roem. & Schult. *(See* E. dulcis (Burm.f.) Henschel)		THI		BND	IND JPN
plantaginoidea W. F. Wight		IND			
quadrangulata (Michx.) Roem. & Schult.				USA	
retroflexa (Poir.) Urb.			MAL		
sphacelata R. Br.				AUS	

ELEOCHARIS R. Br. Cyperaceae *(Cont'd)*	S	P	C	X	F
subtilis Boeck.				TRI	
tetraquetra Nees (Syn. *E. wichurai* Boeck.)					
tuberosa Schult.		JPN			
variegata Presl				MAL	CAB
wichurai Boeck. *(See* **E. tetraquetra** Nees)			JPN		
wolfii A. Gray			IDO		
ELEPHANTOPUS L. Asteraceae *(Compositae)*					
augustifolius Sw.			TRI		
mollis H. B. K. *(See* **E. tomentosus** L.)	FIJ HON MEL NGI PLW	TAI	HAW PR TRI	COL DR GHA JAM MIC PER PHI SAL SEG SUD SUR USA VIE	
scaber L.		PHI	IDO IND MAL	BRA MAU USA VIE	
spicatus Aubl. *(See* **Pseudelephantopus spicatus** (Juss. *ex* Aubl.) C. F. Baker)	PLW		FIJ	PHI	
tomentosus L. (Syn. *E. mollis* H. B. K.)			PHI		
ELEUSINE Gaertn. **Poaceae** *(Gramineae)*					
aegyptiaca Desf.				IND	
africana Kenn-O'byrne		SAF	KEN		
compressa (Forsk.) Asch. & Schwienf. (Syn. *E. flagellifera* Nees)		PK			IND
flagellifera Nees *(See* **E. compressa** (Forsk.) Asch. & Schwienf.)		PK			AFG
indica (L.) Gaertn.	ANG BOR HAW IDO ISR JPN MAL MOZ NGI NIG NZ PHI PR RHO	AUS BRA CEY COL CUB GHA GUA GUI IND IVO JAM KEN MAU MEX	ARG CHN EGY PLW	ANT ARB BND BOL CNK CR DR ECU FIJ HK HON HUN IRA MEL	BUR ETH KOR

ELEUSINE Gaertn. **Poaceae** *(Gramineae) (Cont'd)*	S	P	C	X	F
indica (L.) Gaertn. *(Cont'd)*	SAF TAI THI TNZ VEN ZAM	NIC PER SEG SWZ TRI UGA USA		MIC NET PLE SUR VIE	
tristachya (Lam.) Lam.	MEX	BRA	ARG	CHL	
ELEUTHERANTHERA Poit. *ex* Bosc. **Asteraceae** *(Compositae)*					
ruderalis (Sw.) Sch.-Bip.				JAM PLW	
ELLISIA L. **Hydrophyllaceae**					
nyctelea (L.) L.				USA	
ELODEA Michx. **Hydrocharitaceae**					
canadensis Michx.		AUS ENG GER ITA NET NOR NZ POL PR SWE USA	SPA	ARG BEL CAN CHN CZE EGY FIN HAW HUN IND MAU MEX POR SOV THI	
densa (Planch.) Casp. *(See* **Egeria densa** Planch.*)*		USA			
nuttallii St. John.				BEL	
ELSHOLTZIA Willd. **Lamiaceae** *(Labiatae)*					
cristata Willd. (Syn. *E. patrini* Garcke)					
patrini Garcke *(See* **E. cristata** Willd.*)*				JPN	SOV
ELYMUS L. **Poaceae** *(Gramineae)*					
caput-medusae Boiss. *(See* **Taeniatherum asperum** (Sim.) Nevski*)*			POR USA		
triticoides Buckl.				USA	
ELYTRARIA Michx. **Acanthaceae**					
crenata Vahl (Syn. *E. lyrata* Vahl)					
lyrata Vahl *(See* **E. crenata** Vahl*)*		GHA			
ELYTROPAPPUS Cass. **Asteraceae** *(Compositae)*					
rhinocerotis Less.		SAF			

ELYTROPHORUS Beauv. Poaceae *(Gramineae)*	S	P	C	X	F
articulatus Beauv. (Syn. *E. spicatus* (Willd.) Camus)					
spicatus (Willd.) Camus *(See* **E. articulatus** Beauv.*)*			NIG	CEY	SAF
EMELISTA Rafin. **Caesalpiniaceae** *(Leguminosae)*					
tora (L.) Britton & Rose *(See* **Cassia tora** L.*)*			PR	DR NIC	
EMEX Campdera **Polygonaceae**					
australis Steinh.	AUS SAF		HAW KEN RHO	NZ TAI TRI USA	
spinosa (L.) Campdera		HAW	EGY ISR KEN MOR POR	MAU USA	AUS
EMILIA Cass. **Asteraceae** *(Compositae)*					
coccinea (Sims) Sweet *(See* **E. javanica** (Burm.f.) Merr.*)*			HAW PR TRI	ANT NIG	
javanica (Burm. *f.*) Merr. (Syn. *E. coccinea* (Sims) Sweet) (Syn. *E. sagittata* (Vahl) Dc.)		CEY			
praetermissa Milne-Redhead				NIG	
sagittata (Vahl) Dc. *(See* **E. javanica** (Burm. *f.*) Merr.*)*				CNK JAM	
sonchifolia (L.) Dc. *ex* Wight (Syn. *Senecio sonchifolius* (L.) Moench)	BRA GHA HAW	HON PLW TAI USA	BOR IDO IND PER PHI TRI	ANT AUS CAB CHN COL CR DAH DR ECU FIJ GUA GUI IVO JAM JPN KEN MAL MAU MEL MIC NGI NIG PR SAL SEG SUD SUR THI VIE	HK

ENARTHROCARPUS Labill. **Brassicaceae** *(Cruciferae)*	S	P	C	X	F
lyratus Dc.			EGY		
ENCELIA Adans. **Asteraceae** *(Compositae)*					
mexicana Mart. *ex* Dc.	MEX				
ENICOSTEMA Bl. **Gentianaceae**					
verticillatum (L.) Engl.			TRI		
ENTEROLOBIUM Mart. **Mimosaceae** *(Leguminosae)*					
contortisiliquum (Vell.) Morong			BRA		
cyclocarpum Griseb.			VEN		
ENTEROMORPHA Link **Ulvaceae**					
intestinalis (L.) Grev.			BEL		
ENYDRA Lour. **Asteraceae** *(Compositae)*					
fluctuans Lour.			BND VIE		
sessilis Dc.	BRA				
EPALTES Cass. **Asteraceae** *(Compositae)*					
australis Dc. *(See* **E. cunninghamii** Benth.*)*	TAI				
cunninghamii Benth. (Syn. *E. australis* Dc.)					
EPHEDRA L. **Ephedraceae**					
torreyana S. Wats.			USA		
trifurca Torr.			USA		
EPILOBIUM L. **Onagraceae** *(Syn.* **CHAMAENERION** Adans.*)*					
angustifolium L. (Syn. *Chamaenerion angustifolium* Scop.)			BEL CAN SOV USA		
hirsutum L.			BEL EGY TUR USA		
palustre L.		FIN	BEL TUR		
paniculatum Nutt.			USA		
parviflorum Schreb.			TUR		
pyrricholophum Fr. & Sav.		JPN			
roseum Schreb.			TUR		
spicatum Lam.			TUR		

EQUISETUM L. Equisetaceae	S	P	C	X	F
arvense L.		BEL CAN ENG FIN GER JPN NZ SOV USA YUG	ALK ARG BRA CZE FRA IND IRA MAD MAU NET POL ROM SPA SWE	CHL CHN ICE ITA KOR TUR	
bogotense H. B. K.				BRA CHL	
fluviatile L. (Syn. *E. limosum* L.)				BEL ENG FIN GER USA	
giganteum L. (Syn. *E. pyramidale* Goldm.) (Syn. *E. xylochaetum* Mett.)				BRA CHL PER	
hyemale L.			JPN	USA	
laevigatum A. Br.				USA	
limosum L. *(See* **E. fluviatile** L.*)*				GER NET	
maximum Auct. *(See* **E. telmateia** Ehrh.*)*				SPA	
palustre L.	FIN		GER POL POR	BEL JPN USA	
pratense Ehrh.			SOV	FIN	
pyramidale Goldm. *(See* **E. giganteum** L.*)*				BRA	
ramosissimum Desf. *forma lacustre* Wilcz. & Wirtg. *(See* **E. ramosissimum** Desf. Ssp. Ramosissimum*)*			TAI	BRA	
ramosissimum Desf. Ssp. Ramosissimum (Syn. *E. ramosissimum* Desf. *forma lacustre* Wilcz. & Wirtg.)					
sylvaticum L.				FIN USA	
telmateia Ehrh. (Syn. *E. maximum* Auct.)				USA	
xylochaetum Mett. *(See* **E. giganteum** L.*)*				BRA	
ERAGROSTIS Beauv. **Poaceae** *(Gramineae)*					
aethiopica Chiov.			MOZ	KEN TNZ	SAF
airoides Nees			BRA		

ERAGROSTIS Beauv. **Poaceae** *(Gramineae) (Cont'd)*	S	P	C	X	F
amabilis (L.) Wight & Arn. *(See* **E. tenella** (L.) Beauv. *ex* Roem. & Schult.*)*			HAW TAI	BND IDO TRI USA VIE	CEY
arenicola C. E. Hubb.				MOZ	SAF
aspera (Jacq.) Nees			SAF	BND IVO	IND
atrovirens (Desf.) Trin.		MAL	TAI	THI	BUR IND SAF
bahiensis Schult.		BRA	IDO		
barrelieri Daveau				USA	
blepharostachya Schum.				GHA	
chloromelas Steud.				AUS	HAW SAF
cilianensis (All.) Link *ex* Lutati		AUS SUD	ARG CHN EGY HAW IDO IRQ JPN KEN SAF TAI	BND TNZ TUR USA	CEY IND NZ
ciliaris (L.) R. Br.		MOZ SAL		GHA IND IVO NIC SAF TNZ TRI	AFG AUS
curvula (Schrad.) Nees			AUS LEB	COL	BUR IND NZ SAF
cynosuroides (Retz.) Beauv. *(See* **Desmostachya bipinnata** (L.) Stapf*)*	PK				AFG
diarrhena (Schult.) Steud.		IND	IRQ	PK	
diplachnoides Steud.				SUD	
elongata Jacq.				THI VIE	
fascicularis Trin.				ANG	
ferruginea (Thunb.) Beauv.		JPN	CHN		CHL IND
frankii Steud.				USA	
gangetica (Roxb.) Steud.				IND	BUR
glomerata (Walt.) L. H. Dewey				TRI	

ERAGROSTIS Beauv. **Poaceae** *(Gramineae) (Cont'd)*	S	P	C	X	F
hypnoides (Lam.) B. S. F. Prel.				TRI	
interrupta (Lam.) Doell.				BND IND THI	CEY
japonica (Thunb.) Trin.	IDO	CEY TAI		PHI	AUS IND
leptostachya Steud.				AUS	
lugens Nees			ARG	MEX	BRA
malayana Stapf			MAL		
megastachya (Koel.) Link				ISR	AFG AUS HAW
mexicana (Hornem.) Link				USA	HAW
montana Balansa				VIE	
multicaulis Steud.			JPN		
namaquensis Nees *ex* Schrad.				ANG	
neo-mexicana Vasey			ARG	USA	HAW
nigra Nees *ex* Steud.			IDO		
niwahokori Honda			TAI	KOR	JPN
pectinacea (Michx.) Nees	HAW			THI USA	AUS TAI
pilosa (L.) Beauv.	AFG BRA IDO TAI	IND IRQ ITA KOR SOV	BOT ECU FIJ HUN SEG SUD	ARG CHN CNK COL EGY GHA JOR JPN KEN MEL MEX NEP PHI PK PLW RHO THI TNZ UGA USA	AUS BUR CEY HAW HK ISR SAF
plumosa (Retz.) Link *(See* **E. tenella** (L.) Beauv. *ex* Roem. & Schult.)		IND	TAI		AFG HAW
poaeoides Beauv.			IRQ JPN LEB	AUS USA	HAW IND ISR SAF
pusilla Hack.				ANG	
schimperi Benth. (Syn. *Harpachne schimperi* Hochst.)					

ERAGROSTIS Beauv. Poaceae *(Gramineae) (Cont'd)*	S	P	C	X	F
superba Peyr.				TNZ	IND
tenella (L.) Beauv. *ex* Roem. & Schult. (Syn. *E. amabilis* (L.) Wight & Arn.) (Syn. *E. plumosa* (Retz.) Link)		IND	HAW	BND NGI NIG PLW	AUS BOR CEY MAD SAF THI
tenuifolia Hochst.		ZAM	KEN	AUS TNZ	IND
tremula Hochst.		IND		GHA IVO	AFG BUR SEG
turgida (Schum.) De Wild.				GHA	
unioloides (Retz.) Nees *ex* Steud.	IND	MAL	BOR IDO	CEY FIJ THI	BUR HAW
virescens Presl			ARG SAF	CHL PER	
viscosa (Retz.) Trin.		MAL			IND SAF
xylanica Hack.	IND				
ERECHTITES Rafin. Asteraceae *(Compositae)*					
arguta Dc.		NZ			
atkinsoniae F. Muell.		NZ			
hieracifolia (L.) Raf. *ex* Dc.			HAW IDO JPN	ANT BRA TRI USA VEN	ARG PAN SAL
minima Dc.		NZ			
prenanthoides Dc.				USA	
quadridentata Dc.			AUS	NZ	
scaberula Hook. *f.*				NZ	
valerianaefolia Dc.		IND TAI	BOR HON IDO SOV THI	AUS BRA CHN HAW MAL NGI NZ PLW SAL	
EREMOCARPUS Benth. Euphorbiaceae					
setigerus Benth.				USA	
EREMOCHLOA Buese Poaceae *(Gramineae)*					
ciliaris (L.) Merr.				VIE	

	S	P	C	X	F
EREMOCITRUS Swingle **Rutaceae**					
glauca Swingle				AUS	
EREMODAUCUS Bunge **Apiaceae** *(Umbelliferae)*					
lehmanni Bunge				SOV	
EREMOPHILA R. Br. **Myoporaceae**					
maculata F. Muell.			AUS		
mitchelli Benth.		AUS			
ERICA L. **Ericaceae**					
arborea L.				NZ	
ERIGERON L. **Asteraceae** *(Compositae)*					
acris L.			SOV	TUR	
annuus (L.) Pers.		JPN	CAN CHN IND USA	NZ	
asteroides Roxb.				IND	
bonariensis L. *(See* **Conyza bonariensis** (L.) Cronq.)		BRA	AUS HAW IND JPN KEN RHO SAF	ARG COL JAM NZ TRI URU	AFG
canadensis L. *(See* **Conyza canadensis** (L.) Cronq.)	KOR	HUN IND ITA JPN NZ PHI POL TAI TUN USA	AST AUS CAN CHN GER GRE HAW HON IRQ ISR POR SAF SOV SPA SUR TUR YUG	BEL CHL COL ECU ENG FIJ IRA JAM JOR MAU MEX MOZ NGI NOR PK RHO THI VIE	AFG
crispus Pour. *(See* **E. linifolius** Willd.)	ISR		EGY	MOR	
divaricatus Michx.				USA	
floribundus (H. B. K.) Sch.-Bip.	GHA		AUS JPN KEN SAF	IVO NIG NZ	AFG FIJ RHO
karvinskianus Dc.				JAM	
linifolius Willd. (Syn. *E. crispus* Pour.)		IDO KEN	CHN JPN TAI	IND	AFG RHO

ERIGERON L. Asteraceae *(Compositae) (Cont'd)*	S	P	C	X	F
philadelphicus L.			CAN JPN	USA	
pulchellus Michx.				USA	
speciosus (Lindl.) Dc.				USA	
spiculosus Hook. & Arn.				CHL	
strigosus Muhl. *ex* Willd.			CAN	USA	
sumatrensis Retz.			JPN	CEY PHI	
ERIOCAULON L. Eriocaulaceae					
buergerianum Koern. (Syn. *E. pachypetalum* Hay.)					
cinereum R. Br.	IDO		TAI	CEY	
cinereum R. Br. *var.* sieboldianum (Sieb. & Zucc.) T. Koyama (Syn. *E. formosanum* Hay.)					
decemflorum Maxim. *var.* nipponicum (Maxim.) Nakai (Syn. *E. nipponicum* Maxim.)					
echinulatum Mart.				THI	
formosanum Hay. *(See* **E. cinereum** R. Br. *var.* sieboldianum (Sieb. & Zucc.) T. Koyama)			TAI		
gracile Mart.				VIE	
hexangularis Kunth *(See* **E. sexangulare** L.)				IND	
hondoense Satake			JPN		
luzulaefolium Mart.		IND			
miquelianum Koern.			JPN		
nipponicum Maxim. *(See* **E. decemflorum** Maxim. *var.* nipponicum (Maxim.) Nakai)			JPN		
odoratum Dalz.				THI	
pachypetalum Hay. *(See* **E. buergerianum** Koern.)			TAI		
quinquangulare L.		IND			
robustius (Maxim.) Makino			JPN	KOR	
setaceum L.				IND	VIE
sexangulare L. (Syn. *E. hexangularis* Kunth) (Syn. *E. sieboldianum* Sieb. & Zucc.)					THI
sieboldianum Sieb. & Zucc. *(See* **E. sexangulare** L.)				CAB IND	JPN KOR
truncatum Buch.-Ham. *ex* Mart.		IND		ANT CAB IDO	

ERIOCEREUS Riccob. **Cactaceae**	S	P	C	X	F
martinii (Lab.) Riccob.	AUS				
regelii (Weing.) Backeb.				AUS	
tortuosus Riccob.			AUS		
ERIOCHLOA H. B. K. **Poaceae** *(Gramineae)*					
gracilis (Fourn.) Hitchc.				USA	
pacifica Mez				PER	
polystachya H. B. K. *(See* **E. procera** (Retz.) C. E. Hubb.)				COL TRI	
procera (Retz.) C. E. Hubb. (Syn. *E. polystachya* H. B. K.) (Syn. *E. ramosa* (Retz.) O. Ktze.)			MAL		BOR BUR CEY HAW IND THI
puncata (L.) Desv. *ex* Hamilt.				ANT PER SUR TRI	
ramosa (Retz.) O. Ktze. *(See* **E. procera** (Retz.) C. E. Hubb.)	GUY			PER VIE	IDO
ERIOCHRYSIS Beauv. **Poaceae** *(Gramineae)*					
cayanensis Beauv.		PR	DR		
ERIODICTYON Benth. **Hydrophyllaceae**					
angustifolium Nutt.				USA	
californicum (Hook. & Arn.) Greene				USA	
crassifolium Benth.				USA	
ERIOGONUM Michx. **Polygonaceae**					
deflexum Torr.				USA	
ERLANGEA Sch.-Bip. **Asteraceae** *(Compositae)*					
cordifolia (Benth.) S. Moore			KEN		
laxa S. Moore		ZAM			
marginata (Oliv. & Hiern.) S. Moore			KEN		
ERODIUM L'herit. **Geraniaceae**					
botrys (Cav.) Bertol.		AUS		CHL USA	ISR
cicutarium (L.) L'herit. *ex* W. Ait.	AFG AUS GRE	ITA NZ SWE TUN	ENG FIN GER HAW IRA IRQ ISR JAM LEB PER	ARG CAN CHL EGY JOR MEX NOR PK TAS TUR	HK

ERODIUM L'herit. Geraniaceae (Cont'd)	S	P	C	X	F
cicutarium (L.) L'herit. *ex* W. Ait. *(Cont'd)*			POL SAF SOV SPA USA	YUG	
cygnorum Nees				AUS	
gruinum Soland.			LEB		ISR
malachoides (L.) Willd.			ARG IRQ LEB POR	GRE TUR	
moscatum (L.) L'herit. *ex* Ait.		AUS NZ	COL POR SAF	ARG CHL URU USA VEN	ISR
romanum (L.) Willd.			LEB		ISR
EROPHILA Dc. **Brassicaceae** *(Cruciferae)*					
vulgaris Dc.			SOV		AFG AUS
ERUCA Mill. **Brassicaceae** *(Cruciferae)*					
cappadocica Reut.			IRA		AFG
sativa Mill.	IRA	ARG	SPA	GRE JOR MEX NZ PK USA	AFG AUS ISR
vesicaria Cav.			MOR		
ERUCARIA Gaertn. **Brassicaceae** *(Cruciferae)*					
aleppica Gaertn. (Syn. *E. myagroides* (L.) Hal.)				TUR	
myagroides (L.) Hal. *(See* **E. aleppica** Gaertn.*)*		ISR			
ERUCASTRUM (Dc.) Presl **Brassicaceae** *(Cruciferae)*					
arabicum Fisch. & C. Mey.			KEN	TNZ	
gallicum (Willd.) O. E. Schulz			CAN	USA	
ERVUM L. **Papilionaceae** *(Leguminosae)*					
hirsutum L. *(See* **Vicia hirsuta** (L.) S. F. Gray*)*			SOV		
ERYNGIUM L. **Apiaceae** *(Umbelliferae)*					
campestre L.		CZE	MOR POR	SPA TUR	GRE POL
coronatum Hook. & Arn.			ARG		
creticum Lam.			LEB	ISR	

ERYNGIUM L. Apiaceae *(Umbelliferae)* *(Cont'd)*	S	P	C	X	F
eburneum Decne.				URU	
elegans Cham. & Schlecht.				BRA	
foetidum L.				IDO PR TRI	·ANT DR FIJ HON JAM PAN	CAB
hookeri Walp.				USA	
maritimum L.			SPA		
pandanifolium Cham. & Schlecht.			POR	ARG	
paniculatum Cav. & Dombey			ARG	VEN	
plantagineum F. Muell.				AUS	
rostratum Cav.				AUS CHL	
sanguisorba Cham. & Schlecht.		BRA			
tricuspidatum L.				MOR	
triquetrum Vahl				MOR	
yuccifolium Michx.				USA	
ERYSIMUM L. Brassicaceae *(Cruciferae)*					
alpestre Kotschy *ex* Boiss.				TUR	
asperum (Nutt.) Dc.				USA	
canescens Roth.				TUR	
cheiranthoides L.		FIN	CAN ENG GER SOV	USA	
hieracifolium L. (Syn. *E. strictum* Gaertn.)			CAN		
inconspicuum Macmillan				CAN	
orientale Mill. *(See* **Conringia orientalis** (L.) Dum.)			SPA		
repandum L.			ARG	USA	AFG AUS ISR
strictum Gaertn. *(See* **E. hieracifolium** L.)			SOV		
ERYTHRAEA Borck. **Gentianaceae**					
centaurium Pers.				TUR	
ramosissima Pers.				PK TUR	AFG

	S	P	C	X	F
ERYTHRINA L. **Papilionaceae** *(Leguminosae)*					
herbacea L.				USA	
ERYTHROCOCCA Benth. **Euphorbiaceae**					
oleraccea Prain				CNK	
ERYTHROXYLUM P. Br. **Erythroxylaceae**					
chlorostachys (F. Muell.) Bail.		AUS			
coca Lam.		AUS			
ESCHSCHOLZIA Cham. **Papaveraceae**					
californica Cham.				AUS CHL NZ USA	
EUCALYPTUS L'herit. **Myrtaceae**					
cambageana Maiden		AUS			
ferruginea Schau.		AUS			
gracilis F. Muell.		AUS			
marginata Sm.		AUS			
miniata A. Cunn. *ex* Schau.		AUS			
pilularis J. E. Sm.		AUS			
populnea F. Muell.		AUS			
tetradonta F. Muell.		AUS			
EUCLEA Murr. **Ebenaceae**					
divinorum Hiern.		KEN TNZ			
keniensis R. E. Fries			KEN		
schimperi (A. Dc.) Dandy			KEN		
EUCLIDIUM R. Br. **Brassicaceae** *(Cruciferae)*					
syriacum (L.) R. Br.				USA	
EUGENIA L. **Myrtaceae**					
cumini (L.) Druce *(See* Syzygium cumini (L.) Skeels)	HAW				
jambolana Lam. *(See* Syzygium cumini (L.) Skeels)				USA	
jambos L. *(See* Syzygium jambos (L.) Alst.)			HAW	USA	
myrtoides Poir.				USA	
EUONYMUS L. **Celastraceae**					
atropurpureus Jacq.				USA	CHL

EUPATORIUM L. Asteraceae *(Compositae)*	S	P	C	X	F
adenophorum Spreng. *(See* **Ageratina adenophora** (Spreng.) R.M. King & H. Robinson)	AUS PHI	HAW NZ		USA	
ballotaefolium H. B. K.				BRA	
betonicaeforme Baker				BRA	
bunifolium H. B. K. *(See* **Acanthostyles buniifolius** (Hook. & Arn.) R.M. King & H. Robinson	URU				
cannabinum L.				BEL	
cannabium L.			POR		
capillifolium (Lam.) Small			USA		
compositifolium Walt.				USA	
formosanum Hay.			TAI		
glandulosum H. B. K.			IND	USA	CEY
hirsutum Baker				BRA	
inulaefolium H. B. K. *(See* **Austroeupatorium inulaefolium** (H. B. K.) R.M. King & H. Robinson	PHI				
japonicum Thunb.	NEP		CHN		
liatrideum Dc. (Syn. *E. squarrulosum* Hook. & Arn.)					
macrocephalum Less.		BRA			
macrophyllum L.				JAM PER TRI	
maculatum L.				USA	
microstemon Cass.				PER	
odoratum L. *(See* **Chromolaena odorata** (L.) R. M. King & H. Robinson)	BOR CEY IDO MAL NIG THI	AUS GHA IND TRI	PR SAF	ANT CAB DR HON JAM PAN PER PHI SAL VEN VIE	
pallescens Dc.			IDO		
perfoliatum L.				USA	
pycnocephalum Less.			SAL		
riparium Regel *(See* **Ageratina riparia** (Regel) R. M. King & H. Robinson)	HAW	IDO	AUS	NZ USA	
rugosum Hout. *(See* **Ageratina altissima** (L.) R. M. King & H. Robinson)				USA	
serotinum Michx.				USA	

EUPATORIUM L. Asteraceae *(Compositae)* *(Cont'd)*	S	P	C	X	F
squarrulosum Hook. & Arn. *(See* E. liatrideum Dc.)		BRA			
tremulum Hook. & Arn. *(See* **Raulinoreitzia tremula** (Hook. & Arn.) R.M. King & H. Robinson				ARG	
triste Dc.				JAM	
villosum Sw.				JAM	
EUPHORBIA L. Euphorbiaceae					
acalyphoides Hochst. *ex* Boiss.		SUD			
aegyptiaca Boiss.		EGY SUD			ISR
albomarginata Torr. & Gray			USA		
allepica L.		LEB		TUR	ISR
arguta Soland.			EGY		
atoto Forst. *f.)*				PLW	TAI
brasiliensis Lam.			BRA JAM		
capitellata Engelm.			MEX		
chamaesyce L.			ARG JPN LEB MOR SAF	BRA USA	AFG ISR
corollata L. (Syn. *E. paniculata* Ell.)			USA	MEX	
cotinifolia L.			VEN		
cyparissias L.		CZE	CAN GER SOV	ITA NZ USA	POL TRI
cythophora Murr. *(See* **E. heterophylla** L.)		TAI			
dentata Michx.			MEX		
dracunculoides Lam.		IND	PK		
drummondii Boiss.			AUS		
edulis Lour.			VEN		
eremophila A. Cunn. *ex* Hook.			AUS		
esula L.			CAN GER IRA SOV USA		AFG ARG
exigua L.			ENG POR	GER NZ	ISR POL
falcata L.			MOR		AFG

EUPHORBIA HETEROPHYLLA *(Cont'd)*	S	P	C	X	F
falcata L. *(Cont'd)*			POR		ISR POL
fulgens Karw. *ex* Klotzsch		MEX			
gaillardoti Boiss. & Blanche		LEB			
geniculata Orteg. (*See* **E. prunifolia** Jacq.)	IDO NGI THI	BRA CEY	HAW SAF TRI	IND	
glomerifera (Millsp.) Wheeler		SAL	HAW		PAN
glyptosperma Engelm.			CAN	USA	
graminea Jacq.			IND	HON	
granulata Forsk.			EGY		
helioscopia L.	IND IRA ITA KOR PK POL SWE TUR	ARG AUS CAN CHN EGY ENG GER HUN IRQ JPN LEB MOR NET POR SAF SOV SPA TUN	ARB BEL CHL FIN FRA GRE ICE JOR NOR NZ TAI USA YUG	AFG HK ISR	
heterophylla L. (Syn. *E. cythophora* Murr.) (Syn. *Poinsettia heterophylla* (L.) Small)	FIJ GHA MEX PHI	CUB HON ITA PER UGA **USA**	BOT JAM NIG SAL SUD TUN ZAM	ANT ARB AUS BOL BOR BRA CHN CNK COL CR DR ECU EGY GUA HAW IDO IND JPN MAU MEL MIC MOZ NGI PLW PR RHO SEG SUR VEN	

EUPHORBIA L. Euphorbiaceae *(Cont'd)*	S	P	C	X	F
hirta L. (Syn. *E. pilulifera* L.) (Syn. *Chamaesyce hirta* (L.) Millsp.)	GHA HAW THI TRI	BRA CAB ECU FIJ IND MAU NIG PER PHI PK SAF SAL TAI UGA	ARG COL EGY ETH IDO KEN MAL RHO	ANG ANT ARB AUS BND BOR CEY CHN DR GUA HON IVO JAM MEX MOZ NGI PLW SOV SUD SUR TNZ USA VEN VIE	
hispida Boiss.				VEN	
humifusa Willd. (Syn. *E. pseudochamaesyce* Fisch. & Mey.)				JPN	
hypericifolia L. (Syn. *E. indica* Lam.) (Syn. *E. parviflora* L.) (Syn. *E. reniformis* Bl.)		PER	COL IDO	FIJ IND JAM MAU PK SUR TRI	AFG
hyssopifolia L.			TRI	JAM NIG USA	
inaequilatera Sond.			KEN RHO SAF		
indica Lam. (*See* **E. hypericifolia** L.)			EGY		
lanata Sieber *ex* Spreng.			LEB		
lasiocarpa Klotzsch				JAM TRI	
lathyris L.			AUS	CHL NZ	
leptocaula Boiss.			SOV		
lucida Waldst. & Kit.				USA	
macroclada Boiss. (*See* **E. tinctoria** Boiss.)			LEB		
maculata L.		MEX	ARG JPN USA	NZ	LEB
marginata Pursh				USA	

EUPHORBIA L. Euphorbiaceae *(Cont'd)*	S	P	C	X	F
medicaginea Boiss.			POR	MOR	
micromera Boiss.				USA	
microphylla Heyne *ex* Roth		IND		BND	
milii Desm. (Syn. *E. splendens* Bojer *ex* Hook.)					
oerstediana Boiss.	TRI				
ovalifolia Engelm. *ex* Klotzsch & Garcke				CHL	
palustris L.				TUR	
paniculata Ell. *(See* **E. corollata** L.)	BRA				
papillosa St. Hil.			ARG	BRA	
parviflora L. *(See* **E. hypericifolia** L.)			IDO		
patagonica Hieron.			ARG		
peplus L.	CHL	ITA JOR POL SWE TUN	AUS EGY ENG HAW HUN IRQ ISR LEB MOR POR SAF SOV SPA	ARB ARG BRA CAN GER GRE IRA JAM MAU MEX NOR NZ PER RHO SUD TUR USA	CHN
pilulifera L. *(See* **E. hirta** L.)		BRA CAB IND PK	EGY	JAM PHI VIE	ARG CHN HK
pinea L.				SPA	
platyphyllos L.				CHL NZ	
polycnemoides Hochst. *ex* Boiss.				GHA	
preslii Guss.			AUS		
prostrata W. Ait.		HUN IND PK THI	EGY HAW HON IRQ KEN PHI RHO SAF TNZ TRI UGA	ANG ARG CNK COL DAH GHA GUI IDO IVO JAM MAU MEL	AUS

EUPHORBIA L. Euphorbiaceae *(Cont'd)*	S	P	C	X	F
prostrata W. Ait. *(Cont'd)*				MEX MOZ NGI NGR NIG PLW SEG SUR USA ZAM	
prunifolia Jacq. (Syn. *E. geniculata* Orteg.)	IDO PHI THI	BRA IND MEL NGI	CEY EGY HAW ISR JAM JOR	ETH KEN MAU NEP SAF SUD TRI VIE	ARG
pseudochamaesyce Fisch. & Mey. *(See* **E. humifusa** Willd.)			JPN		
pterococca Brot.			POR		
reinwardtiana Steud. (Syn. *E. serrulata* Rein.) (Syn. *E. vachellii* Hook. & Arn.)					
reniformis Bl. *(See* **E. hypericifolia** L.)			IDO		
rossica P. Smirnow			SOV		
sanguinea Hochst. & Steud.			KEN		
scordifolia Jacq.				GHA	
segetalis L.			POR		AUS
serpens H. B. K.	MEX	·	ARG	URU	
serpyllifolia Pers.			CAN	MEX	
serrata L.				USA	
serrula Engelm.				USA	
serrulata Rein. *(See* **E. reinwardtiana** Steud.)				PHI	
splendens Bojer *ex* Hook. *(See* **E. milii** Desm.)				VEN	
stricta L.				TUR	
sulcata De Lens *ex* Loisel.				MOR	
supina Raf.		JPN		NZ	USA
tarapacana Phil.				CHL	
terracina L.		AUS			ISR
thymifolia L.		IND ITA PK	CHN HAW HON IDO IRQ	ANT BOR CAB CEY CNK	HK

EUPHORBIA L. Euphorbiaceae *(Cont'd)*	S	P	C	X	F
thymifolia L. *(Cont'd)*			JPN TAI THI TRI	CR FIJ GHA GUY IVO JAM MAU MEX NGI PHI SEG SUR VIE	
tinctoria Boiss. (Syn. *E. macroclada* Boiss.)					
turcomanica Boiss.			IRA		AFG
vachellii Hook. & Arn. *(See* **E. reinwardtiana** Steud.)			TAI		
vermiculata Rafin.				USA	
virgata Waldst. & Kit.			SOV	CAN	
EUPHRASIA L. Scrophulariaceae					
odontites L. *(See* **Bartsia odontites** (L.) Huds.)				SOV	
EURYOPS Cass. Asteraceae *(Compositae)*					
floribundus N. E. Br.			SAF		
EUXOLUS Rafin. Amaranthaceae					
blitum Gren. *(See* **Amaranthus lividus** L.)			JPN	KOR	
EVAX Gaertn. Asteraceae *(Compositae)*					
multicaulis Dc.				USA	
EVOLVULUS L. Convolvulaceae					
alsinoides (L.) L. (Syn. *E. nummularius* L.)				GHA IND VEN	AFG
nummularius L. *(See* **E. alsinoides** (L.) L.)	IND	PR		BRA GHA	
tenuis Mart. *ex* Choisy				VEN	
EXOTHECA Anderss. Poaceae *(Gramineae)*					
chevalieri A. Camus				VIE	
FACELIS Cass. Asteraceae *(Compositae)*					
apiculata Cass.				USA	
retusa (Lam.) Sch.-Bip.			SAF		ARG

FAGOPYRUM Mill. Polygonaceae	S	P	C	X	F
esculentum Moench (Syn. *F. sagittatum* Gilib.)				USA	
sagittatum Gilib. *(See* **F. esculentum** Moench)				CAN	
tataricum (L.) Gaertn. (Syn. *Polygonum tataricum* L.)			CAN SOV	USA	
FAGRAEA Thunb. **Loganiaceae**					
racemosa Jack *ex* Wall.				MAL	
FAGUS L. **Fagaceae**					
grandifolia Ehrh.				USA	
FALCARIA Fabr. **Apiaceae** *(Umbelliferae)*					
rivini Host *(See* **F. vulgaris** Bernh.)				SOV TUR	
sioides (Wib.) Asch. *(See* **F. vulgaris** Bernh.)			IRA		
vulgaris Bernh. (Syn. *F. rivini* Host) (Syn. *F. sioides* (Wib.) Asch.) (Syn. *Prionitis falcaria* Dum.)		LEB		SOV	ISR POL
FARFUGIUM Lindl. **Asteraceae** *(Compositae)* *(Syn.LIGULARIA* Cass.)					
japonicum (L.) Kitam. (Syn. *Ligularia tussilginea* (Burm.) Makino)					
FARSETIA Turra **Brassicaceae** *(Cruciferae)*					
aegyptia Turra (Syn. *F. edgeworthii* Hook. *f.*) & Thoms.)					
edgeworthii Hook. *f.*) & Thoms. *(See* **F. aegyptia** Turra)				PK	
jacquemontii Hook. *f.*) & Thoms.				PK	AFG
stenoptera Hochst.	SUD				
FATOUA Gaudich. **Moraceae**					
villosa (Thunb.) Nakai	JPN				
FEDIA Gaertn. **Valerianaceae**					
caput-bovis Pomel				MOR	
cornucopiae (L.) Gaertn. (Syn. *F. scorpioides* Dufr.)			MOR		AUS POR
scorpioides Dufr. *(See* **F. cornucopiae** (L.) Gaertn.)			MOR		POR
FERULA L. **Apiaceae** *(Umbelliferae)*					
communis L. (Syn. *F. nodiflora* L.)					
nodiflora L. *(See* **F. communis** L.)			SPA	TUR	

FESTUCA L. Poaceae *(Gramineae)*	S	P	C	X	F
arundinacea Schreb.		AUS			AFG ISR NZ SAF
megalura Nutt.				USA	
myuros L. (Syn. *F. sciuroides* (Kunth) Roth)				USA	
octoflora Walt.				USA	
ovina L.			SPA		
rubra L.			FIN		
sciuroides (Kunth) Roth *(See* **F. myuros** L.*)*			CHL		
FEVILLEA L. **Cucurbitaceae**					
cordifolia L.				JAM	
FIBIGIA Medic. **Brassicaceae** *(Cruciferae)*					
clypeata (L.) Medic.			LEB		
FICARIA Hall **Ranunculaceae**					
grandiflora Robert				TUR	
FICUS L. **Moraceae**					
aurantiaca Griff. (Syn. *F. callicarpa* Miq.)					
callicarpa Miq. *(See* **F. aurantiaca** Griff.*)*				PHI	
fistulosa Reinw. *ex* Bl.				IDO	
hirta Vahl			IDO	IND	
pumila L.			AUS		
retusa L.			HAW		
villosa Bl.				IDO	
FILAGO L. **Asteraceae** *(Compositae)*					
arvensis L.			SOV		AFG ISR
gallica L.				MOR NZ	ISR POR
germanica L.				MOR	AFG ISR POR
minima Fries			SOV		
spathulata Presl					AFG ISR POR

FILIPENDULA L. Rosaceae	S	P	C	X	F
ulmaria (L.) Maxim.				BEL	
FIMBRISTYLIS Vahl **Cyperaceae**					
acuminata Vahl	IND		MAL		
aestivalis (Retz.) Vahl	BOR IND	TAI	IDO JPN		
annua (All.) Roem. & Schult. *(See F. dichotoma (L.) Vahl)*			COL IDO JPN PHI	FIJ PER	
arvensis Vahl *(See F. ferruginea (L.) Vahl)*				FIJ	
autumnalis (L.) Roem. & Schult.		BRA		JPN USA	
barbata (Rottb.) Benth. *(See Bulbostylis barbata (Rottb.) C. B. Clarke)*			JPN PHI	IDO	
bis-umbellata (Forsk.) Bub.			EGY		
complanata (Retz.) Link	PHI		IDO	FIJ	
cymosa R. Br. (Syn. *F. spathacea* Roth)					
dichotoma (L.) Vahl (Syn. *F. annua* (All.) Roem. & Schult.) (Syn. *F. diphylla* (Retz.) Vahl)	NIG	FIJ PR	HAW IRQ JPN MAL PHI TAI	GHA NGI PK PLW THI	AFG CHN HK ISR KOR
diphylla (Retz.) Vahl *(See F. dichotoma (L.) Vahl)*	IND	BRA HAW	TRI	PHI USA VIE	
ferruginea (L.) Vahl (Syn. *F. arvensis* Vahl)	IND			IDO PHI THI	
globulosa (Retz.) Kunth	BOR		IND	BND MAL PHI	IDO
koidzumiana Ohwi			TAI		
littoralis Gaudich. (Syn. *F. miliacea* (L.) Vahl)	PHI	JPN			
miliacea (L.) Vahl *(See F. littoralis Gaudich.)*	BOR CEY GUY IDO IND MAL SUR TAI THI TRI	AUS JPN KOR PHI	USA	ANT BND BUR CAB CHN CUB DR FIJ HAW IVO NIG PER VIE	HK

FIMBRISTYLIS Vahl Cyperaceae *(Cont'd)*	S	P	C	X	F
monostachyos (L.) Hassk. *(See F. ovata (Burm. f.) Kern)*				BND PHI	FIJ IDO
ovata (Burm.f.) Kern (Syn. *F. monostachyos* (L.) Hassk.)					
quinquangularis (Vahl) Kunth	NIG			CEY VIE	IDO
schoenoides (Retz.) Vahl	IND		MAL		
spathacea Roth *(See F. cymosa R. Br.)*		MAL		FIJ	
squarrosa Vahl			JPN		
subbispicata Nees & Mey.			JPN		
tenera Roem. & Schult.				PK	
tetragona R. Br.	IND				
thonningiana Boeck.				NIG	
tristachya R. Br.			JPN	NIG	
FLACOURTIA L'herit. **Flacourtiaceae**					
indica (Burm. *f.*) Merr.				RHO	IDO
FLAGELLARIA L. **Flagellariaceae**					
indica L.				VIE	
FLAVERIA Juss. **Asteraceae** *(Compositae)*					
australasica Hook.			AUS IND RHO		
bidentis (L.) O. Ktze.			ARG SAF	CHL PER	
contrayerba Pers.			SAF		
trinervia Mohr				VEN	
FLEMINGIA Roxb. *ex* W. T. Aiton **Papilionaceae** *(Leguminosae)*					
lineata (L.) O. Ktze. *(See Moghania lineata (L.) O. Ktze.)*				IDO	
strobilifera (L.) R. Br. *(See Moghania strobilifera (L.) St. Hil. ex O. Ktze.)*		TRI		ANT JAM MEL	IDO
FLEURYA Gaudich. **Urticaceae**					
aestuans (L.) Gaudich. *(See Laportea aestuans (L.) Chew)*			NIG PR TRI	ANT DR GHA NIC	IDO
interrupta (L.) Gaudich. *(See Laportea interrupta (L.) Chew)*				PHI THI	IDO
FLOURENSIA Dc. **Asteraceae** *(Compositae)*					
cernua Dc.				USA	

FOENICULUM Mill. Apiaceae *(Umbelliferae)*	S	P	C	X	F
vulgare Mill.		MEX NZ	ARG AUS HAW SPA	CHL MOR URU USA VEN	AFG ISR
FORESTIERA Poir. Oleaceae					
acuminata (Michx.) Poir.				USA	
neo-mexicana Gray				USA	
pubescens Nutt.				USA	
FORRESTIA Rich. Commelinaceae					
hispida A. Rich.				PHI	
FRAGARIA L. Rosaceae					
indica Andr. (Syn. *Duchesnea chrysantha* Miq.) (Syn. *D. indica* (Andr.) Focke)				MAU	
vesca L.				USA	
virginiana Decne.				USA	
FRANKENIA L. Frankeniaceae					
hirsuta L.				TUR	
salina (Mol.) I. M. Johnston				CHL	
FRANSERIA Cav. Asteraceae *(Compositae)*					
acanthicarpa (Hook.) Coville				USA	
confertiflora (Dc.) Rydb.		MEX		USA	
discolor Nutt.			USA		
strigulosa Rydb.			HAW	USA	
tenuifolia Harv. & Gray		AUS			
tomentosa A. Gray				USA	
FRAXINUS L. Oleaceae					
americana L.				USA	
caroliniana Mill.				USA	
latifolia Benth.				USA	
nigra Marsh.				USA	
oregona Nutt. (Syn. *F. pensylvanica ssp. oregona* (Nutt.) G. Miller)				CHL	
pensylvanica Marsh.				USA	
pensylvanica ssp. oregona (Nutt.) G. Miller (*See* **F. oregona** Nutt.)					
quadrangulata Michx.				USA	
tomentosa Michx. *f.*				USA	

FRAXINUS L. Oleaceae *(Cont'd)*	S	P	C	X	F
velutina Torr.				USA	
FROELICHIA Moench **Amaranthaceae**					
tomentosa Moq.			ARG		
FUCHSIA L. **Onagraceae**					
magellanica Lam.			HAW		
FUIRENA Rottb. **Cyperaceae**					
glomerata Lam.			IND TAI THI		
umbellata Rottb.	BOR	MAL		CAB VIE	IDO
FUMARIA L. **Fumariaceae**					
agraria Lag.		TUN	ARG MOR POR		ISR
asepala Boiss.			LEB		ISR
capreolata L.		AUS	MOR POR	ARG BRA NZ PER	ISR
densiflora Dc.		AUS	LEB MOR	ISR	
indica Pugsl.				IND	AUS
media Loisel.				CHL	
micrantha Lag.		AUS	LEB		
muralis Sond.			POR SAF	NZ	AUS
officinalis L.	ARG ENG	ALG AST AUS CZE FIN NOR NZ SPA SWE	FRA GER GRE HUN ISR LEB MOR POL POR SAF SOV	BEL CAN CHL DEN EGY ICE IRA IRQ ITA JOR MAU NET SUD TUR URU USA YUG	
parviflora Lam.		IND TUN	LEB POR	PK	AUS ISR
schleicheri Soyer-Willem.			SOV		
vaillantii Loisel.			IRA SOV	CHL TUR	POL

	S	P	C	X	F
FUNASTRUM Fourn. **Asclepiadaceae**					
clausum (Jacq.) Schltr.				JAM	
FURCRAEA Vent. **Agavaceae**					
tuberosa (Mill.) Ait.			PR		
GAGEA Salisb. **Liliaceae**					
arvensis Kar. & Kir. *ex* Ledeb.				TUR	
GAILLARDIA Fouger. **Asteraceae** *(Compositae)*					
pulchella Fouger.			HAW	USA	
GALACTIA P. Br. **Papilionaceae** *(Leguminosae)*					
tenuiflora (Willd.) Wight & Arn.			TAI	GHA	
GALACTITES Moench **Asteraceae** *(Compositae)*					
tomentosa (L.) Moench			MOR SPA		POR
GALEGA L. **Papilionaceae** *(Leguminosae)*					
officinalis L.				ARG CHL NZ TUR USA	SOV
GALEOPSIS L. **Lamiaceae** *(Labiatae)*					
angustifolia Ehrh. *ex* Hoffm.				GER	
bifida Boenn. *(See* **G. tetrahit** L.)	FIN			SOV	
intermedia Vill. *(See* **G. ladanum** L.)				SOV	
ladanum L. (Syn. *G. intermedia* Vill.) (Syn. *G. segetum* S. F. Gray)				GER SOV	
pubescens Bess.				GER	
segetum S. F. Gray *(See* **G. ladanum** L.)				GER	
speciosa Mill.	FIN	NOR	ENG GER	ICE SOV	
tetrahit L. (Syn. *G. bifida* Boenn.)		ALK CZE NOR TUN	CAN ENG FIN GER SPA USA	NZ SOV	
versicolor Curt.				SOV	
GALINSOGA Ruiz & Pav. **Asteraceae** *(Compositae)*					
caracasana Sch.-Bip.				COL	
ciliata (Raf.) Blake			CAN COL GER	NZ USA	JPN

GALINSOGA Ruiz & Pav. Asteraceae *(Compositae)* *(Cont'd)*	S	P	C	X	F
ciliata (Raf.) Blake *(Cont'd)*			SAL		
parviflora Cav.	ANG GER MEX SAF TNZ UGA	ARG AUS BEL BRA COL ETH HAW ISR MOZ NGI PER POL RHO TAI ZAM	CAN ENG IDO JPN SOV USA	CHL DR IND ITA KEN LEB NET NZ PHI SAL VEN	AFG YUG
urticaefolia Benth.				ECU	
GALIUM L. Rubiaceae					
aparine L.	ENG GER ITA JPN	BEL FRA GRE NZ SWE TUN	ARG CHN SPA USA	ALK CAN CHL FIN HUN ICE NOR PK POR SOV TUR URU YUG	AFG AUS ETH HK ISR KOR POL
asprellum Michx.				USA	
boreale L.			FIN	USA	AFG
gracilens (A. Gray) Makino			JPN		
mollugo L.			FIN SOV SPA USA	BEL NZ	AFG
palustre L.			FIN POR	BEL NZ	
parisiense L.				NZ POR	
pseudo-asprellum Makino			JPN		
rotundifolium L.				TUR	
saccharatum All.			SPA		POR
spurium L.	TNZ	ETH KEN	GER SOV		AFG ISR POL
trachyspermum A. Gray			JPN		
tricorne Stokes	IRA		GER IRQ ISR LEB	CHL MOR SOV TUR	AFG AUS POL POR

GALIUM L. Rubiaceae *(Cont'd)*	S	P	C	X	F
tricorne Stokes *(Cont'd)*			SPA		
trifidum L.			JPN		
uliginosum L.			FIN		
vaillantii Dc.				FIN	
verum L.			FIN JPN	ITA MEX NZ USA	AFG SOV
viscosum Vahl			MOR		
GALPHIMIA Cav. **Malpighiaceae**					
brasiliensis (L.) A. Juss.				BRA	
GAMOCHAETA Wedd. **Asteraceae** *(Compositae)*					
americana Wedd.				ARG	
GARRYA Dougl. *ex* Lindl. **Garryaceae**					
elliptica Dougl.				USA	
flavescens S. Wats.				USA	
fremontii Torr.				USA	
GASTRIDIUM Beauv. **Poaceae** *(Gramineae)*					
ventricosum (Gouan) Schinz & Thell.			POR	CHL NZ USA	HAW ISR
GASTROLOBIUM R. Br. **Papilionaceae** *(Leguminosae)*					
callistachys Meissn.		AUS			
grandiflorum F. Muell.			AUS		
parviflorum Benth.		AUS			
villosum Benth.		AUS			
GAUDINIA Beauv. **Poaceae** *(Gramineae)*					
fragilis (L.) Beauv.			POR		ISR
GAULTHERIA L. **Ericaceae**					
shallon Pursh				USA	
GAURA L. **Onagraceae**					
biennis L.				USA	
coccinea (Pursh) Nutt.				USA	
odorata Lag.				USA	
parviflora Dougl.			AUS		
sinuata Nutt. *ex* Ser.				USA	
villosa Torr.				USA	

	S	P	C	X	F
GAYLUSSACIA H. B. K. Ericaceae					
frondosa (L.) T. & G.				USA	
GEISSASPIS Wight & Arn. Papilionaceae *(Leguminosae)*					
cristata Wight & Arn.				VIE	
GELSEMIUM Juss. Loganiaceae					
sempervirens (L.) Ait. *f.*				USA	
GENISTA L. Papilionaceae *(Leguminosae)*					
linifolia L.				AUS	
tinctoria L.				USA	
GEOFFRAEA L. Papilionaceae *(Leguminosae)*					
decorticans (Gill. *ex* Hook. & Arn.) Burkart			ARG		
GERANIUM L. Geraniaceae					
arabicum Forsk.			KEN		
carolinianum L.			HAW JPN	USA	
columbinum L.				USA	
core-core Steud.			CHL		
dissectum L.			ENG GER IRQ POR SPA	NZ USA	AUS ISR POL
maculatum L.				USA	
molle L.		AUS	ENG GER POR SOV SPA	NZ USA	AFG CHL POL
pilosum Cav.				AUS USA	
pratense L.			ENG GER	USA	
pusillum Burm. *f.*				FIN GER ICE USA	
robertianum L.			POR	CHL TUR	
rotundifolium L.				SPA TUR	
simense A. Rich.		ETH			
thunbergii Sieb. & Zucc.			JPN		
tuberosum L.		TUN	ISR LEB	TUR	

GEROPOGON L. Asteraceae *(Compositae)*	S	P	C	X	F
glaber L. (Syn. *Tragopogon hybridus* L.)			MOR		
GEUM L. **Rosaceae**					
aleppicum Jacq.				USA	
macrophyllum Willd.				USA	
urbanum L.				TUR	
GISEKIA L. **Aizoaceae**					
africana (Lour.) O. Ktze.	TNZ		RHO		
pharnacioides L.				IDO PK VIE	
GLADIOLUS L. **Iridaceae**					
byzantinus Mill.		LIB TUN			
cuspidatus Jacq.				NZ	AUS
segetum Ker-Gawl.		TUN	IRA ISR LEB POR SPA	ITA	
GLANDULARIA G. F. Gmel. **Verbenaceae**					
dissecta (Willd. *ex* Spreng.) Schnack & Covas)				ARG	
laciniata (L.) Schnack & Covas			ARG		
peruviana (L.) Small			ARG		
GLASTARIA Boiss. **Brassicaceae** *(Cruciferae)*					
deflexa Boiss. (Syn. *Texiera glastifolia* (Dc.) Jaub. & Spach)					LEB
GLAUCIUM Mill. **Papaveraceae**					
corniculatum Curt.				SPA TUR	
flavum Crantz (Syn. *G. luteum* Scop.)					EGY
luteum Scop. *(See* **G. flavum** Crantz)			SPA		
GLECHOMA L. **Lamiaceae** *(Labiatae)*					
hederacea L.			CAN FIN GER JPN SOV USA	NZ	POL
GLECHON Spreng. **Lamiaceae** *(Labiatae)*					
ciliata Benth.				BRA	

	S	P	C	X	F
GLEDITSCHIA L. **Mimosaceae** *(Leguminosae)*					
triacanthos L.				USA	
GLEICHENIA Sm. **Gleicheniaceae**					
laevigata (Willd.) Hook.			IDO		
linearis (Burm.) Clarke (Syn. *Dicranopteris linearis* (Burm.) Und.)		CEY IDO MAL		DR FIJ GHA	
GLINUS L. **Aizoaceae**					
dahomensis A. Cheval.				ANG	
lotoides L. (Syn. *Mollugo hirta* Thunb.) (Syn. *M. lotoides* Wight & Arn. *ex* Clarke)			EGY IDO	GHA	
oppositifolius (L.) A. Dc. (Syn. *Mollugo oppositifolia* L.) (Syn. *M. spergula* L.) (Syn. *M. verticillata* Roxb. *non* L.)			IND	GHA THI	IDO
GLIRICIDIA H. B. K. **Papilionaceae** *(Leguminosae)*					
sepium (Jacq.) Steud.				JAM	
GLOBBA L. **Zingiberaceae**					
parviflora Presl				PHI	
GLOCHIDION J. R. & G. Forst. **Euphorbiaceae**					
litorale Bl.				VIE	
GLORIOSA L. **Liliaceae**					
superba L.				NIG	
GLYCERIA R. Br. **Poaceae** *(Gramineae)*					
declinata Breb.				POR	
fluitans (L.) R. Br.				BEL GER USA	
maxima (Hartm.) Holmb.	NZ		AUS	BEL ENG GER NET	
GLYCINE L. **Papilionaceae** *(Leguminosae)*					
hedysaroides Willd.				GHA	
soja (L.) Sieb. & Zucc.			JPN		CAB
GLYCYRRHIZA Willd. **Papilionaceae** *(Leguminosae)*					
acanthocarpa (Lindl.) J. M. Black			AUS		
astragalina Gill.			ARG		
echinata L.			YUG		ISR
foetida Desf.			MOR		

GLYCYRRHIZA L. Papilionaceae *(Leguminosae) (Cont'd)*	S	P	C	X	F
glabra L. (Syn. *G. glandulifera* W. & K.)		IRA TUR	IRQ SAF	SOV	AFG AUS ISR
glandulifera W. & K. *(See* **G. glabra** L.)				SOV	
lepidota (Nutt.) Pursh				USA	
GNAPHALIUM L. Asteraceae *(Compositae)*					
affine D. Don *(See* **G. luteo-album** L. *var.* affine (D. Don) Koster)		JPN		NEP	
americanum Mill. *(See* **G. purpureum** L.)				BRA MEX	
calviceps Fern.				USA	
cheiranthifolium Lam.				ARG BRA	
falcatum Lam.				USA	
indicum L.		TAI	IDO PR	IND MAU	
japonicum Thunb.			AUS HAW JPN	NZ	
luteo-album L. *var.* affine (D. Don) Koster (Syn. *G. affine* D. Don) (Syn. *G. multiceps* Wall. *ex* Dc.)			EGY KEN POR RHO SAF	ANG AUS BND NZ PK	AFG ISR
macounii Greene				USA	
multiceps Wall. *ex* Dc. *(See* **G. luteo-album** L. *var.* affine (D. Don) Koster)		JPN		TAI	
obtusifolium L.				USA	
pensylvanicum Willd.				USA	
peregrinum Fern.				TNZ USA	
pulvinatum Delile				EGY	
purpureum L. (Syn. *G. americanum* Mill.) (Syn. *G. spicatum* Lam.)		PAR	HAW IND JPN TAI	AUS BRA CHL NZ USA VEN	.
spicatum Lam. *(See* **G. purpureum** L.)			ARG AUS COL PR		
sylvaticum L.				SOV	
uliginosum L.		FIN	ENG GER POR SOV	USA	POL

GNAPHALIUM L. Asteraceae *(Compositae) (Cont'd)*	S	P	C	X	F
undulatum L.			SAF		
GODETIA Spach Onagraceae					
tenuifolia (Cav.) Spach				CHL	
GOEBELIA Bunge *ex* Boiss. Papilionaceae *(Leguminosae)*					
alopecuroides Bunge *ex* Boiss.			IRA		
pachycarpa Bunge *ex* Boiss.			IRA		
GOMPHOCARPUS R. Br. Asclepiadaceae					
fruticosus (L.) R. Br. *ex* W. T. Ait. (Syn. *Asclepias fruticosa* L.)					EGY
physocarpus E. Mey.				HAW	USA
GOMPHRENA L. Amaranthaceae					
celosioides Mart.	TAI THI		AUS IND RHO SAF	GHA	
decumbens Jacq.				IND	
dispersa Standl.			PR		
globosa L.				GHA PHI	
GONIOPTERIS Presl Aspidiaceae					
prolifera Fee *(See* **Dryopteris vivipara** (Raddi) C. Chr.)			TAI		
GONOLOBUS Michx. Asclepiadaceae					
carolinensis (Jacq.) Schultes				USA	
gonocarpos (Walt.) Perry				USA	
stenanthus (Standl.) Woodson				HON	
GONOSTEGIA Turcz. Urticaceae					
hirta (Bl.) Miq.			PHI TAI		
GONZALAGUNIA Ruiz & Pav. Rubiaceae					
spicata (Lam.) G. Maza			PR		
GOSSYPIUM L. Malvaceae					
tomentosum Nutt.				USA	
GOUANIA Jacq. Rhamnaceae					
lupuloides Urb.				JAM	
GRANGEA Adans. Asteraceae *(Compositae)*					
maderaspatana (L.) Poir.				IND THI	IDO

GRATIOLA L. Scrophulariaceae	S	P	C	X	F
japonica Miq.			JPN		
GREVILLEA R. Br. Proteaceae					
banksii R. Br.			HAW	USA	
robusta A. Cunn. *ex* R. Br.			HAW		
GRINDELIA Willd. Asteraceae *(Compositae)*					
brachystephana Griseb.				ARG	
rubricaulis Dc.				CHL	
squarrosa (Pursh) Dunal				USA	AUS
GUADUA Kunth Poaceae *(Gramineae)*					
trinii Rupr.				ARG	
GUAIACUM L. Zygophyllaceae					
officinale L.				JAM	
GUAREA L. Meliaceae					
trichilioides L.				BRA	
GUETTARDA L. Rubiaceae					
platypoda Dc.				BRA	
GUIZOTIA Cass. Asteraceae *(Compositae)*					
villosa Sch.-Bip. *ex* A. Rich.	ETH				
GUNDELIA L. Asteraceae *(Compositae)*					
tournefortii L.			IRA LEB	ISR	
GUTIERREZIA Lag. Asteraceae *(Compositae)*					
dracunculoides (Dc.) Blake			USA		
microcephala (Dc.) Gray				USA	
sarothrae (Pursh) Britt. & Rusby				USA	
texana Torr. & Gray				USA	
GYMNOCLADUS Lam. Caesalpinaceae *(Leguminosae)*					
dioicus (L.) K. Koch				USA	
GYMNOCORONIS Dc. Asteraceae *(Compositae)*					
spilanthoides Dc.				ARG	
GYMNOPETALUM Arn. Cucurbitaceae					
leucostictum Miq.				IDO	
GYNANDROPSIS Dc. Capparidaceae *(Syn. PEDICELLARIA Schrank)*					
gynandra (L.) Briq. *(See Cleome gynandra L.)*	IND TNZ	THI	EGY HAW KEN	CAB CHN FIJ	MAU

GYNANDROPSIS Dc. Capparidaceae *(Cont'd)*	S	P	C	X	F
gynandra (L.) Briq. *(Cont'd)*			NIG RHO SAF TAI	GHA IDO MAL MOZ PR SUD USA	
pentaphylla (L.) Dc. (See **Cleome gynandra** L.)			KEN	FIJ IND MAU PHI	CAB HAW
GYNERIUM Humb. & Bonpl. **Poaceae** *(Gramineae)*					
saggittatum (Aubl.) Beauv.			PR		HON
GYNURA Cass. **Asteraceae** *(Compositae)*					
crepidioides Benth. (See **Crassocephalum crepidioides** (Benth.) S. Moore)	PHI		IDO IND	CEY	
pseudo-china (L.) Dc.				VIE	
GYPSOPHILA L. **Caryophyllaceae**					
muralis L.			SOV		
porrigens (L.) Boiss.			TUR		
HACKELIA Opiz **Boraginaceae**					
floribunda (Lehm.) I. M. Johnston			USA		
HACKELOCHLOA O. Ktze. **Poaceae** *(Gramineae)*					
granularis (L.) O. Ktze.		MEX		GHA IVO NIC	BUR IND SAF
HAKEA Schrad. **Proteaceae**					
gibbosa (Sm.) Cav.		SAF		NZ	
sericea Schrad.		SAF		NZ	
suaveolens R. Br.			SAF	NZ	
tenuifolia Britten			SAF		
HALOGETON C. A. Mey. **Chenopodiaceae**					
glomeratus (Bieb.) C. A. Mey.			CAN IRA USA		AFG TRI
HAMAMELIS L. **Hamamelidaceae**					
macrophylla Pursh				USA	
virginiana L.				CHL USA	
HAMELIA Jacq. **Rubiaceae**					
erecta Jacq. (See **H. patens** Jacq.)				TRI	

HAMELIA Jacq. **Rubiaceae** *(Cont'd)*	S	P	C	X	F
patens Jacq. (Syn. *H. erecta* Jacq.)				BRA CR GUA HON PAN SAL VEN	
HAPLOPAPPUS Cass. **Asteraceae** *(Compositae)*					
arborescens (A. Gray) H. M. Hall				USA	
bloomeri A. Gray				USA	
laricifolius A. Gray				USA	
pluriflorus (T. & G.) H. M. Hall				USA	
tenuisectus (Greene) Blake *ex* Benson			USA		
HAPLOPHYLLUM A. Juss. **Rutaceae**					
buxbaumii (Poir.) Boiss.			LEB		
HARPACHNE Hochst. *ex* A. Rich. **Poaceae** *(Gramineae)*					
schimperi Hochst. *(See* **Eragrostis schimperi** Benth.)			KEN		
HARPAGOPHYTUM Dc. *ex* Meissn. **Pedaliaceae**					
procumbens Dc.			SAF		
HEDEOMA Pers. **Lamiaceae** *(Labiatae)*					
pulegioides (L.) Pers.				USA	
HEDERA L. **Araliaceae**					
helix L.				USA	
HEDYCHIUM Koen. **Zingiberaceae**					
coronarium Koen.		PR		MAU	GUA SAL
HEDYOTIS L. **Rubiaceae**					
auricularia L. (Syn. *Oldenlandia auricularia* (L.) F. V. M.)	CEY		IDO		
biflora (L.) Lam. (Syn. *Oldenlandia biflora* L.)			HAW PHI	PLW	
corymbosa (L.) Lam. (Syn. *Oldenlandia corymbosa* L.)			CNK GHA	BND CAB CAM CHN DAH GUI IDO IND IVO JAM JPN LIB MAL MLI NEP	

HEDYOTIS L. Rubiaceae *(Cont'd)*	S	P	C	X	F
corymbosa (L.) Lam. *(Cont'd)*				NGI NGR NIG PHI RHO SEG SUD SUR THI UGA VIE	
diffusa Willd. (Syn. *Oldenlandia diffusa* (Willd.) Roxb.)			JPN TAI		
herbacea L. (Syn. *Oldenlandia herbacea* (L.) Roxb.)					
lindleyana Hook.			JPN		
lineata Roxb.				IND	IRA
pinifolia (Wall. *ex* G. Don) K. Schum. (Syn. *Oldenlandia pinifolia* Wall. *ex* G. Don)					
pterita Bl. (Syn. *Oldenlandia pterita* (Bl.) Miq.)					
tenelliflora Bl.			TAI		IDO
umbellata (L.) Lam. (Syn. *Oldenlandia umbellata* L.)					
HEDYPNOIS Scop. **Asteraceae** *(Compositae)*					
cretica (L.) Willd.			ARG ISR MOR	CHL	AUS PER
HELENIUM L. **Asteraceae** *(Compositae)*					
amarum (Raf.) H. Rock			USA	AUS	
aromaticum Bailey				CHL	
autumnale L.				USA	
hoopesii Gray				USA	
nudiflorum Nutt.				USA	
tenuifolium Nutt.			AUS		
HELEOCHLOA Host *ex* Roem. **Poaceae** *(Gramineae)* *(See* **CRYPSIS** Ait.*)*					
schoenoides (L.) Host *(See* **Crypsis schoenoides** (L.) Lam.*)*				USA	
HELIANTHEMUM Hall. **Asteraceae** *(Compositae)*					
salicifolium (L.) Mill.				TUR	
vulgare Gaertn.				TUR	
HELIANTHUS L. **Asteraceae** *(Compositae)*					
annuus L.	MEX		ARG USA	BRA CAB	ISR

HELIANTHUS L. Asteraceae *(Compositae) (Cont'd)*	S	P	C	X	F
annuus L. *(Cont'd)*				DR PER PHI TUR URU VEN	
californicus Dc.				USA	
ciliaris Dc.				USA	
grosse-serratus Martins				USA	
maximiliani Schrad.				USA	
petiolaris Nutt.				USA	
tuberosus L.				DR JPN USA VEN	NZ
HELICONIA L. **Musaceae**					
bihai (L.) L.				NIC	
cannoidea A. Rich.				PER	
psittacorum L. *f.*		TRI		BRA	
HELICTERES L. **Sterculiaceae**					
jamaicensis Jacq.				JAM	
HELIOTROPIUM L. **Heliotropiaceae**					
amplexicaule Vahl	MAU	AUS		ARG IDO USA	
angiospermum Murr. (See **H. parviflorum** L.)				COL PER	HON
anomalum Hook. & Arn.				USA	
bovei Boiss.		LEB			
bracteatum R. Br.				THI	
curassavicum L.			IND IRA PR	ARG CHL PER USA VEN	
eduardii Martelli			KEN	TNZ	
eichwaldi Steud.		IND		PK	AFG
europaeum L.	IRA	AUS ISR ITA JOR SPA SUD YUG	EGY HUN IND IRQ MOR POR	LEB NGI SOV TUR USA	
filiforme H. B. K.				VEN	

HELIOTROPIUM L. Boraginaceae *(Cont'd)*	S	P	C	X	F
fruticosum L.				MEX	
indicum L. (Syn. *Tiaridium indicum* (L.) Lehm.)	IDO	IND TAI THI	KEN PR TRI	ANT BND BOR BRA CAB CHN COL CR CUB DAH GHA GUI HON JAM LIB MAL MAU MLI NIG PAN PHI PK SAL SEG TNZ USA VEN VIE	HK MIC
ovalifolium Forsk.			IND KEN RHO	ANG SUD	
parviflorum L. (Syn. *H. angiospermum* Murr.)				JAM	
procumbens Mill.			TRI	VEN	HON TAI
scabrum Retz.				IND	
steudneri Vatke			KEN		
strigosum (L.) Willd.				GHA IND NIG	
sudanicum F. W. Andrews		SUD			
supinum L.			EGY POR	IND PK	ISR
undulatifolium Turrill			KEN		
HELMINTHIA Juss. **Asteraceae** *(Compositae)*					
echioides (L.) Gaertn. *(See* **Picris echioides** L.)		YUG	SPA	GRE URU	AFG ISR
HELMINTHOSTACHYS Kaulfuss **Ophioglossaceae**					
zeylanica (L.) Hook.				PHI	
HEMARTHRIA R. Br. **Poaceae** *(Gramineae)*					
altissima (Poir.) Stapf & C. E. Hubb.				MAU	BUR IND

HEMARTHRIA R. Br. Poaceae *(Gramineae) (Cont'd)*	S	P	C	X	F
altissima (Poir.) Stapf & C. E. Hubb. *(Cont'd)*					SAF
compressa (L. *f.*) R. Br. (Syn. *Rottboellia compressa* L. *f.*)			TAI	IND	
japonica (Hack) Roshev. *(See* **H. sibirica** (Gand.) Ohwi)			JPN		
longiflora (Hook. *f.*) A. Camus				VIE	BUR MAL
protensa Nees *ex* Steud.				BND	BUR IND
sibirica (Gand.) Ohwi (Syn. *H. japonica* (Hack.) Roshev.)					
HEMEROCALLIS L. **Liliaceae**					
fulva (L.) L.				USA	
HEMIDIODIA K. Schum. **Rubiaceae**					
ocimifolia (Willd.) K. Schum.		MEX	PR	SUR	IDO
HEMIGRAPHIS Nees **Acanthaceae**					
hirta T. Anders.				IND	
primulaefolia Villar				PHI	
HEMISTEPTA Bunge **Asteraceae** *(Compositae)*					
lyrata Bunge (Syn. *Saussurea affinis* Spreng. *ex* Dc.)		TAI	JPN		
HEMITELIA R. Br. **Cyatheaceae**					
latebrosa Mett. *(See* **Alsophila latebrosa** Wall.)			IDO		
HEMIZONIA Dc. **Asteraceae** *(Compositae)*					
congesta Dc.				USA	
HEMIZYGIA Briq. **Lamiaceae** *(Labiatae)*					
welwitschii (Rolfe) Ashby				ANG	
HERACLEUM L. **Apiaceae** *(Umbelliferae)*					
maximum Bartr.				USA	
sphondylium L.			ENG GER	TUR USA	
HERNIARIA L. **Caryophyllaceae**					
cinerea Dc.				CHL	
HERPETICA Rafin. **Caesalpiniaceae** *(Leguminosae)* *(See* **CASSIA** L.)					
alata Rafin. *(See* **Cassia alata** L.)		PR	DR		
HESPERIS L. **Brassicaceae** *(Cruciferae)*					
matronalis L.				USA	

	S	P	C	X	F
HESPEROCNIDE Torr. **Urticaceae**					
sandwicensis Wedd.			HAW	USA	
HETERANTHERA Ruiz & Pav. **Pontederiaceae**					
dubia (Jacq.) Macm.			USA		
limosa (Sw.) Willd.				USA	
reniformis Ruiz & Pav.				USA	
HETERODENDRUM Desf. **Sapindaceae**					
oleaefolium Desf.			AUS		
HETEROPAPPUS Less. **Asteraceae** *(Compositae)*					
hispidus (Thunb.) Less.			TAI		
HETEROPOGON Pers. **Poaceae** *(Gramineae)*					
contortus (L.) Beauv. *ex* Roem. & Schult. (Syn. *Andropogon contortus* L.)			AUS SAF	FIJ GHA IND MEL PER	AFG BUR HAW IDO
HETEROPTERIS Kunth **Malpighiaceae** *(Syn.BANISTERIA* L.)					
purpurea (L.) Kunth (Syn. *Banisteria purpurea* L.)					
HETEROTHECA Cass. **Asteraceae** *(Compositae)*					
grandiflora Nutt.			HAW	USA	
subaxillaris (Lam.) Britt. & Rusby				USA	
HETEROTRICHUM Dc. **Melastomataceae**					
cymosum (Wendl.) Urb.			PR		
HIBISCUS L. **Malvaceae**					
abelmoschus L. (Syn. *Abelmoschus moschatus* Medic.)				PLW	
articulatus Hochst. *ex* A. Rich.				ANG	
aspera Hook. *f.* *(See* **H. cannabinus** L.)		-		GHA IVO	
cannabinus L. (Syn. *H. aspera* Hook. *f.*)	ZAM		KEN RHO	CAB	
elatus Sw.				JAM	
esculentus L.			SUD	IND NIG TUR	
ficulneus L.	AUS			IND SUD	
lasiocarpus Cav.				USA	
mastersianus Hiern.			KEN		
micranthus L. *f.*				GHA	

HIBISCUS L. Malvaceae *(Cont'd)*	S	P	C	X	F
micranthus L. *f.* *(Cont'd)*				IND	
obtusilobus Garcke		SUD			
palustris L.				USA	
panduraeformis Burm. *f.*				IND	
rosa-sinensis L.				IND	
rugosus Roxb. *ex* Steud.				IND	
sabdariffa L.				JAM	
tetraphyllus Roxb.				IND	
tiliaceus L.			HAW		CAB
trionum L.	AFG IRA	AUS ITA KOR LEB SAF SUD SWZ	EGY ETH HUN KEN POL POR RHO YUG	BOT CAN GHA IND IRQ ISR JOR JPN NZ PK SEG SOV USA ZAM	CHN HK
vitifolius L.				GHA IND	
HIERACIUM L. Asteraceae *(Compositae)*					
aurantiacum L.			CAN	NZ USA	
bauhini Schwaeg. *ex* Schrank			SOV		
florentinum All.				CAN USA	
floribundum Wimm. & Grab.				USA	
pilosella L.				CAN GER USA	
praealtum Gochnat				USA	
pratense Tausch				USA	
umbellatum L.			FIN		
vulgatum Fries				USA	
HIEROCHLOE R. Br. Poaceae *(Gramineae)*					
borealis R. & S.				SOV	
odorata (L.) Beauv. (Syn. *Holcus odoratus* L.)			JPN SOV	USA	

	S	P	C	X	F
HILLIA Jacq. **Rubiaceae**					
parasitica Jacq.			PR		
HIPPEASTRUM Herb. **Amaryllidaceae**					
puniceum (Lam.) O. Ktze.				JAM	
HIPPURIS L. **Hippuridaceae**					
vulgaris L.			IRA	ENG	
HIRSCHFELDIA Moench **Brassicaceae** *(Cruciferae)*					
incana (Lowe) Lag.-Foss.	AUS		ARG LEB MOR POR	GRE USA	ISR
HOFFMANNSEGGIA Cav. **Caesalpiniaceae** *(Leguminosae)*					
densiflora Benth.				USA	
falcaria Cav.			ARG		
HOLCUS L. **Poaceae** *(Gramineae)*					
halepensis Pers. (See **Sorghum halepense** (L.) Pers.)				SOV	
lanatus L.	NZ		COL POR SPA	BRA CHL USA	ARG AUS HAW IND SAF
mollis L.			USA	CHL NZ	
odoratus L. (See **Hierochloe odorata** (L.) Beauv.)				SOV	
HOLOCALYX Micheli **Caesalpiniaceae** *(Leguminosae)*					
glaziovii Taub.				BRA	
HOLOSTEUM L. **Caryophyllaceae**					
umbellatum L.				USA	
HOMERIA Vent. **Iridaceae**					
breyniana (L.) Lewis				AUS NZ	
miniata Sweet		AUS			
HOMOLEPIS Chase **Poaceae** *(Gramineae)*					
aturensis (H. B. K.) Chase				PER	
HORDEUM L. **Poaceae** *(Gramineae)*					
berteroanum E. Desv. *ex* C. Gay				CHL	
brachyantherum Nevski				USA	HAW
bulbosum L.			IRA LEB	JOR TUR	AFG AUS HAW ISR

HORDEUM L. Poaceae *(Gramineae) (Cont'd)*	S	P	C	X	F
glaucum Steud.			IRQ		IND
hystrix Roth			AUS	NZ USA	
jubatum L.	ALK		CAN USA	CHL NZ	
leporinum Link		AUS		ARG USA	AFG IND
marinum Huds.			LEB	NZ TUR	
murinum L.	AFG NZ	ITA TUN	ARG AUS GRE HAW HUN IRA IRQ JOR LEB MOR POR SAF SPA	CHL EGY ENG GER ISR JPN POL TAS TUR USA YUG	IND
pusillum Nutt.			USA		
secalinum Schreb.			POR		AFG
spontanum Koch				IRA JOR	
vulgare L.			ARG		AFG HAW IND
HOTTONIA Boerh. *ex* L. **Primulaceae**					
palustris L.				BEL ENG GER NET	
HOUTTUYNIA Thunb. **Saururaceae**					
cordata Thunb.			CHN JPN TAI		
HULTHEMIA Bl. *ex* Miq. **Papilionaceae** *(Leguminosae)*					
persica Bornm.			IRA		
HUMULUS L. **Cannabaceae**					
japonicus Sieb. & Zucc.			JPN		
lupulus L.			JPN	NZ	
HYBANTHUS Jacq. **Violaceae**					
enneaspermus F. Muell.				GHA MOZ SUD	IDO
humilis Standl.				MEX	

	S	P	C	X	F
HYBANTHUS Jacq. **Violaceae** *(Cont'd)*					
parviflorus Baill.				ARG	
HYDRANGEA L. **Hydrangeaceae**					
arborescens L.				USA	
quercifolia Bartr.				USA	
HYDRILLA L. C. Rich. **Hydrocharitaceae**					
verticillata (L. *f.*) Royle	IND THI USA	AUS CAB PK VIE	IDO IRA LEB PHI	BND BOR CEY CHN ENG FIJ JPN KEN MAU NEP NZ TNZ UGA	KOR
HYDROCHARIS L. **Hydrocharitaceae**					
dubia (Bl.) Backer			JPN		
morsus-ranae L.			JPN POR	BEL ENG FRA GER	AUS ISR
HYDROCHLOA Beauv. **Poaceae** *(Gramineae)*					
carolinensis Beauv.				USA	
HYDROCOTYLE L. **Apiaceae** *(Umbelliferae)*					
americana L.				USA	
asiatica L. *(See* **Centella asiatica** (L.) Urb.*)*	SAF		JPN	BND FIJ IND MAU	
bonariensis Lam. *(See* **H. umbellata** L.*)*			ARG	BRA CHL MAU PER	
exigua (Urb.) Malme				BRA	
japonica Makino				IND	
formosana Masamune			TAI		
laxiflora Dc.			AUS		
leucocephala Cham. & Schlecht.			ARG		
maritima Honda			JPN		
mexicana Cham. & Schlecht.			SAL		
novae-zeelandiae Dc.			HAW		
poeppigi Dc.				CHL	

HYDROCOTYLE L. Apiaceae *(Umbelliferae) (Cont'd)*	S	P	C	X	F
ramiflora Maxim.			JPN		
ranunculoides L. *f.*				COL	ISR
rotundifolia Roxb. *(See* **H. sibthorpioides** Lam.)				IND VIE	JPN
sibthorpioides Lam. (Syn. *H. rotundifolia* Roxb.)			HAW IDO JPN	USA	
tripartita R. Br. *ex* A. Rich.			AUS		
umbellata L. (Syn. *H. bonariensis* Lam.)				MAU MEX PER SUR USA VEN	
verticillata Thunb.			HAW	USA	
HYDRODICTYON Roth **Hydrodictyaceae**					
reticulatum Lagerh.			JPN	USA	
HYDROLEA L. **Hydrophyllaceae**					
glabra Schum. & Thonn. (Syn. *H. guineensis* Choi.)					
graminifolia A. W. Benn.		NIG			
guineensis Choi. *(See* **H. glabra** Schum. & Thonn.)		NIG			
macrosepala A. W. Benn.		NIG			
quadrivalis Walt.				USA	
spinosa L.				TRI	IDO
uniflora Raf.				USA	
zeylanica (L.) Vahl				CEY	BRA IDO IND PHI
HYDROTRIDA Willd. **Scrophulariaceae**					
caroliniana Small *(See* **Bacopa caroliniana** (Walt.) Robins.)				USA	
HYGROPHILA R. Br. **Acanthaceae**					
angustifolia R. Br.		CAB			
aristata Nees				BND CEY IND THI	
chevalieri Benoist				NIG	
phlomoides Nees		CAB IND			
pobeguini Benoist	NIG				

	S	P	C	X	F
HYGROPHILA R. Br. **Acanthaceae** *(Cont'd)*					
quadrivalvis Nees				THI	
spinosa T. Anders. (Syn. *Asteracantha longifolia* (L.) Nees)					
HYGRORYZA Nees **Poaceae** *(Gramineae)*					
aristata (Retz.) Nees				BND IND	CEY
HYMENACHNE Beauv. **Poaceae** *(Gramineae)*					
amplexicaulis (Rudge) Nees	SUR		IDO	TRI	
indica Buese *(See* **Sacciolepis indica** (L.) A. Chase)			IDO JPN		
myurus (Lam.) Beauv. *(See* **Sacciolepis myurus** (Lam.) A. Chase)				BND MAL THI	
HYMENOCARDIA Wall. *ex* Lindl. **Euphorbiaceae**					
acida Tul.				ANG	
HYMENOCARPOS Savi **Papilionaceae** *(Leguminosae)*					
circinnata Savi			LEB		ISR
HYMENOCLEA Torr. & Gray *ex* Torr. **Asteraceae** *(Compositae)*					
monogyra Torr. & Gray				USA	
HYMENOXYS Cass. **Asteraceae** *(Compositae)*					
odorata Dc.			USA		
richardsoni (Hook.) Cock.			USA		
tweediei Hook. & Arn.			ARG		
HYOSCYAMUS L. **Solanaceae**					
niger L.			AUS ENG GER IRA	NZ SOV USA	AFG BRA
reticulatus L.			LEB		AFG ISR
HYPARRHENIA Anderss. **Poaceae** *(Gramineae)*					
gazensis (Rendle) Stapf		MOZ			
hirta (L.) Stapf (Syn. *Andropogon hirtus* L.)					EGY
rufa (Nees) Stapf	COL	HON MOZ	HAW SUD	ANG BOT BRA CNK DAH ECU GHA GUA MAU NIG RHO	BUR IND

HYPARRHENIA Anderss. Poaceae *(Gramineae) (Cont'd)*	S	P	C	X	F
rufa (Nees) Stapf *(Cont'd)*				SAL SEG THI TRI USA	
HYPECOUM L. **Papaveraceae**					
grandiflorum Benth.		GRE	POR SPA	TUR	ISR
procumbens L.				TUR	
trilobum Trautv.				PK	
HYPERICUM L. **Hypericaceae**					
androsaemum L.		NZ		AUS	SAF
angustifolium Lam.				TUR	
brasiliense Choisy				BRA	
crispum L. (Syn. *H. triquetrifolium* Turra)		TUN	LEB	GRE TUR	ISR
elodes Huds.				POR	
erectum Thunb.				JPN	
japonicum Thunb.			TAI	IND NZ	AUS
laxum (Bl.) Koidz.				JPN	
maculatum Walt.				FIN	
montbretti Spach				TUR	
perforatum L.	AUS USA	HUN ITA NZ SWE TUR	CAN HAW IRA POL SAF SOV	AFG BEL CHL IRQ JPN PK SUD	TRI
punctatum L.				USA	
quadrangulum L.			SOV		
tetrapterum Fries				AUS NZ	
triquetrifolium Turra (*See* **H. crispum** L.)			LEB		
HYPOCHOERIS L. **Asteraceae** *(Compositae)*					
brasiliensis Griseb.			BRA	ARG	
glabra L.		KEN		AUS BRA CHL NZ USA	
radicata L.		CHL NZ	ARG AUS	BND USA	

HYPOCHOERIS L. Asteraceae *(Compositae) (Cont'd)*	S	P	C	X	F
radicata L. *(Cont'd)*			CAN COL HAW JPN SAF		
tweediei (Hook. & Arn.) Cabrera				ARG	
HYPOESTES Soland. *ex* R. Br. **Acanthaceae**					
verticillaris R. Br.				KEN	
HYPOXIS L. **Amaryllidaceae**					
decumbens L.				ARG	BRA
HYPTIS Jacq. **Lamiaceae** *(Labiatae)*					
atrorubens Poit.			PR	TRI	
brevipes Poit.		MAL	BOR PHI TAI	PER VIE	IDO
capitata Jacq.		AUS	PR TRI	COL HON IDO JAM NGI PAN PER PHI SAL	
gaudichaudii Benth.		BRA			
lanceolata Poir.	NIG			GHA	
mutabilis (Rich.) Briq.				PER	
pectinata (L.) Poit.		FIJ	HAW PR	GHA JAM PLW TRI	IDO
spicigera Lam.				GHA	
suaveolens (L.) Poit.	BRA	PR TNZ	AUS IND MIC PHI TAI THI	CAB CEY CNK COL CR DAH FIJ GHA IDO JAM LEB MAL MAU MEL MEX NET NGI NIC PAN SAL SEG USA	CHN HK

HYPTIS Jacq. Lamiaceae *(Labiatae)* *(Cont'd)*	S	P	C	X	F
suaveolens (L.) Poit. *(Cont'd)*				VEN VIE	
verticillata Jacq.				HON JAM PAN SAL	
IBATIA Decne. Asclepiadaceae					
maritima Decne.			PR		
IBERIS L. Brassicaceae *(Cruciferae)*					
amara L.				SPA	
IBICELLA Van Eselt. Martyniaceae					
lutea (Lindl.) Van Eselt.				ARG AUS SAF	
ICACINA A. Juss. Icacinaceae					
senegalensis Juss.		GHA			
IFLOGA Cass. Asteraceae *(Compositae)*					
fontanesii Cass.				PK	AFG
ILEX L. Aquifoliaceae					
ambigua (Michx.) Torr.				USA	
decidua Walt.				USA	
glabra (L.) Gray				USA	
opaca Ait.				USA	
vomitoria Ait.				USA	
ILYSANTHES Rafin. Scrophulariaceae (*See* LINDERNIA All.)					
antipoda (L.) Merr. (*See* Lindernia ruelloides (Colsm.) Pennell)	IDO		PHI TAI		
ciliata O. Ktze. (Syn. *Bonnaya brachiata* Link & Otto) (*See* Lindernia ciliata (Colsm.) Pennell)				IDO	
parviflora Benth.				IND	
serrata (Thunb.) Makino (*See* Vandellia anagallis (Burm.) Yamazaki *var.* verbenaefolia (Colsm.) Yamazaki)			TAI		IDO
veronicaefolia Urban				IND	
IMPATIENS L. Balsaminaceae					
balsamina L.			PR	MEX USA VIE	
capensis Meerb.				USA	
chinensis L.				VIE	

	S	P	C	X	F
IMPATIENS L. Balsaminaceae *(Cont'd)*					
pallida Nutt.				USA	
platypetala Lindl.			IDO		
poilanei Tardieu				VIE	
sultani Hook. *f.*				VEN	
IMPERATA Cyr. Poaceae *(Gramineae)*					
arundinacea Cyr. *var.* indica Anderss. *(See* **I. cylindrica** (L.) Beauv. *var.* major (Nees) C. E. Hubb.)			CHN SPA	BND IND	
brasiliensis Trin.		TRI	ARG	BOL BRA	
cylindrica (L.) Beauv. *var.* major (Nees) C. E. Hubb. (Syn. *I. arundinacea* Cyr. *var.* indica Anderss.)	AUS BOR CEL CEY DAH GHA IDO IND IRQ KEN MAL NGI NIG PHI TAI THI TNZ UGA	ARB CAB JPN MAD MOZ PK	CHN EGY IRA ISR SPA	AFG BND FIJ FRA GUI HAW IVO JOR LIB MAU MEL MLI NEP NZ PLW SAF SEG USA VIE VOL	CNK KOR PR
exaltata Brongn.				PHI	
tenuis Hack.				BOL	
INDIGOFERA L. Papilionaceae *(Leguminosae)*					
anil L. *(See* **I. suffruticosa** Mill.)				BRA	HAW
arrecta Hochst. *ex* A. Rich.			IDO	ANG	
australis Willd.			AUS		
bracteola Dc.			GHA		
colutea (Burm. *f.*) Lam. (Syn. *I. viscosa* Lam.)					EGY
endecaphylla Jacq. *(See* **I. spicata** Forsk.)				IND MAU	PR
enneaphylla L. *(See* **I. linnaei** Ali)			IND	MAU	AUS IDO
glandulosa Willd.			IND		IDO
hendecaphylla Jacq. *(See* **I. spicata** Forsk.)			KEN		
hirsuta L.		GHA	COL NIG TAI	ANG BRA IND	AUS

INDIGOFERA L. Papilionaceae *(Leguminosae) (Cont'd)*	S	P	C	X	F
hirsuta L. *(Cont'd)*				NGI PHI SUR	
linifolia (L. *f.*) Retz.			IND		AFG IDO
linnaei Ali (Syn. *I. enneaphylla* L.)					
macrophylla Schum. & Thonn.			GHA		
mucronata Willd. *ex* Spreng.			PER		
parviflora Heyne			IND		
pseudo-tinctoria Matsum.		JPN			
spicata Forsk. (Syn. *I. endecaphylla* Jacq.) (Syn. *I. hendecaphylla* Jacq.)			KEN	TRI	PR
subulata Vahl				GHA NIG	
suffruticosa Mill. (Syn. *I. anil* L.)	PLW		HAW PR TRI	BOL COL CR DR FIJ MEX NGI PER PHI	IDO
sumatrana Gaertn. *(See* **I. tinctoria** L.)				IND	
tinctoria L. (Syn. *I. sumatrana* Gaertn.)			PR	PHI	CAB
tomentosa L.			IDO		
trita L. *f.*			IND	ANG	
viscidissima Bak.				ANG	
viscosa Lam. *(See* **I. colutea** (Burm. *f.*) Lam.)				IND	AFG AUS IDO
INULA L. Asteraceae *(Compositae)*					
britannica L.			SOV		CHN
dysenterica L.				TUR	
graveolens Desf.		AUS	MOR SAF		ISR
helenium L.				TUR USA	
heterolepis Boiss.				TUR	
indica L.		IND			
viscosa Dryand.			MOR SPA		

	S	P	C	X	F
IONDRABA Reichb. **Brassicaceae** *(Cruciferae)*					
auriculata (L.) Schulz (Syn. *Biscutella auriculata* L.)			POR		
IONIDIUM Vent. **Violaceae**					
suffruticosum Ging.				IND	
IONOXALIS Small **Oxalidaceae**					
martiana (Zucc.) Small *(See* **Oxalis martiana** Zucc.)			PR	ANT	
IPOMOEA L. **Convolvulaceae** *(Syn. CALONYCTION* Choisy)					
acuminata Roem. & Schult.				URU	
alba L. (Syn. *Calonyction aculeatum* (L.) House)			HAW	PLW	AUS
amoena Choisy				GHA	
angulata Mart. *ex* Choisy *(See* **I. hederifolia** L.)				AUS	
angustifolia Jacq.				VIE	
aquatica Forsk. (Syn. *I. reptans* (L.) Poir.)	IND MOZ THI	CAB DAH PHI	HAW IDO NIG	AUS BND BOR CNK COL FIJ GHA IVO KEN MAL MEL MIC RHO SEG SUD SUR TNZ TRI TUN TUR USA VIE ZAM	CHN HK
asperifolia Hallier *f.*				ANG GHA	
barbigera Sweet				USA	
batatas (L.) Lam.			TAI	CNK HON	CAB IDO
blepharosepala Hochst. *ex* A. Rich.	SUD				
cairica (L.) Sweet			HAW TAI	BRA GHA URU USA	ARG AUS COL
calobra W. Hill & F. Muell.			AUS		
caloneura Meissn.				COL	

IPOMOEA L. Convolvulaceae *(Cont'd)*	S	P	C	X	F
cardiosepala Meissn.				SUD	
chryseides Ker-Gawl.				VIE	
coccinea L.				MAU USA VEN	
congesta R. Br. (Syn. *I. indica* (Burm.) Merr.)			HAW TAI	AUS NZ PLW USA	
coptica (L.) Roth *ex* Roem. & Schult. (Syn. *I. dissecta* Willd.)					
cordofana Choisy		SUD			
coscinosperma Hochst. *ex* Choisy			SAF		
crassicaulis (Benth.) B. L. Robinson *(See* **I. fistulosa** Mart. *ex* Choisy)			IDO		
crassifolia Cav.				PER	
cymosa Roem. & Schult.				IND	
cynanchifolia C. B. Clarke			RHO		
digitata L. (Syn. *I. mauritiana* Jacq.)					
dissecta Willd. *(See* **I. coptica** (L.) Roth *ex* Roem. & Schult.)		PR		JAM MEX	COL
eriocarpa R. Br. (Syn. *I. hispida* Roem. & Schult. *non* Zuccagni)				PK	
fistulosa Mart. *ex* Choisy (Syn. *I. crassicaulis* (Benth.) B. L. Robinson)				ARG BRA USA	
gossypioides Parodi				ARG	
gracilis R. Br.		THI	TAI	AUS PLW	
grandiflora (Choisy) Hallier *f. non* Lam. *(See* **I. tuba** (Schlectend.) G. Don)				URU	
hardwikii Hemsl.		TAI			
hederacea (L.) Jacq. (Syn. *Convolvulus hederaceus* L.)		USA		BRA GHA IND PHI	
hederifolia L. (Syn. *I. angulata* Mart. *ex* Choisy)				DR	
hirsutula Jacq. *f.*				USA	
hispida Roem. & Schult. *non* Zuccagni *(See* **I. eriocarpa** R. Br.)				GHA IND	IDO
indica (Burm.) Merr. *(See* **I. congesta** R. Br.)			AUS		HAW
indivisa H. Hallier				ARG	

IPOMOEA L. Convolvulaceae *(Cont'd)*	S	P	C	X	F
indivisa H. Hallier *(Cont'd)*				URU	
involucrata Beauv.				GHA NIG	CNK
lacunosa L.			JPN		
leari Paxt.				GHA	
mauritiana Jacq. *(See* **I. digitata** L.)				GHA	
maxima (L. *f.*) G. Don *ex* Sweet (Syn. *I. sepiaria* Koen. *ex* Roxb.)					
muelleri Benth.			AUS		
muricata (L.) Jacq. (Syn. *Calonyction muricatum* G. Don)					
nil (L.) Roth		IND		BRA HON MEX TRI	
obscura (L.) Ker-Gawl.				HAW TAI	USA	IDO
pandurata (L.) G. F. W. Mey.				USA	
pes-caprae (L.) Sweet	THI		HAW PR TAI TRI	FIJ GHA IND MEX PLW USA	IDO
pes-tigridis L.			IND PHI		IDO
plebeia R. Br.			AUS	ANG	
polyantha Roem. & Schult.			PR	MEX	
purpurea (L.) Roth		BRA MEX SAF USA	ANG ARG AUS	CHL HON PER VEN	
quamoclit L. (Syn. *Quamoclit pinnata* (Desv.) Boj.)				USA VIE	
quinquefolia L. *(See* **Merremia quinquefolia** (L.) Hall. *f.)*			PR		USA
reptans (L.) Poir. *(See* **I. aquatica** Forsk.)	NIG	IND PHI		BND SUD .	AUS HAW IDO THI
sepiaria Koen. *ex* Roxb. *(See* **I. maxima** (L. *f.*) G. Don *ex* Sweet)				IND	
setifera Poir.				SUR	
stolonifera (Cyr.) Gmel.			TAI	USA	
tiliacea (Willd.) Choisy	COL		PR	DR JAM	

IPOMOEA L. Convolvulaceae *(Cont'd)*	S	P	C	X	F
tiliacea (Willd.) Choisy *(Cont'd)*				NIC	
trichocarpa Ell.				USA	
trifida (H. B. K.) G. Don			MEX	HON	
triloba L.	AUS PHI	CUB HAW HON	ARG IDO JAM	CAB COL CR ECU IVO MIC NEP NGI PLW PR SAL , SEG THI USA	
tuba (Schlectend.) G. Don (Syn. *I. grandiflora* (Choisy) Hallier *f. non* Lam.)					
tuboides Deg. & Ooststr.				HAW	
IRESINE P. Br. **Amaranthaceae**					
calea Standl.				SAL	
celosia L.			PR TRI	DR HON JAM NIC SAL	
rhizomatosa Standl.				USA	
IRIS L. **Iridaceae**					
foetidissima L.		NZ			
missouriensis Nutt.				USA	
pseudacorus L.			ARG IRA POR SPA	BEL ENG FRA GER NZ USA	
versicolor L.				USA	
ISACHNE R. Br. **Poaceae** *(Gramineae)*					
australis R. Br. *(See* **I. globosa** (Thunb.) O. Ktze.)	CEY			BND MAL THI VIE	AUS
chevalieri A. Camus				VIE	
dispar Trin.				IND	
globosa (Thunb.) O. Ktze. (Syn. *I. australis* R. Br.)	CEY		MAL	IDO THI	AUS IND JPN

ISACHNE R. Br. **Poaceae** *(Gramineae)* *(Cont'd)*	S	P	C	X	F
kunthiana (Wight & Arn.) Miq. (Syn. *I. schmidii* Hack.)		CEY			IND
myosotis Nees				VIE	
schmidii Hack. (*See* **I. kunthiana** (Wight & Arn.) Miq.)			TAI		
ISATIS L. **Brassicaceae** *(Cruciferae)*					
tinctoria L.		SPA		CHL TUR USA	AFG
ISCHAEMUM L. **Poaceae** *(Gramineae)*					
afrum (J. F. Gmel.) Dandy	SUD				
aristatum L.	THI	MAL		CAB IND MAU PLW TRI VIE	IDO
brachyatherum (Hochst.) Fenzl *ex* Hack.				SUD	
ciliare Retz. (*See* **I. indicum** (Houtt.) Merr.)				VIE	
crassipes (Steud.) Thell.				JPN	
indicum (Houtt.) Merr. (Syn. *I. ciliare* Retz.)					
muticum L.	BOR MAL			NGI NIC	CEY
pilosum (Trimen) Wight				IND	
rugosum Salisb.	BRA CEY FIJ GHA GUI IND MAD MEL SUR THI	LIB PER SEG	BOR PHI	BND CAB COL CUB DR GUY IDO JAM MAL TRI VIE	BUR CHN
timorense Kunth		MAL	BOR IDO	IND NGI TRI	BUR CEY
ISCHNOSIPHON Koern. **Marantaceae**					
leucophoeus Koern.				PER	
ISEILEMA Anderss. **Poaceae** *(Gramineae)*					
vaginiflorum Domin	AUS				
ISOCARPHA R. Br. **Asteraceae** *(Compositae)*					
bilbergiana Less.				TRI	

	S	P	C	X	F
ISOTOMA Lindl. **Campanulaceae** (See **LAURENTIA** Michx. *ex* Adans.)					
longiflora (L.) Presl (See **Laurentia longiflora** (L.) Peterm.)			PR TRI	BRA DR HON MAU MEX NGI SAL USA	FIJ HAW IDO
IVA L. **Asteraceae** *(Compositae)*					
axillaris Pursh.			CAN	AUS USA	
ciliata Willd.				USA	
xanthifolia Nutt.			CAN USA		
IXERIS Cass. **Asteraceae** *(Compositae)*					
dentata (Thunb.) Nakai (Syn. *Lactuca dentata* (Thunb.) Robins.)			JPN TAI		
japonica (Burm.) Nakai (Syn. *Lactuca debilis* (Thunb.) Benth.)			JPN		
laevigata (Bl.) Sch.-Bip.			TAI		
polycephala Cass.			JPN		
stolonifera A. Gray (Syn. *Lactuca stolonifera* (A. Gray) Benth.)			JPN		
IXIOLIRION Fisch. **Amaryllidaceae**					
montanum Herb.			IRA LEB		AFG ISR
IXOPHORUS Schlecht. **Poaceae** *(Gramineae)*					
unisetus (Presl) Schlecht.				HON SAL	CR
JABOROSA Juss. **Solanaceae**					
integrifolia Lam.			ARG		
runcinata Lam.		ARG			
JACARANDA Juss. **Bignoniaceae**					
oxyphylla Cham.				BRA	
JACQUEMONTIA Choisy **Convolvulaceae**					
martii Choisy				BRA	
pentantha (Jacq.) G. Don			PR		
pycnocephala Benth.				MEX	
sandwicensis A. Gray			HAW	USA	
tamnifolia Griseb.	GHA			SUR USA	

JASMINUM L. Oleaceae	S	P	C	X	F
azoricum L.			PR	DR	
fruticans L.				TUR	
subtriplinerve Bl.			TAI		
JATROPHA L. Euphorbiaceae *(Syn. ADENOROPIUM* Pohl)					
curcas L. (Syn. *Curcas curcas* (L.) Britton)			PR	BRA FIJ HON IND JAM PAN SAL	AUS IDO
glandulifera Roxb.				IND	
gossypifolia L. (Syn. *Adenoropium gossypifolium* (L.) Pohl)			IND	AUS BRA JAM TRI	IDO PR
JOSEPHINIA Vent. Pedaliaceae					
eugenia F. Muell.				AUS	
JUGLANS L. Juglandaceae					
cinerea L.				USA	
microcarpa Berlandier				USA	
nigra L.				CHL USA	
JUNCELLUS C. B. Clarke **Cyperaceae**					
alopecuroides (Rottb.) C. B. Clarke *(See* **Cyperus alopecuroides** Rottb.)				EGY SUD	
nipponicus (Fr. & Sav.) C. B. Clarke *(See* **Cyperus nipponicus** Fr. & Sav. *var.* nipponicus)				KOR	
JUNCUS L. **Juncaceae**					
acutus L.				ARG AUS NZ	ISR
alatus Fr. & Sav.			JPN		
articulatus L.			AUS POR	BEL ITA NZ	AFG
balticus Willd.				USA	
bufonius L.		AUS BRA	EGY FIN GER JPN POR SOV	CHL COL ITA NZ USA	AFG
compressus Jacq.			GER		
conglomeratus L.				BEL	

JUNCUS L. Juncaceae *(Cont'd)*	S	P	C	X	F
conglomeratus L. *(Cont'd)*				GER	
effusus L.			GER JPN POR	BEL ENG KOR NZ TUR USA	
filiformus L.			FIN	ENG	
gerardi Loisel.				TUR USA	
glaucus Sibth.			LEB	TUR	AFG ISR POR
inflexus L.				BEL ENG	
koidzumii Satake			JPN		
lampocarpus Ehrh.				TUR	
leschenaultii J. Gay *ex* Laharpe		JPN	KOR TAI		
marginatus Rostk.				USA	
maritimus Lam.			IRQ	NZ TUR	IRA ISR
microcephalus H. B. K.		BRA			
papillosus Fr. & Sav.			JPN		
phaeocephalus Engelm.				USA	
polyanthemus Buchem.				AUS	
prismatocarpus R. Br.	IND	TAI	JPN	NZ VIE	
procerus E. Mey.				CHL NZ	
roemerianus Scheele				USA	
rugosus Steud.			POR		
tenuis Willd.			JPN	NZ USA	
yokoscensis (Fr. & Sav.) Satake			JPN		
JUNIPERUS L. Cupressaceae					
ashei Buchholz				USA	
communis L.				USA	
deppeana Steud.				USA	
horizontalis Moench				USA	
monosperma (Engelm.) Sarg.				USA	
occidentalis Hook.				USA	

JUNIPERUS L. Cupressaceae *(Cont'd)*	S	P	C	X	F
osteosperma (Torr.) Little				USA	
pinchotii Sudw.				USA	
scopulorum Sarg.				USA	
silicicola (Small) Bailey				USA	
virginiana L.				USA	
JUSSIAEA L. Onagraceae *(See* **LUDWIGIA** L.)					
affinis Dc. *(See* **Ludwigia affinis** (Dc.) Hara)				SUR	
angustifolia Lam. *(See* **Ludwigia octovalvis** (Jacq.) Raven *ssp.* octovalvis)			IDO PR		
californica (Wats.) Jeps.				USA	
decurrens (Walt.) Dc. *(See* **Ludwigia decurrens** Walt.)			JPN USA	ANT BND TRI	
diffusa Forsk. *(See* **Ludwigia adscendens** (L.) Hara)				USA	
erecta L. *(See* **Ludwigia erecta** (L.) Hara)	THI	NIG	PHI PR TAI	COL FIJ MEX PER SUR	GHA IDO
grandiflora L. *(See* **Ludwigia peruviana** (L.) Hara)				MEX	
leptocarpa Nutt. *(See* **Ludwigia leptocarpa** (Nutt.) Hara)		GHA		SUR	
linifolia Vahl *(See* **Ludwigia hyssopifolia** (G. Don) Exell)	BOR MAL TRI	IDO NIG THI	COL PHI	BRA CHN HK PLW VIE	NGI
longifolia Dc. *(See* **Ludwigia longifolia** (Dc.) Hara)		BRA			
michauxiana Fern.				USA	
perennis Brenan *(See* **Ludwigia perennis** L.)				IND	
peruviana L. *(See* **Ludwigia peruviana** (L.) Hara)		COL		USA	IDO
prostrata Lev. *(See* **Ludwigia prostrata** Roxb.)				JPN	
repens L. *(See* **Ludwigia adscendens** (L.) Hara)	PLW	GHA IND THI USA	ARG AUS IDO MAL PHI	BND CAB CEY CHL CHN COL EGY JPN MAU	ANG BOT ISR MOZ

JUSSIAEA L. Onagraceae *(Cont'd)*	S	P	C	X	F
repens L. *(Cont'd)*				PR RHO VIE	
stenorraphe Brenan *(See* **Ludwigia stenorraphe** (Brenan) Hara*)*		NIG			
suffruticosa L. *(See* **Ludwigia octovalvis** (Jacq.) Raven *ssp.* octovalvis*)*	NIG TRI	HAW IND	IDO PR TAI	AUS BRA CHN CR FIJ MAL MAU NIC PHI PLW SAL THI	SEG
uruguayensis Camb. *(See* **Ludwigia uruguayensis** (Camb.) Hara*)*		BRA			
JUSTICIA L. **Acanthaceae**					
americana (L.) Vahl				USA	
diffusa Willd.				IND	
exigua S. Moore				KEN TNZ	
flava Kurz			GHA UGA		
insularis T. Anders.			GHA	NIG	
pectoralis Jacq.			TRI		
procumbens L.			CHN JPN TAI	IND	IDO
prostrata Schlecht. *ex* Nees				IND	
simplex D. Don.		IND			IDO
striata (Klotzsch) Bullock				KEN	
KALANCHOE Adans. **Crassulaceae** *(Syn.BRYOPHYLLUM* Salisb.*)*					
pinnata (Lam.) Pers. (Syn. *Bryophyllum pinnatum* (Lam.) Kurz)					
verticillata Elliot (Syn. *Bryophyllum tubiflorum* Harvey)					
KALIMERIS Cass. **Asteraceae** *(Compositae)*					
pinnatifolia (Maxim.) Kitam.			JPN		
yomena Kitam.			JPN		
KALLSTROEMIA Scop. **Zygophyllaceae**					
californica (Wats.) Vail				USA	
caribaea Urb.		PR			

	S	P	C	X	F
KALLSTROEMIA Scop. Zygophyllaceae *(Cont'd)*					
grandiflora Torr.				USA	
hirsutissima Vail				USA	
maxima Gray	HON MEX TRI	COL GUA SAL	JAM PR	CR DR ECU GHA USA VEN	
pubescens (G. Don) Dandy			COL	GHA	
KALMIA L. Ericaceae					
angustifolia L.				USA	
latifolia L.				USA	
polifolia Wang.				USA	
KARWIMSKIA Zucc. Rhamnaceae					
humboldtiana (Roem. & Schult.) Zucc.		MEX		USA	
KICKXIA Bl. Apocynaceae					
elatine (L.) Dum.			ENG	NZ USA	AUS
lanigera Hand.-Mazz.				POR	AFG
spuria (L.) Dum.			ENG		AUS POR
KIRKIA Oliv. Simaroubaceae					
acuminata Oliv.				RHO	
KNAUTIA L. Dipsacaceae					
arvensis Coult.			FIN POL SPA	PR USA	
sylvatica Duby				TUR	
KOCHIA Roth Chenopodiaceae					
aphylla R. Br.				AUS	
indica Wight				PK	ARG
scoparia (L.) Schrad.			ARG	CAN USA	AFG
villosa Lindl.				AUS	
KOEBERLINIA Zucc. Simaroubaceae					
spinosa Zucc.				USA	
KOELERIA Pers. Poaceae *(Gramineae)*					
phleoides (Vill.) Pers.			ARG POR	GRE	AFG AUS BRA ISR SAF

KOHAUTIA Cham. & Schlect. **Rubiaceae**	S	P	C	X	F
grandiflora Dc.				GHA	
senegalensis Cham. & Schlecht.				GHA NIG	
KOHLRAUSCHIA Kunth **Caryophyllaceae**					
prolifera (L.) Kunth			AUS		
KOLOBOPETALUM Engl. **Menispermaceae**					
chevalieri (Hutchinson & Dalziel) Troupin				CNK	
KOPSIA Bl. **Apocynaceae**					
ramosa Dum.				SOV	
KOSTELETZKYA Presl **Malvaceae**					
virginica (L.) Presl				USA	
KRIGIA Schreb. **Asteraceae** *(Compositae)*					
virginica (L.) Willd.				USA	
KUMMEROWIA Schindl. **Papilionaceae** *(Leguminosae)*					
stipulacea (Maxim.) Makino			JPN		
striata (Thunb.) Schindl. (Syn. *Microlespedeza striata* Makino)			JPN TAI		
KUNDMANNIA Scop. **Apiaceae** *(Umbelliferae)*					
sicula Dc.			MOR		
KYLLINGA Rottb. **Cyperaceae**					
aurata (Nees) Nees			KEN		
brevifolia Rottb. *(See* **Cyperus brevifolius** (Rottb.) Hassk.*)*		TAI TRI	IDO	BOR CHN CR FIJ NGI PHI USA VIE	HAW
cylindrica Nees *(See* **Cyperus sesquiflorus** (Torr.) Mattf. & Kuk. *var.* subtriceps (Nees) T. Koyama*)*			KEN	TNZ	
erecta Schum.			KEN SAF	GHA TNZ	
monocephala Rottb. *(See* **Cyperus kyllingia** Endl.*)*		PHI	IDO TRI	AUS BOR CAB FIJ IND MAU VIE	HAW
odorata Vahl			TRI	FIJ	
polyphylla Thou. *ex* Link		MAU			
pumila Michx.			TRI		

	S	P	C	X	F
LABURNUM Fabr. **Papilionaceae** *(Leguminosae)*					
anagyroides Medic.				TUR	
LACHNANTHES Ell. **Haemodoraceae**					
tinctoria Ell.				USA	
LACTUCA L. **Asteraceae** *(Compositae)*					
biennis (Moench) Fern.				CAN USA	
canadensis L.				USA	
capensis Thunb.		ETH	KEN	GHA	
debilis (Thunb.) Benth. *(See* **Ixeris japonica** (Burm.) Nakai)				CHN JPN	
dentata (Thunb.) Robins. *(See* **Ixeris dentata** (Thunb.) Nakai)			JPN		
dissecta D. Don				PK	AFG
formosana Maxim.			TAI		
indica L.			CHN JPN TAI	MAU	NGI
indica L. *var.* laciniata (O. Ktze.) Hara (Syn. *L. laciniata* (Houtt.) Makino)					
intybacea Jacq.		PR	COL VEN		
laciniata (Houtt.) Makino *(See* **L. indica** L. *var.* laciniata (O. Ktze.) Hara)			JPN		
muralis E. Mey.				USA	
orientalis Boiss.			IRA LEB		AFG
pulchella Dc.			CAN USA		
runcinata Dc.				IND	
saligna L.			AUS LEB	MOR USA	AFG ISR
scariola L.		LEB	CAN HAW IRA SOV	BRA CHL ISR TUR USA	AFG AUS SAF
serriola L.			ARG AUS CAN EGY LEB SAF USA	CHL IRA IRQ MOR NZ	
stolonifera (A. Gray) Benth. *(See* **Ixeris stolonifera** A. Gray)		JPN			
taraxacifolia Schum. & Thonn.			GHA	KEN	

LACTUCA L. Asteraceae *(Compositae) (Cont'd)*	S	P	C	X	F
tatarica (L.) C. A. Mey. (Syn. *Sonchus tartaricus* L.)				SOV	
versicolor Sch.-Bip.				CHN	
viminea J. & C. Presl			MOR SOV	TUR	AFG ISR
LAFOENSIA Vand. **Lythraceae**					
pacari St. Hil.				BRA	
LAGAROSIPHON Harv. **Hydrocharitaceae**					
major (Ridley) Moss		NZ			
roxburghii Benth.		IND		BND	
LAGASCEA Cav. **Asteraceae** *(Compositae)*					
mollis Cav.			IND	MEX USA	
LAGGERA Sch.-Bip. *ex* Hochst. **Asteraceae** *(Compositae)*					
aurita Sch.-Bip. *ex* C. B. Clarke (Syn. *Blumea aurita* Dc.)					
LAGONYCHIUM M. Bieb. **Mimosaceae** *(Leguminosae)*					
farctum (Banks & Sol.) Bobrov.			IRQ		AFG
LAGURUS L. **Poaceae** *(Gramineae)*					
ovatus L.			SAF	CHL NZ	AUS ISR
LALLEMANTIA Fisch. & Mey. **Lamiaceae** *(Labiatae)*					
iberica (M. B.) Fisch. & Mey.				SOV	
LAMARCKIA Moench **Poaceae** *(Gramineae)*					
aurea (L.) Moench				CHL USA	AUS HAW IND ISR SAF
LAMIUM L. **Lamiaceae** *(Labiatae)*					
album L.			ENG SOV	CHN GER NZ	AFG POL
amplexicaule L.	POL	AST AUS BEL BUL SWE TUN	AFG ARG CHN EGY ENG GER HUN IRA ISR JPN LEB MOR NET POR	CAN CHL COL ECU FIN FRA GRE ICE IRQ ITA JOR KOR NEP NOR	

LAMIUM L. Lamiaceae *(Labiatae)* *(Cont'd)*	S	P	C	X	F
amplexicaule L. *(Cont'd)*			SOV SPA USA YUG	NZ PK TUR	
macrodon Boiss.				TUR	
maculatum L.			SOV	USA	
purpureum L.			ENG GER SOV SPA	FIN ICE ITA NZ TUR USA	POL POR
LANTANA L. Verbenaceae					
aculeata L. *(See* **L. camara** L.)		IDO MAL		BOR CEY DR IND NIC THI	
camara L. (Syn. *L. aculeata* L.)	BOR FIJ GHA HAW IDO MEL NGI NIG PHI PLW	AUS IND KEN MAD MOZ NZ RHO SAF TNZ TRI TUR UGA	MAL PLE PR	ANT ARB BOL BRA CAB CAM CHN CR GUA GUI HON IVO JAM LIB MAU NIC PAN SEG THI USA VIE ZAM	HK ISR MIC
crocea Jacq.				JAM	
cujabensis Shau.				COL	
involucrata L.				ANT JAM TRI	PR
montevidensis Briq.				AUS	
moritziana Otto & Dietr.				PER	COL
ovata Hayek				USA	
salvifolia Jacq.				GHA	
trifolia L.		UGA	KEN	HON PER SAL TRI	COL IDO

LAPORTEA Gaudich. Urticaceae	S	P	C	X	F
aestuans (L.) Chew (Syn. *Fleurya aestuans* (L.) Gaudich.)					DR
gigas Wedd.			AUS		
interrupta (L.) Chew (Syn. *Fleurya interrupta* (L.) Gaudich.)					
LAPPULA Moench **Boraginaceae**					
echinata Gilib.			CAN SAF SOV	USA	AFG AUS
occidentalis (S. Wats.) Greene				USA	
redowskii (Hornem.) Greene				CAN	
LAPSANA L. **Asteraceae** *(Compositae)*					
apogonoides Maxim.		JPN			
communis L.	FIN		ENG GER HAW IRA	CHL ICE NZ TUR USA	AUS JPN POL SOV
humilis (Thunb.) Makino			JPN		
stellata L.			LEB		
LARIX Adans. **Pinaceae**					
laricina (Duroi) K. Koch				USA	
occidentalis Nutt.				USA	
LARREA Cav. **Zygophyllaceae**					
divaricata Cav.		MEX			ARG
tridentata (Dc.) Coville				USA	
LASIACIS Hitchc. **Poaceae** *(Gramineae)*					
ligulata Hitchc. & Chase				PER	
ruscifolia (H. B. K.) Hitchc.			PR	DR	
LASTREA Bory **Thelypteridaceae** *(See* **THELYPTERIS** Schmidel)					
gracilescens (Bl.) Moore *(See* Thelypteris gracilescens (Bl.) Ching)				PHI	
leucolepis Presl *(See* Macrothelypteris polypodioides (Hook.) Holtt.)				PHI	
LATHYRUS L. **Papilionaceae** *(Leguminosae)*					
angulatus L.			POR		
annuus L.			IRA IRQ POR	TUR	
aphaca L.		IND	IRA MOR	GRE PK	ISR

LATHYRUS L. Papilionaceae *(Leguminosae) (Cont'd)*	S	P	C	X	F
aphaca L. *(Cont'd)*			POR SPA	TUR	
articulatus L.			POR	MOR	
blepharicarpus Boiss.			LEB		
cicera L.			MOR POR	GRE TUR	
erectus Lag.				TUR	
heterophyllus L.			POR		
hirsutus L.			POR SPA	GRE ITA	
latifolius L.			POR	NZ	AUS
nissolia L.				TUR	
ochrus Dc.		TUN	MOR POR		
ophaca L.		IND			
palustris L.				USA	
pratensis L.			ENG SOV	NZ USA	AFG FIN
quadrimarginatus Bory & Chaub.				MOR	
sativus L.			MOR POR SPA	ARG IND TUR	AFG AUS
sphaericus Retz.			POR		AFG
sylvestris L.			SPA	USA	
tingitanus L.				MOR	
tuberosus L.		YUG	GER SOV SPA		POL
LATUA Phil. Solanaceae					
pubiflora Baill.				CHL	
LAUNAEA Cass. Asteraceae *(Compositae)*					
asplenifolia Hook. *f.*		IND			
nudicaulis Hook. *f.*			MOR	IND PK	AFG ISR
pinnatifida Cass. (*See* **L. sarmentosa** (Willd.) Sch.-Bip. *ex* O. Ktze.)				VIE	
sarmentosa (Willd.) Sch.-Bip. *ex* O. Ktze (Syn. *L. pinnatifida* Cass.)					
LAURENTIA Michx. *ex* Adans. Campanulaceae *(Syn.ISOTOMA Lindl.)*					
longiflora (L.) Peterm. (Syn. *Isotoma longiflora* (L.) Presl)			HAW USA		

LAVANDULA L. Lamiaceae *(Labiatae)*	S	P	C	X	F
spica Cav.			SPA		
stoechas L.			SPA	AUS	
LAVATERA L. Malvaceae					
cretica L.			MOR POR	NZ	
punctata All.			LEB		
thuringiaca L.			SOV		
trimestris L.			ISR MOR POR		
LEANDRA Raddi Melastomataceae					
longicoma Cogn.			PER		
LEDUM L. Ericaceae					
groenlandicum Oeder			USA		
LEERSIA Sw. Poaceae *(Gramineae)*					
hexandra Sw.	BRA GUY IDO PHI	AUS MAD MAL NIG SUR	ARG BOR THI USA	BND CAB CEY CUB IND TNZ TRI VEN VIE	BUR CHN GHA HK ISR JAM PR SAF
japonica Honda *ex* Honda			JPN		
oryzoides (L.) Sw.			JPN POR	USA	
LEGOUSIA Durand Campanulaceae					
hybrida Delarb.			ENG		
LEMNA L. Lemnaceae					
gibba L.			POR SPA		ISR SAF
minor L.			CAB GER MOR POL SPA SWE THI USA	AUS JPN NZ PHI POR	ARG AST BEL BOR CAN CHN CNK COL ECU EGY ENG FIN HAW HON IDO IND IRA IRQ	HK

LEMNA L. Lemnaceae *(Cont'd)*	S	P	C	X	F
minor L. *(Cont'd)*				ISR JAM JOR KEN KOR LEB MEL MEX NET PK SAF UGA VIE	
oligorrhiza (Hegelm.) Kurz *(See* **Spirodela oligorrhiza** (Kurz) Hegelm.)				FIJ	AUS
paucicostata Hegelm. *ex* Engelm. *(See* **L. perpusilla** Torr.)		JPN	TAI	IDO IND	KOR MAD
perpusilla Torr. (Syn. *L. paucicostata* Hegelm. *ex* Engelm.)					
polyrhiza L. *(See* **Spirodela polyrhiza** (L.) Schleid.)			JPN	IND THI	ISR
trisulca L.			JPN	BEL BND ENG GER USA	IND ISR
LEONOTIS R. Br. **Lamiaceae** *(Labiatae)*					
africana Th. & Dur.				GHA	
leonurus (L.) R. Br.				AUS	
mollissima Guercke			KEN		
nepetaefolia (L.) R. Br.		CAB	COL HAW KEN PR	ANT AUS BRA DR JAM TRI	IDO
LEONTICE L. **Berberidaceae**					
leontopetalum L.		JOR	LEB	TUR	
LEONTODON L. **Asteraceae** *(Compositae)*					
autumnalis L.		NOR	CAN ENG GER SOV SPA	FIN USA	POL
hispidus L.			GER		
nudicaule (L.) Banks				CHL USA	
LEONURUS L. **Lamiaceae** *(Labiatae)*					
cardiaca L.				USA	
sibiricus L.		BRA	HAW	ARG	IDO

LEONURUS L. Lamiaceae *(Labiatae) (Cont'd)*	S	P	C	X	F
sibiricus L. *(Cont'd)*		GUA	JPN PR TAI	BND DR IND MAU VEN	PK
villosus Desf.			SOV		
LEPIDIUM L. **Brassicaceae** *(Cruciferae)*					
africanum (Burm. *f.*) Dc.			SAF		
auriculatum Regel & Koern.			HAW		
bipinnatifidum Desv.		COL		CHL	
bonariense L.			ARG SAF	AUS BRA URU	
campestre (L.) R. Br.			CAN GER USA	NZ TUR	AUS POL
chalepense L.		LEB	ARG		ISR
densiflorum Schrad.			CAN	NZ USA	
draba L. *(See* **Cardaria draba** (L.) Desv.*)*		ARG	CAN IRA LEB POR SAF SOV SPA USA	AUS GRE IRQ TUR	AFG ISR
graminifolium L.				SPA	
hyssopifolium Desv.			AUS		
latifolium L.			LEB	TUR USA	
nitidum Nutt.				USA	
perfoliatum L.			CAN LEB SOV	TUR USA	AFG JPN
repens Boiss.				USA	
ruderale L.			SOV	BRA MAU NZ	AFG ISR POL
sagittulatum Thell.				AUS	
sativum L.			EGY ENG LEB	PK	AFG AUS
schinzii Thell.		SAF			
virginicum L.		MEX VEN	CHN JPN LEB PR	AUS BRA CAN DR	COL

LEPIDIUM L. Lamiaceae *(Labiatae)* (Cont'd)	S	P	C	X	F
virginicum L. *(Cont'd)*				USA	NZ
LEPILAENA J. Drum. *ex* Harv. **Zannichelliaceae**					
biloculata Kirk				AUS	
LEPTADENIA R. Br. **Asclepiadaceae**					
reticulata Wight				IND	
LEPTILON Rafin. **Asteraceae** *(Compositae)*					
pusillum (Nutt.) Britton			PR	MEX	
LEPTOCARPUS R. Br. **Restionaceae**					
disjunctus Mast.				VIE	
LEPTOCHLOA Beauv. **Poaceae** *(Gramineae)*					
chinensis (L.) Nees	IDO PHI SWZ THI	IND JPN MAL	CEY TAI	CHN NGI VIE	BOR BUR HK KOR
coerulescens Steud.	NIG				GHA
fascicularis (Lam.) A. Gray				CHL JAM USA	
filiformis (Lam.) Beauv. *(See* **L. panicea** (Retz.) Ohwi)	GUA MEX PER USA	ARG COL ECU HON JAM PHI TAI VIE		ANT BND CHN DR GHA IDO IND NIC PR SAL TRI VEN	HK
panicea (Retz.) Ohwi (Syn. *L. filiformis* (Lam.) Beauv.)	MOZ PHI			CHN CUB PR USA	CEY IND MAL SAF
scabra Nees	GUY	SUR	TRI		
uninervia (Presl) Hitchc. & A. Chase	PER			CHL USA	
virgata (L.) Beauv.				ANT PER SAL TRI	HAW
LEPTOSPERMUM J. R. & G. Forst. **Myrtaceae**					
scoparium Forst.		NZ			
LEPTOTAENIA Nutt. *ex* Torr. & Gray **Apiaceae** *(Umbelliferae)*					
dissecta Nutt.				USA	

	S	P	C	X	F
LEPTURUS R. Br. **Poaceae** *(Gramineae)*					
cylindricus (Willd.) Trin.				CHL	
LESPEDEZA Mich. **Papilionaceae** *(Leguminosae)*					
cuneata (Dum.-Cours.) G. Don			JPN TAI	USA	AFG
pilosa (Thunb.) Sieb. & Zucc.			JPN		
stipulacea Maxim. *(See* **L. striata** (Thunb.) Hook. & Arn.)				USA	
striata (Thunb.) Hook. & Arn. (Syn. *L. stipulacea* Maxim.)				USA	
violacea (L.) Pers.				USA	
LESQUERELLA S. Wats **Brassicaceae** *(Cruciferae)*					
gordoni (Gray) Wats.				USA	
LEUCAENA Benth. **Mimosaceae** *(Leguminosae)*					
glauca Benth. *(See* **L. leucocephala** (Lam.) De Wit)	PLW	GHA	HAW IDO TRI	CNK FIJ JAM NGI USA	CAB
leucocephala (Lam.) De Wit (Syn. *L. glauca* Benth.)	NGI	HAW		AUS PLW USA	
LEUCANTHEMUM L. **Asteraceae** *(Compositae)*					
myconis (L.) Giraud	POR			MOR	
paludosum Poir.				MOR	
segetum (L.) Stankov				POR	
vulgare Lam.			SOV	GRE	
LEUCAS R. Br. **Lamiaceae** *(Labiatae)*					
aspera (Willd.) Link				BND IND MAU PHI VIE	IDO
cephalotes Spreng.				IND	
decurvata Baker *ex* Hiern.				ANG	
glabrata R. Br.		TNZ			
javanica Benth.				PHI	IDO
lanata Benth.				IND	
lavandulaefolia J. E. Sm. (Syn. *L. linifolia* Spreng.)		IND	IDO	PHI	
linifolia Spreng. *(See* **L. lavandulaefolia** J. E. Sm.)				IND	
martinicensis R. Br.	ZAM	RHO	KEN SAF	BRA	

LEUCAS R. Br. **Lamiaceae** *(Labiatae)* *(Cont'd)*	S	P	C	X	F
mollissima Wall.			TAI		
neufliseana Courb.			KEN		
urticaefolia R. Br.		SUD		IND	
LEUCOJUM L. **Amaryllidaceae**					
aestivum L.				TUR	
LIBOCEDRUS Endl. **Cupressaceae**					
decurrens Torr.				USA	
LIGULARIA Cass. **Asteraceae** *(Compositae)* *(See* **FARFUGIUM** Lindl.)					
tussilginea (Burm.) Makino *(See* **Farfugium japonicum** (L.) Kitam.)			JPN		
LIGUSTRUM L. **Oleaceae**					
lucidum Ait.			AUS		
LIMEUM L. **Aizoaceae**					
linifolium Fenzl				SAF	
LIMNANTHEMUM S. G. Gmel. **Gentianaceae** *(See* **NYMPHOIDES** Hill)					
cristatum (Roxb.) Griseb. *(See* **Nymphoides cristata** (Roxb.) O. Ktze.)				IND	
indicum (L.) Griseb. *(See* **Nymphoides indica** (L.) O. Ktze.)				IDO VIE	
LIMNOBIUM Rich. **Hydrocharitaceae**					
boscii Rich (Syn. *L. spongia* (Bosc) Steud.)					
spongia (Bosc) Steud. *(See* **L. boscii** Rich)				USA	
LIMNOCHARIS H. B. K. **Butomaceae**					
emarginata Humb. & Bonpl.				THI	
flava (L.) Buch.	CEY IDO MAL		BOR	VIE	THI
LIMNOPHILA R. Br. **Scrophulariaceae**					
aromatica (Lam.) Merr.			JPN		IDO
conferta Benth.		IND		CAB CEY	
dasyantha Skan				NIG	
gratioloides R. Br.		IND			
heterophylla Benth.	IND		THI		
indica (L.) Druce			JPN	NIG	
micrantha Benth.		IND			

LIMNOPHILA R. Br. Scrophulariaceae *(Cont'd)*	S	P	C	X	F
sessiliflora Bl.			JPN		IDO
LINARIA Mill. Scrophulariaceae					
aucheri Boiss.			LEB		
biebersteinii Bess.			SOV		
canadensis Dum.-Cours.			BRA CHL USA		
commutata Bernh. *ex* Reichb.			MOR		
corifolia Desf.			TUR		
dalmatica Mill.			CAN	USA	
elatine Mill. (Syn. *Elatinoides elatine* Well.)					
latifolia Desf.			MOR		
minor Desf.			SOV	ITA USA	
spuria Mill.			MOR		
supina Desf.			SPA		
triphylla Mill.		TUN	MOR		
vulgaris Mill.			CAN ENG SOV USA	ITA NZ	ALK AUS FIN POL
LINDERNIA All. Scrophulariaceae *(Syn. ILYSANTHES Rafin.)*					
anagallis (Burm. *f.*) Pennell			IDO		
angustifolia (Benth.) Wettst.		CAB	JPN	KOR	IDO
antipoda (L.) Merr. *(See L. ruelloides (Colsm.) Pennell)*			TAI	CEY	
ciliata (Colsm.) Pennell (Syn. *Bonnaya brachiata* Link & Otto) (Syn. *Ilysanthes ciliata* O. Ktze.)				IND	
cordifolia (Colsm.) Merr.			IDO TAI	CEY	
crustacea (L.) F. Muell.	TRI	MAL	CHN IDO JPN	SUR VIE	
diffusa Wettst.				CNK NIG	
dubia (L.) Pennell			JPN POR	SPA	
hyssopioides Haines				IND	
parviflora (Roxb.) Haines				IND	

LINDERNIA All. Scrophulariaceae *(Cont'd)*	S	P	C	X	F
procumbens (Krock.) Philcox (Syn. *L. pyxidaria* All.)	HUN JPN KOR	ITA POL	TAI	CAB CHN IDO NEP PHI SOV THI VIE	
pyxidaria All. *(See* **L. procumbens** (Krock.) Philcox)		JPN KOR TAI	IDO		
ruelloides (Colsm.) Pennell (Syn. *L. antipoda* (L.) Merr.) (Syn. *Ilysanthes antipoda* (L.) Merr.)					
LINUM L. **Linaceae**					
anatolicum Boiss.				TUR	
austriacum L.				TUR	
flavum L.				TUR	
neo-mexicanum Greene				USA	
peyroni Post			LEB		ISR
usitatissimum L.				AUS CHL IRA TUR USA	AFG
LIPOCARPHA R. Br. **Cyperaceae**					
argentea R. Br. *ex* Nees *(See* **L. chinensis** (Osb.) Kern)				MAL	
chinensis (Osb.) Kern (Syn. *L. argentea* R. Br. *ex* Nees			IDO		
microcephala (R. Br.) Kunth			JPN	KOR	
LIPPIA L. **Verbenaceae**					
alba N. E. Br. *ex* Britton & Wilson			PR	BRA	
asperifolia Rich.			KEN		
cuneifolia (Torr.) Steud.				USA	
filiformis Schrad.				CHL	
geminata H. B. K.				JAM	
helleri Britton			PR		
javanica Spreng.			KEN		
nodiflora L. *(See* **Phyla nodiflora** (L.) Greene)		IND	IRQ PHI PR TAI	CAB CHL NGI PK USA VIE	AFG AUS IDO
reptans H. B. K.			PR	CHL	

LIPPIA L. Verbenaceae *(Cont'd)*	S	P	C	X	F
reptans H. B. K. *(Cont'd)*				SAL	
ukambensis Vatke				KEN	
LIQUIDAMBAR L. **Hamamelidaceae**					
styraciflua L.				CHL USA	
LIRIODENDRON L. **Magnoliaceae**					
tulipifera L.				USA	
LISAEA Boiss. **Apiaceae** *(Umbelliferae)*					
heterocarpa Boiss.			IRA		
syrica Boiss.			LEB		ISR
LISIANTHUS L. **Gentianaceae**					
alatus Aubl.				PER	
LITHOCARPUS Bl. **Fagaceae**					
densiflora Rehd.				USA	
LITHOSPERMUM L. **Boraginaceae**					
arvense L.		ARG AUS	ENG GER SOV SPA USA	CHN FIN GRE ICE TUR	AFG ISR POL POR
officinale L.			ARG	CAN USA	AFG
ruderale Dougl. *ex* Lehm.				USA	
zollingeri A. Dc.				CHN	
LITSEA Lam. **Lauraceae**					
citrata Bl. *(See* **L. cubeba** (Lour.) Pers.*)*				IND	
cubeba (Lour.) Pers. (Syn. *L. citrata* Bl.)					
glutinosa (Lour.) C. B. Rob.				MAU	
LOBELIA L. **Lobeliaceae**					
affinis Wall.			TAI		
angulata Forst. *f.* (Syn. *Pratia nummularia* (Lam.) A. Br. & Aschers.)					
cardinalis L.				USA	
chinensis Lour. (Syn. *L. radicans* Thunb.)		JPN			IDO TAI
cliffortiana L.		MAU			
inflata L.				USA	
pratioides Benth.				AUS	

LOBELIA L. Lobeliaceae (Cont'd)	S	P	C	X	F
purpurascens R. Br.				AUS	FIJ
radicans Thunb. *(See* **L. chinensis** Lour.*)*			IDO JPN TAI		
syphilitica L.				USA	
trigona Roxb.				IND	
trinitensis Griseb.				TRI	
LOCHNERA Reichb. **Apocynaceae**					
pusilla K. Schum.		IND			
LOLIUM L. Poaceae *(Gramineae)*					
italicum A. Br. *(See* **L. multiflorum** Lam.*)*				ITA	AUS HAW
linicolum A. Br. *(See* **L. perenne** L.*)*				SOV	
multiflorum Lam. *(Syn. L. italicum* A. Br.*)*	AFG BRA POL	ARG HUN ITA PER TUN URU	CAN EGY GRE JPN LEB MOR SAF USA	AUS BEL CHL COL ECU ENG FIJ FRA GER IND IRA IRQ JOR KEN NEP NET NZ POR RHO SPA SUR YUG	HAW
perenne L. *(Syn. L. linicolum* A. Br.*)*		NZ	CAN EGY IRA SPA	ARG CHL ITA JOR TUR	FIN HAW IND ISR SAF TAI
persicum Boiss. & Hohen.				CAN SOV	
remotum Schrank			JPN SOV		AFG IND POL
rigidum Gaudich.	TUN	AUS	IRQ LEB MOR	ISR POR	HAW IND SAF
temulentum L.	CHN ETH PHI	SPA TUN	ARG AUS CEY COL	AST BRA CHL ENG	AFG HAW POL

LOLIUM L. Poaceae *(Gramineae) (Cont'd)*	S	P	C	X	F
temulentum L. *(Cont'd)*			EGY FRA GER GRE IRA IRQ JOR JPN KEN LEB MOR SAF SOV USA	IND ISR ITA KOR NZ POR TUR URU VEN	
LOMATIUM Rafin. **Apiaceae** *(Umbelliferae)*					
leptocarpum (Nutt.) Coult. & Rose				USA	
LOMOPLIS Rafin. **Mimosaceae** *(Leguminosae)*					
ceratonia (L.) Raf.			PR		
LONCHOCARPUS Kunth **Papilionaceae** *(Leguminosae)*					
capassa Rolfe				RHO	
LONICERA L. **Caprifoliaceae**					
hirsuta Eaton				USA	
japonica Thunb.		NZ	ARG USA	CHL	AFG AUS
sempervirens L.				USA	AUS
subspicata Hook. & Arn.				USA	
tatarica L.				USA	
LOPEZIA Cav. **Onagraceae**					
mexicana Jacq.		MEX			
LOPHIOCARPUS Turcz. **Chenopodiaceae**					
guyanensis Micheli				VIE	
LORANTHUS L. **Loranthaceae**					
elasticus Desr.				IND	
longiflorus Desr.				IND	
pulverulentus Wall.				IND	
LOTUS L. **Papilionaceae** *(Leguminosae)*					
arabicus L.			EGY		
corniculatus L.			EGY JPN SOV SPA TAI	NZ TUR USA	
hispidus Desf.				POR	
scoparius (Nutt.) Oxley				USA	

LOTUS L. Papilionaceae *(Leguminosae) (Cont'd)*	S	P	C	X	F
tenuis Waldst. & Kit.				USA	
LUDWIGIA L. Onagraceae (*Syn. JUSSIAEA* L.)					
adscendens (L.) Hara (Syn. *Jussiaea diffusa* Forsk.) (Syn. *J. repens* L.)	PHI	CAB		ARG BND BOR CHL CHN COL DAH GHA HON IDO IND JAM JOR JPN KEN MAU MEX NGI PER PR RHO SEG THI TNZ TUR UGA USA VIE	ANG BOT HK MOZ
affinis (Dc.) Hara (Syn. *Jussiaea affinis* Dc.)					
decurrens Walt. (Syn. *Jussiaea decurrens* (Walt.) Dc.)			JPN		
erecta (L.) Hara (Syn. *Jussiaea erecta* L.)			DR		
hyssopifolia (G. Don) Exell (Syn. *Jussiaea linifolia* Vahl)		BOR	PHI THI	CAB CHN CNK COL DAH EGY GHA HON IDO IRQ NEP NGI NIG PLW SEG SUD SUR VIE	
leptocarpa (Nutt.) Hara (Syn. *Jussiaea leptocarpa* Nutt.)					
longifolia (Dc.) Hara (Syn. *Jussiaea longifolia* Dc.)					

LUDWIGIA L. Onagraceae *(Cont'd)*	S	P	C	X	F
octovalvis (Jacq.) Raven *ssp.* octovalvis (Syn. *Jussiaea angustifolia* Lam.) (Syn. *J. suffruticosa* L.)		BOR HAW		AUS BOT BRA CHN CNK COL DAH GHA HON IDO IND IRQ JAM JOR JPN KEN MAU MIC NGI PER PHI PLW RHO SUD SUR TAI THI TNZ UGA VIE ZAM	HK
palustris (L.) Ell.				USA	
parviflora Roxb. *(See* **L. perennis** L.)		IND	TAI		AFG IDO
peploides (Kunth) Raven				AUS NZ	
peploides (Kunth) Raven *ssp.* stipulacea (Ohwi) Raven (Syn. *L. stipulacea* Ohwi)					
perennis L. (Syn. *L. parviflora* Roxb.) (Syn. *Jussiaea perennis* Brenan)			PHI	CEY	
peruviana (L.) Hara (Syn. *Jussiaea grandiflora* L.) (Syn. *J. peruviana* L.)				DR	
prostrata Roxb. (Syn. *Jussiaea prostrata* Lev.)	JPN	TAI			IDO
stenorraphe (Brenan) Hara (Syn. *Jussiaea stenorraphe* Brenan)					
stipulacea Ohwi *(See* **L. peploides** (Kunth) Raven *ssp.* stipulacea (Ohwi) Raven)				TAI	
uruguayensis (Camb.) Hara (Syn. *Jussiaea uruguayensis* Camb.)					
LUEHEA Willd. **Tiliaceae**					
tarapotina Macbride				PER	

LUFFA Mill. Cucurbitaceae	S	P	C	X	F
aegytiaca Mill.		GHA		AUS	
cylindrica (L.) M. Roem.			PR	HON PLW	CAB IDO ISR
operculata Cogn.				PER	
LUPINUS L. Papilionaceae *(Leguminosae)*					
albus L.			SPA		
angustifolius L.				MOR SPA	
arboreus Sims				CHL NZ	
argenteus Pursh				USA	
caudatus Kell.				USA	
hirsutus L.			SPA	MOR	
kingii S. Wats.				USA	
laxiflorus Dougl.				USA	
leucophyllus Dougl.				USA	
luteus L.			IDO MOR TAI		
macrocarpus Hook. & Arn.				CHL	
perennis L.				USA	
pilosus Murr.			MOR		
pusillus Pursh				USA	
rivularis Dougl.				USA	
sericeus Pursh				USA	
LUZIOLA Juss. Poaceae *(Gramineae)*					
spruceana Benth. *ex* Doell.		SUR			
LUZULA Dc. Juncaceae					
campestris (L.) Dc.			SOV	FIJ JPN	AUS
LYCHNIS L. Caryophyllaceae					
alba Mill. (Syn. *Melandrium album* (Mill.) Garcke)			CAN USA	SOV	AUS
dioica L.				CAN USA	
flos-cuculi L.			ENG	GER USA	
LYCIUM L. Solanaceae					
barbarum L.				PK	AFG

LYCIUM L. Solanaceae *(Cont'd)*	S	P	C	X	F
barbarum L. *(Cont'd)*				TUR	ISR
europaeum L.				TUR	
ferocissimum Miers		AUS		NZ	
halimifolium Mill.				CHL TUR USA	
LYCOPERSICON Mill. **Solanaceae**					
esculentum Mill. *(See* **L. lycopersicum** (L.) Karsten)			TAI	HON USA	
lycopersicum (L.) Karsten (Syn. *L. esculentum* Mill.) (Syn. *Solanum lycopersicum* L.)					
peruvianum Mill.				PER	
pimpinellifolium Mill.				PER	
LYCOPODIUM L. **Lycopodiaceae**					
cernuum L.			IDO PR	DR	
LYCOPSIS L. **Boraginaceae**					
arvensis L. (Syn. *Anchusa arvensis* Bieb.)			ARG GER SOV SPA	USA	AFG AUS POL POR
LYCOPUS L. **Lamiaceae** *(Labiatae)*					
americanus Muhl.				USA	
asper Greene				USA	
europaeus L.			POR	BEL CHN TUR USA	
lucidus Turcz.				JPN	
parviflorus Maxim. *(See* **L. uniflorus** Michx.)				JPN	
uniflorus Michx. (Syn. *L. parviflorus* Maxim.)				USA	
LYGODESMIA D. Don **Asteraceae** *(Compositae)*					
juncea (Pursh) D. Don				USA	
LYGODIUM Sw. **Schizaeaceae**					
circinnatum Sw.			IDO		
flexuosum (L.) Sw.		MAL		PHI VIE	IDO
japonicum (Thunb.) Sw.			TAI	PHI	
polymorphum (Cav.) H. B. K.				HON SAL	

LYGODIUM Sw. Schizaeaceae *(Cont'd)*	S	P	C	X	F
scandens (L.) Sw.		MAL		VIE	
LYSIMACHIA L. **Primulaceae**					
clethroides Duby			JPN		
fortunei Maxim.			JPN		
japonica Thunb.			JPN		
mauritiana Lam.			JPN		
nummularia L.				BEL NZ USA	
punctata L.				TUR	
thyrsiflora L.				BEL	
vulgaris L.		POR		BEL NZ TUR	
LYTHRUM L. **Lythraceae**					
acutangulum Lag.				MOR	
anceps (Kohne) Makino			JPN		
hyssopifolia L.			ARG MOR POR	BRA CHL NZ URU USA	AUS ISR
junceum Banks & Soland.			MOR	NZ POR	
salicaria L.			FIN IRA POR	BEL JPN NZ TUR USA	AFG AUS ISR
MACARANGA Thou. **Euphorbiaceae**					
harveyana Muell.-Arg.	PLW				
triloba (Reinw. *ex* Bl.) Muell.-Arg.		MAL			
MACLEAYA R. Br. *ex* G. Don **Papaveraceae**					
cordata (Willd.) R. Br. *ex* G. Don			JPN		
MACLURA Nutt. **Moraceae**					
pomifera (Rap.) Schneid.				CHL USA	
MACROTHELYPTERIS Ching **Thelypteridaceae**					
polypodioides (Hook.) Holtt. (Syn. *Lastrea leucolepis* Presl)					
MACROZAMIA Miq. **Cycadaceae**					
communis (L.) Johnson			AUS		

MACROZAMIA Miq. Cycadaceae *(Cont'd)*	S	P	C	X	F
spiralis Miq.				CHL	
MADIA Molina **Asteraceae** *(Compositae)*					
elegans D. Don				USA	
glomerata Hook.				USA	
sativa Molina			ARG	CHL NZ USA	AUS
MAESA Forsk. **Myrsinaceae**					
tenera Mez.			TAI		
MALACHIUM Fries **Caryophyllaceae**					
aquaticum (L.) Fries *(See* **Stellaria aquatica** (L.) Scop.)	JPN		SOV		
MALACHRA L. **Malvaceae**					
alceaefolia Jacq. *(See* **M. capitata** (L.) L.)			HAW	BRA GUA HON JAM PER TRI	
capitata (L.) L. (Syn. *M. alceaefolia* Jacq.)			PR	DR IND PER PHI USA	
fasciata Jacq.			PHI	HON TRI	
MALCOLMIA R. Br. **Brassicaceae** *(Cruciferae)*					
africana R. Br.	IRA			PK	
conringioides Boiss. *(See* **M. exacoides** Spreng.)			LEB		
exacoides Spreng. (Syn. *M. coringioides* Boiss.)			LEB		
MALOPE L. **Malvaceae**					
trifida Cav.				MOR	
MALVA L. **Malvaceae**					
crispa L.			SOV		
hispanica L.			MOR POR		
moschata L.			CAN	USA	AUS
neglecta Wallr.	NZ		CAN LEB SOV USA	TUR	ISR POL
nicaeensis All.			ISR LEB	CHL NZ	AFG AUS

MALVA L. Malvaceae *(Cont'd)*	S	P	C	X	F
nicaeensis All. *(Cont'd)*				USA	
parviflora L.		AUS HAW	ARG EGY IRA IRQ SAF	BRA MEX MOR NZ PK URU USA	AFG ISR
pusilla Sm.			CAN SOV		POL
rotundifolia L.			ENG IRA IRQ LEB	AUS CHL COL TUR USA	POL
sinensis Cav. *(See* **M. sylvestris** L.)			SOV		
sylvestris L. (Syn. *M. sinensis* Cav.) (Syn. *M. vulgaris* S. F. Gray)		COL	POR SOV SPA	AUS GRE NZ TUR	AFG ARG ISR POL
verticillata L.	KEN			AUS TNZ	
vulgaris S. F. Gray *(See* **M. sylvestris** L.)					ITA
MALVASTRUM A. Gray **Malvaceae**					
americanum (L.) Torr. (Syn. *M. spicatum* (L.) A. Gray)					
coromandelianum (L.) Garcke		PHI	ARG AUS HAW TAI	BRA CAB HON IND MAU NIG	IDO
peruvianum (L.) A. Gray			COL	PER	
scabrum A. Gray				PER	
scoparium A. Gray				PER	
spicatum (L.) A. Gray *(See* **M. americanum** (L.) Torr.)			AUS		
MALVAVISCUS Adans. **Malvaceae**					
arboreus Cav.				PER	
MAMMILLARIA Haw. **Cactaceae**					
vivipara (Nutt.) Haw.				USA	
MANDRAGORA L. **Solanaceae**					
autumnalis Bertol.			MOR SPA		

MANETTIA L. Rubiaceae	S	P	C	X	F
hispida Poepp. & Endl.				PER	
MANGIFERA L. **Anacardiaceae**					
indica L.				JAM	
MANILKARA Adans. **Sapotaceae**					
zapota (L.) Van Royen (Syn. *Sapota achras* Mill.)					
MARISCUS Vahl **Cyperaceae**					
albescens Gaud.				VIE	
coloratus Nees				GHA	
cyperinus (Retz.) Vahl *(See* **Cyperus cyperinus** (Retz.) Valck. Sur.)			TAI		PHI
dregeanus Kunth				MAU	
ferax (L. C. Rich.) C. B. Clarke *(See* **Cyperus odoratus** L.)			TRI		
ligularis Th. & H. Dur.			TRI	GHA SUR	
mutisii H. B. K. (Syn. *Cyperus mutisii* (H. B. K.) Griseb.)					
rufus H. B. K. (Syn. *Cyperus ligularis* L.)					
sieberianus Nees *(See* **Cyperus cyperoides** (L.) O. Ktze.)				MAU	
umbellatus Vahl	GHA	NIG		CNK IVO	
ustulatus C. B. Clarke		NZ			
MARRUBIUM L. **Lamiaceae** *(Labiatae)*					
cuneatum Soland.			LEB		
parviflorum Fisch. & Mey.				TUR	
radiatum Delile *ex* Benth.			LEB		
supinam (Steph. *ex* Willd.) Hu				CHN	
vulgare L.		AUS	ARG IRA LEB SOV	CHL NZ PER URU USA VEN	AFG ISR
MARSDENIA R. Br. **Asclepiadaceae**					
rostrata R. Br.				AUS	
MARSILEA L. **Marsileaceae**					
aegyptiaca Willd.				EGY	
crenata Presl	PHI			IDO THI	

MARSILEA L. Marsileaceae *(Cont'd)*	S	P	C	X	F
drummondii A. Br.				AUS	
minuta L.				CAB IND MAL	
mucronata A. Br. *(See* **M. vestita** Hook. & Grev.*)*				USA	
quadrifolia L.	TAI	CAB CEY IND ITA JPN	POR SPA	AFG BND BOR CHN COL IDO IRA IRQ KOR NEP PK THI TUR VIE	USA
vestita Hook. & Grev. (Syn. *M. mucronata* A. Br.)					
MARSYPIANTHES Mart. *ex* Benth. **Lamiaceae** *(Labiatae)*					
chamaedrys (Vahl) O. Ktze.			BRA	MEX TRI	
MARTYNIA L. **Martyniaceae**					
annua L.			AUS IND PR	NIC	
MASCAGNIA Bert. **Malpighiaceae**					
pubiflora Griseb.	BRA				
MATRICARIA L. **Asteraceae** *(Compositae)*					
chamomilla L.	AFG ENG GER NET POL	AST BEL FRA GRE HUN ITA SPA SWE TUN	CAN EGY IRQ SOV	ARG BRA CHL CR FIN HAW IRA ISR JOR JPN NOR NZ PER TUR YUG	POR URU
discoidea Dc.				CHL FIN SOV TUR	AUS POL
inodora L. (Syn. *Chrysanthemum inodorum* L.)		TUN	SOV SPA	FIN GER ICE NZ	

MATRICARIA L. Asteraceae *(Compositae) (Cont'd)*	S	P	C	X	F
inodora L. *(Cont'd)*				TUR	
maritima L.			CAN GER	ICE USA	
matricarioides (Less.) Porter			CAN ENG FIN GER HAW	CHL ICE NZ	AUS USA
recutita L.			ENG		
suaveolens L.			IRA SOV		
MATTHIOLA R. Br. **Brassicaceae** *(Cruciferae)*					
parviflora R. Br.			MOR		
MAZUS Lour. **Scrophulariaceae**					
japonicus (Thunb.) O. Ktze. *(See* **M. pumilus** (Burm. *f.*) Steen.)	TAI		JPN	NEP	AFG IDO
miquelii Makino (Syn. *M. stolonifer* Makino)					JPN
pumilus (Burm. *f.*) Steen. (Syn. *M. japonicus* (Thunb.) O. Ktze.) (Syn. *M. rugosus* Lour.)	TAI		CHN JPN	IDO JAM NEP PK THI	AFG
rugosus Lour. *(See* **M. pumilus** (Burm. *f.*) Steen.)			CHN JPN	PK	AFG
stolonifer Makino *(See* **M. miquelii** Makino)			CHN TAI		
MECARDONIA Ruiz & Pav. **Scrophulariaceae**					
dianthera (Sw.) Pennell	TRI			IND	
MEDICAGO L. **Papilionaceae** *(Leguminosae)*					
agrestis Ten.			SPA		
arabica All.			ARG POR	NZ PER URU USA	AUS
ciliaris (L.) Krock		TUN	EGY ISR MOR POR		
dentatus Pers.		IND		MAU	
denticulata Willd.		AUS IND PK	JPN SAF	CHL CHN MAU MOR	AFG MEX
disciformis Dc.				TUR	
falcata L.			SOV SPA	ITA	AUS ISR

MEDICAGO L. Papilionaceae *(Leguminosae) (Cont'd)*	S	P	C	X	F
granatensis Willd.				CHL	
hispida Gaertn.		MEX	ARG CHN COL EGY HAW IRQ JPN MOR SAF SPA	CHL GRE IRA PER POR URU VEN	ISR
intertexta (L.) Mill.				MOR	
lupulina L.		IND	ARG CAN ENG HAW IRA POR SOV USA	CHL COL ITA JPN MAU NZ SPA TUR	AFG AUS ISR
maculata Sibth.			SPA	CHL	AUS
minima (L.) Bartal.			ARG AUS CHN	CHL NZ TUR USA	AFG ISR
obscura Retz.			POR		
orbicularis (L.) Bartal.			MOR POR SPA	CHL TUR	AUS ISR
polymorpha L.			AUS	IND NZ POR USA	
sativa L.		AUS	ENG IRA TAI	IND NZ TUR USA	AFG
scutellata (L.) Mill.			POR	TUR	
tribuloides Desr.			POR		AUS
turbinata (L.) Willd.			POR		
MELALEUCA L. Myrtaceae					
lanceolata R. Br. *ex* Benth. *(See* M. leucadendra (L.) L.)		AUS			
leucadendra (L.) L. (Syn. *M. lanceolata* R. Br. *ex* Benth.)					
MELAMPODIUM L. Asteraceae *(Compositae)*					
arvense Robinson	MEX				
divaricatum Dc.	SAL				

MELANDRIUM Roehl Caryophyllaceae	S	P	C	X	F
album (Mill.) Garcke *(See* **Lychnis alba** Mill.)				GER SOV	
noctiflorum (L.) Fries			SOV	GER	
MELANTHERA Rohr. **Asteraceae** *(Compositae)*					
confusa Britton			PR		
nivea Small				PAN TRI	
scandens (Schum. & Thonn.) Brenan		GHA			
MELASTOMA L. **Melastomataceae**					
affine D. Don (Syn. *M. polyanthum* Bl.)					
candidum D. Don				VIE	
decemfidum Roxb. *(See* **M. sanguineum** Sims)		HAW			
malabathricum L.	BOR HAW IDO MAL MEL	CEY		CAB IND MAU MIC PHI PLW TAI THI USA VIE	AUS FIJ
polyanthum Bl. *(See* **M. affine** D. Don)				VIE	
sanguineum Sims (Syn. *M. decemfidum* Roxb.)					
MELIA L. **Meliaceae**					
azedarach L.			TAI	HON PAN USA	
MELIANTHUS L. **Melianthaceae**					
comosus Vahl				AUS	
MELILOTUS Mill. **Papilionaceae** *(Leguminosae)*					
alba Desr.		IND	ARG SAF SOV	AUS BRA CHL COL MAU NZ PER PK TUR USA	ISR POL VEN
altissima Thuill.				USA	
indica (L.) All.	AFG ARG AUS	ARB IND ITA	HAW IRQ ISR	CHL IRA JOR	

MELILOTUS Mill. Papilionaceae *(Leguminosae)* *(Cont'd)*	S	P	C	X	F
indica (L.) All. *(Cont'd)*	EGY MEX		POR SAF TUN	KEN MAU MOR NEP NZ PER PK RHO THI UGA URU USA	
messanensis (L.) All.			POR	GRE	
officialis (L.) Lam.			CHN IRA SOV SPA	GRE ITA NZ PER TUR USA	AFG POL
segetalis Ser.				MOR	
sicula Vitm.			EGY		
sulcata Desf.		TUN	MOR	GRE	
MELINIS Beauv. **Poaceae** *(Gramineae)*					
minutiflora Beauv.		BRA	HAW	COL VEN	AUS CR IND SAF
MELOCHIA L. **Sterculiaceae**					
concatenata L. (Syn. *M. corchorifolia* L.)				CAB CHN PHI USA VIE	
corchorifolia L. (*See* **M. concatenata** L.)	FIJ THI	MAL	PHI TAI	AUS BND GHA IND NIG USA VIE	IDO
lupulina Sw.				PER	
melissaefolia Benth.				GHA	
mollis Triana & Planch.				NIG	
pyramidata L.			COL PR SAL	DR MAU	IDO
tomentosa L.			PR	JAM	
villosa (Mill.) Fawc. & Rendle			PR	COL	
MELOTHRIA L. **Cucurbitaceae**					
fluminensis Gardn.				SUR	

MELOTHRIA L. Cucurbitaceae *(Cont'd)*	S	P	C	X	F
formosana Hayata			TAI		
guadalupensis Cogn.			TRI	HON PAN	
heterophylla (Lour.) Cogn.			TAI		
japonica (Thunb.) Maxim.			JPN		
maderaspatana (L.) Cogn.				GHA	IDO
mucronata (Bl.) Cogn.			TAI		
tridactyla Hook. *f.*				NIG	
MENISPERMUM L. **Menispermaceae**					
canadense L.				USA	
MENTHA L. **Lamiaceae** *(Labiatae)*					
aquatica L. (Syn. *M. citrata* Ehrh.)			IRA POR	BEL ENG TUR	
arvensis L. (Syn. *M. austriaca* Jacq.)			CHN ENG FIN GER SOV	ITA JPN TUR USA	IDO POL
austriaca Jacq. *(See* **M. arvensis** L.*)*			SOV		
citrata Ehrh. *(See* **M. aquatica** L.*)*				CHL	
gentilis L.				USA	
longifolia (L.) Huds.			IRA	TUR	AFG
microphylla C. Koch			EGY LEB		
piperita L.				CHL USA	AUS
pulegium L.			MOR	CHL NZ TUR URU	AUS ISR
rotundifolia (L.) Huds.			POR	NZ	
satureiodes R. Br.			AUS		
spicata L.				NZ PER USA VEN	
tomentosa Urv.				TUR	
viridis L.				JAM	
MENTZELIA L. **Loasaceae**					
albicaulis Dougl.				USA	

MENTZELIA L. Loasaceae *(Cont'd)*	S	P	C	X	F
aspera L.				VEN	
MENYANTHES L. **Gentianaceae**					
trifoliata L.				USA	
MERCURIALIS L. **Euphorbiaceae**					
annua L.		TUN	ENG GER ISR LEB POR SPA	GRE ITA TUR	AUS
MERREMIA Dennst. **Convolvulaceae**					
aegyptia (L.) Urb.		VEN	HAW	GHA PER THI	
angustifolia Hall. *f.*				GHA	
cissoides Hall.				SUR	
dissecta (Jacq.) Hall. *f.*				GHA PLW	
distillatoria (Blanco) Merr.				PHI	
emarginata (Burm. *f.*) Hall. *f.*		SUD		IND	IDO
gemella (Burm. *f.*) Hall. *f.*				PHI	IDO
hastata Hall. *f.* *(See* **M. tridentata** (L.) Hall. *f. ssp.* hastata (Hall.f.) Ooststr.)				PHI	
hederacea (Burm. *f.*) Hall. *f.*				GHA	
peltata (L.) Merr.				PLW	
quinquefolia (L.) Hall. *f.* (Syn. *Ipomoea quinquefolia* L.)		AUS		HON	JAM
tridentata (L.) Hall. *f.*			PHI		
tridentata (L.) Hall. *f. ssp.* hastata (Hall. F.) Ooststr. (Syn. *M. hastata* Hall. *f.*)					
unbellata (L.) Hall. *f.*			TRI	GHA HON PAN PLW SAL	IDO JAM
METAPLEXIS R. Br. **Asclepiadaceae**					
japonica (Thunb.) Makino			JPN		
MICONIA Ruiz & Pav. **Melastomataceae**					
chamissois Naud.				BRA	
laevigata Dc.				JAM	
lateriflora Cogn.				PER	
nervosa Triana				PER	

MICONIA Ruiz & Pav. Melastomataceae *(Cont'd)*	S	P	C	X	F
stenostachya (Schr. & Mart.) Dc.			TRI		
MICROCARPAEA R. Br. Scrophulariaceae					
minima (Koen.) Merr.			JPN		
MICROCOCCA Benth. Euphorbiaceae					
mercurialis Benth.				IND IVO NIG	
MICROCYSTIS Lemmerman Chroococcaceae					
toxica Stephens	SAF				
MICROLAENA R. Br. Poaceae *(Gramineae)*					
stipoides (Labill.) R. Br.			HAW		
MICROLESPEDEZA Makino Papilionaceae *(Leguminosae)*					
striata Makino *(See* **Kummerowia striata** (Thunb.) Schindl.*)*			JPN		
MICROMELUM Bl. Rutaceae					
pubescens Bl.			IDO		
MICROMERIA Benth. Lamiaceae *(Labiatae)*					
brownei (Sw.) Benth.			JAM		
viminea (L.) Urb.			JAM		
MICROSORIUM Link Polypodiaceae					
heterophyllum (L.) A. D. Hawkes (Syn. *Polypodium exiguum* Hew.)					
MICROTEA Sw. Phytolaccaceae					
debilis Sw.				SUR TRI	
MICROTRICHIA Dc. Asteraceae *(Compositae)*					
perrotteti Dc.				NIG	
MIKANIA Willd. Asteraceae *(Compositae)*					
congesta Dc.		MAL	PR		
cordata (Burm. *f.*) B. L. Robins.	IDO MAL		BOR IND	GHA GUI IVO LIB NIG PLW SEG	ANG CEY ETH NGI PHI SAF TAI THI
cordifolia (L. *f.*) Willd.			PR	ARG DR PER	
micrantha H. B. K.	FIJ IND PLW		DR	BND BOR CEY	MAU MEL MEX

MIKANIA Willd. Asteraceae *(Compositae)* *(Cont'd)*	S	P	C	X	F
micrantha H. B. K. *(Cont'd)*				IDO JAM MAL PHI THI TRI	
scandens (L.) Willd.	CEY MAL	GHA IDO	IND	CNK MAU PHI PK USA	
MIMOSA L. Mimosaceae *(Leguminosae)*					
biuncifera Benth.				USA	
casta L.				TRI	
flavescens Splitg.				COL	
invisa Mart.	BOR FIJ MAL MEL NGI PHI PLW TAI	AUS IDO	MAU NIG	CEY HAW IND VIE	ARG CAB
pigra L.		MAD MEX		CR GHA GUA HON SAL	MAU SAF
pudica L.	BOR IDO MAL MEX NGI PHI	CEY CUB FIJ IND PER THI TRI	HAW NIG PLE PLW PR TAI	ANT AUS BND BOL BRA CHN COL DR GHA HON JAM MAU MEL NIC USA VEN VIE	CAB CR
sensitiva L.				BRA	
sepiaria Benth.				BRA	
somnians Humb. & Bonpl. *ex* Willd.			COL		
MIMULUS L. Scrophulariaceae					
nepalensis Benth.			JPN		
orbicularis Wall.			THI		
MIRABILIS L. Nyctaginaceae					
hirsuta (Pursh) Macmillan				USA	

MIRABILIS L. Nyctaginaceae *(Cont'd)*	S	P	C	X	F
jalapa L.			ARG HAW PR	BRA CHL COL DR FIJ GHA HON JAM USA VEN	IDO
linearis (Pursh) Heimerl				USA	
nyctaginea (Michx.) Macm.				USA	
MISCANTHUS Anderss. **Poaceae** *(Gramineae)*					
condensatus Hack. *(See* **M. sinensis** Anderss. *var.* condensatus (Hack.) Makino)			JPN		
floridulus (Labill.) Warb.			JPN		
japonicus Anderss.	PLW			FIJ USA	
sinensis Anderss.			JPN TAI	CHN NZ	AUS
sinensis Anderss. *var.* condensatus (Hack.) Makino (Syn. *M. condensatus* Hack.)					
MITRACARPUM Zucc. **Rubiaceae**					
frigidus K. Schum.				ARG	
scabrum Zucc. (Syn. *M. verticillatum* Vatke)	NIG			GHA IVO	
verticillatum Vatke *(See* **M. scabrum** Zucc.)				NIG	
MODIOLA Moench **Malvaceae**					
caroliniana (L.) G. Don			ARG AUS HAW	CHL NZ URU USA	
MODIOLASTRUM K. Schum. **Malvaceae**					
malvifolium K. Schum.				ARG	
MOGHANIA St. Hil. **Papilionaceae** *(Leguminosae)*					
lineata (L.) O. Ktze. (Syn. *Flemingia lineata* (L.) O. Ktze.)					
strobilifera (L.) St. Hil. *ex* O. Ktze. (Syn. *Flemingia strobilifera* (L.) R. Br.)					
MOLLUGO L. **Molluginaceae**					
berteriana Ser.		PR			
cerviana Ser.				IND	
hirta Thunb. *(See* **Glinus lotoides** L.)				IND PHI PK	

MOLLUGO L. Molluginaceae *(Cont'd)*	S	P	C	X	F
lotoides Wight & Arn. *ex* Clarke *(See* **Glinus lotoides** L.)				IND PHI	
nudicaulis Lam.		GHA		IVO MOZ	PR
oppositifolia L. *(See* **Glinus oppositifolius** (L.) A. Dc.)				IND VIE	
pentaphylla L. (Syn. *M. stricta* L.)		IDO MAL	JPN PHI TAI THI	IND VIE	
spergula L. *(See* **Glinus oppositifolius** (L.) A. Dc.)				IND	
stricta L. *(See* **M. pentaphylla** L.)		JPN	TAI	KOR	
verticillata Roxb. *non* L. *(See* **Glinus oppositifolius** (L.) A. Dc.)	JPN		COL USA	BRA CHL GRE HON NIC	AUS
MOLTKIA Lehm. **Boraginaceae**					
coerulea Lehm.				TUR	
MOLUCELLA L. **Lamiaceae** *(Labiatae)*					
laevis L.	ISR	LEB		TUR	AUS
MOMORDICA L. **Cucurbitaceae**					
balsamina L.		SUD		USA	ISR
charantia L.	MIC	AUS COL ECU IDO SUR USA	GHA GUA HAW PLW TRI	ANT BOL BRA CAB CNK CR DAH DR FIJ HON JAM JPN MEL MEX NEP NGI NIG PAN PER PR RHO SAL SUD THI	CHN HK KOR
tuberosa (Roxb.) Cogn.	SUD			IND	
MONARDA L. **Lamiaceae** *(Labiatae)*					
citriodora L.				USA	

MONARDA L. Lamiaceae *(Labiatae)* *(Cont'd)*	S	P	C	X	F
fistulosa L.				USA	
punctata L.				USA	
MONERMA Beauv. **Poaceae** *(Gramineae)*					
cylindrica (Willd.) Coss. & Dur.			POR	NZ	AUS SAF
filiformis Trin.				CHL	
MONIERA P. Br. **Scrophulariaceae** *(See* **BACOPA** Aubl.)					
cuneifolia Michx. *(See* **Bacopa cuneifolia** (Michx.) Wettst.)				IND	
MONNINA Ruiz & Pav. **Polygonaceae**					
amplibracteata Ferr.				PER	
MONOCHORIA Presl **Pontederiaceae**					
hastata (L.) Solms	BOR FIJ	CAB IND PHI	IDO	BND CHN MAL	
korsakowii Regel & Maack		JPN			
vaginalis (Burn. *f.*) Presl	BOR IDO JPN KOR PHI TAI	CEY HAW IND THI	MAL	AUS BND CAB CHN VIE	HK
MONOCOSMIA Fenzl **Portulacaceae**					
arrigioloides Fenzl				CHL	
MONOLEPIS Schrad. **Chenopodiaceae**					
nuttalliana (Roem. & Schult.) Greene				USA	
MONOTAGMA K. Schum. **Marantaceae**					
plurispicatum (Koern.) Schum. K.				PER	
MONTANOA Cerv. *ex* Llave & Lex. **Asteraceae** *(Compositae)*					
hibiscifolia (Benth) Sch.-Bip.				USA	
MONTIA L. **Portulacaceae**					
perfoliata (Donn) Howell			ENG	NZ USA	
MONTRICHARDIA Cruger **Araceae**					
arborescens Schott				SUR	
MORINDA L. **Rubiaceae**					
citrifolia L.				PLW	
royoc L.				JAM	

	S	P	C	X	F
MORRENIA Lindl. **Asclepiadaceae**					
odorata Lindl.				USA	
MORUS L. **Moraceae**					
alba L. (Syn. *M. multicaulis* Perr.)				DR TUR USA	AFG ISR
australis Poir.			TAI		
multicaulis Perr. *(See* **M. alba** L.)				VEN	
nigra L.				TUR USA	ISR
rubra L.				USA	
MOSCHOSMA Reichb. **Lamiaceae** *(Labiatae)*					
polystachyun (L.) Benth.				IND PHI	IDO
MOSLA Buch.-Ham. *ex* Maxim. **Lamiaceae** *(Labiatae)*					
dianthera (Ham.) Maxim.			JPN		
punctulata (J. F. Gmel.) Nakai			JPN		
MUCUNA Adans. **Papilionaceae** *(Leguminosae)*					
coriacea Baker		MOZ			
gigantea (Willd.) Dc.				MAL PLW	
pruriens (L.) Dc.	MEX MOZ	JAM MAD	GUA KEN MIC TNZ	CNK COL DAH DR ECU GHA IDO MAU NIG PLW PR RHO SAL SEG SUD THI	
sloanei Fawc. & Rendle				GHA	
MUEHLENBECKIA Meissn. **Polygonaceae**					
complexa (A. Cunn.) Meissn.		NZ			
cunninghamii F. Muell.				AUS	
diclina Druce				AUS	
sagittifolia Meissn.		ARG		URU	
MUHLENBERGIA Schreb. **Poaceae** *(Gramineae)*					
asperifolia (Nees & Mey.) Parodi				USA	

MUHLENBERGIA Schreb. Poaceae *(Gramineae) (Cont'd)*	S	P	C	X	F
frondosa (Poir.) Fern.				USA	
japonica Steud.			JPN		
mexicana (L.) Trin.				USA	
schreberi G. F. Gmel			USA		
MULGEDIUM Cass. Asteraceae *(Compositae)*					
tataricum Dc.			SOV		AFG
MURDANNIA Royle **Commelinaceae**					
nudiflora (L.) Brenan (Syn. *Aneilema nudiflorum* (L.) Wall.) (Syn. *Commelina nudiflora* L.)		IND	TRI	BND CHN GHA HAW MEL MEX THI USA	
MUSA L. **Musaceae**					
acuminata Colla (Syn. *M. malaccensis* Ridl.)				MAL	
malaccensis Ridl. *(See* **M. acuminata** Colla*)*				MAL	
x **paradisiaca** L.				JAM	ISR
MUSCARI Mill. **Liliaceae**					
botryoides (L.) Mill.				USA	
comosum (L.) Mill.		POR	LEB MOR SPA	JOR TUR	
racemosum (L.) Mill.		TUN	IRA LEB SPA	USA	AFG ISR
MYAGRUM L. **Brassicaceae** *(Cruciferae)*					
perfoliatum L.			LEB	AUS TUR	
MYCELIS Cass. Asteraceae *(Compositae)*					
muralis Dum.			ENG		
MYOPORUM Banks & Soland. *ex* Forst. *f.* **Myoporaceae**					
acuminatum R. Br.				AUS	
deserti A. Cunn. *ex* Benth.				AUS	
MYOSOTIS L. **Boraginaceae**					
arvensis Lam.			ENG GER SOV	FIN ITA	POL
caespitosa Schultz				TUR	
collina Hoffm.				GER	

MYOSOTIS L. Boraginaceae *(Cont'd)*	S	P	C	X	F
lappula L.			SOV		
micrantha Pall. *ex* Lehm.			GER SOV		
palustris Lam.			SPA	BEL	
sicula Guss.				TUR	
sylvatica Hoffm.			SOV		
MYRICA L. Myricaceae					
californica Cham. & Schlecht.				USA	
cerifera L.				USA	CHL
faya Dryand.		HAW		USA	
pensylvanica Loisel.				USA	
MYRIOPHYLLUM L. Haloragidaceae					
alterniflorum Dc.		NOR USA		ARG CAN ENG FIN GER IND POL POR SWE YUG	
brasiliense Camb.		HAW NZ RHO USA	ARG AUS JPN	CAB IDO IND THI VIE	
elatinoides Gaudich.		NZ	ARG	AUS	TNZ
exalbescens Fern.			USA	CAN POR	
heterophyllum Michx.			USA		
indicum Willd.	IND				
propinquum A. Cunn.		AUS		NZ	
spicatum L.	SAF USA	ENG GER IND	JPN PHI POR	ALG ARB AST AUS BEL CAB CAN CHN GRE HUN IRA ISR ITA JOR NET NOR PK POL	KOR ZAM

MYRIOPHYLLUM L. Haloragidaceae *(Cont'd)*	S	P	C	X	F
spicatum L. *(Cont'd)*				RHO SWE VIE YUG	
verrucosum Lindl.		AUS			
verticillatum L.		GER USA	CHN JPN	AFG AST AUS BEL BOR BRA CAN ENG HUN IND IRA ITA KOR LEB NET NOR PK POL POR SWE YUG	
MYRISTICA Gronov. **Myristicaceae**					
fragrans Houtt.				JAM	
NAJAS L. **Najadaceae**					
flexilis (Willd.) Rostk. & Schmidt			USA		
graminea Del.		AUS ITA PK USA	JPN PHI TAI	AFG CAB CHN EGY IDO IRA JOR KOR MEL SUD THI TNZ UGA VIE	IND
guadelupensis (Spreng.) Magnus			USA		
interrupta K. Schum.				MOZ RHO	
major All. *(See* **N. marina** L.*)*			JPN		
marina L. (Syn. *N. major* All.)				AUS CHN CNK GER HUN IND IRA ISR ITA	KOR

NAJAS L. Najadaceae *(Cont'd)*	S	P	C	X	F
marina L. *(Cont'd)*				JAM	
				JOR	
				JPN	
				NOR	
				PK	
				POL	
				RHO	
				SAF	
				THI	
				TNZ	
				UGA	
				USA	
minor All.	MAD	JPN	AFG	KOR	
		USA	CHN	RHO	
			EGY		
			GER		
			GRE		
			HUN		
			IDO		
			IND		
			IRA		
			IRQ		
			ISR		
			ITA		
			JOR		
			MAL		
			MAU		
			PK		
			POL		
			POR		
			THI		
pectinata Magnus			BOT	EGY	
				GHA	
				RHO	
				SUD	
				TNZ	
				USA	
NAMA L. **Hydrophyllaceae**					
jamaicense L.			ARG		
NARCISSUS L. **Amaryllidaceae**					
pseudo-narcissus L.			SPA		
tazetta L.			POR		AFG
			SPA		ISR
NARDUS L. **Poaceae** *(Gramineae)*					
stricta L.			SOV	ITA	
				NZ	
NARTHECIUM Huds. **Liliaceae**					
ossifragum (L.) Huds.			AUS		
NASSELLA E. Desv. **Poaceae** *(Gramineae)*					
trichotoma (Nees) Hack.			AUS	SAF	
			NZ		
NASTURTIUM R. Br. **Brassicaceae** *(Cruciferae)*					
fontanum Asch. *(See* **N. officinale** R. Br.)			CHN	JAM	

NASTURTIUM R. Br. Brassicaceae *(Cruciferae) (Cont'd)*	S	P	C	X	F
indicum (L.) Dc. *(See* **Roripppa indica** (L.) Hiern)			PHI TAI	IND PK	IDO
madagascariense Dc.				IND	
montanum Wall.				CHN	
officinale R. Br. (Syn. *N. fontanum* Asch.)			IRA JPN POR USA	AUS BEL ENG VEN	ISR
palustre (Pollich) Dc. *(See* **Rorippa islandica** (Oeder) Borb.)			JPN	COL KOR	AUS
sublyratum (Miq.) Fr. & Sav. *(See* **Rorippa dubia** (Pers.) Hara)			TAI		
NAVARRETIA Ruiz & Pav. **Polemoniaceae**					
intertexta (Benth.) Hook.				USA	
squarrosa (Esch.) Hook. & Arn.			AUS	NZ USA	
NECHAMANDRA Planch. **Hydrocharitaceae**					
alternifolia (Roxb.) Thw. (Syn. *Vallisneria alternifolia* Roxb.)				IND	
NELUMBIUM Juss. **Nelumbonaceae** *(See* **NELUMBO** Adans.)					
nuciferum Gaertn. *(See* **Nelumbo nucifera** Gaertn.)			IRA		
speciosum Willd. *(See* **Nelumbo nucifera** Gaertn.)			IRA	IND	
NELUMBO Adans. **Nelumbonaceae** *(Syn. NELUMBIUM* Juss.)					
lutea (Willd.) Pers. *(See* **N. pentaphylla** (Walt.) Fern.)			USA	GUA IND	
nucifera Gaertn. (Syn. *Nelumbium nuciferum* Gaertn.) (Syn. *N. speciosum* Willd.)		CAB PR VIE	IRA LAO	AUS CHN HAW IDO IND ITA JPN KOR MAL NEP PHI THI	
pentaphylla (Walt.) Fern. (Syn. *N. lutea* (Willd.) Pers.)					
NEPETA L. **Lamiaceae** *(Labiatae)*					
cataria L.			CAN SOV	NZ PER USA	AFG POL
hederacea (L.) Trev.			CAN CHN		

NEPETA L. Lamiaceae *(Labiatae) (Cont'd)*	S	P	C	X	F
hederacea (L.) Trev. *(Cont'd)*			IRA		
pilinux P. H. Davis				TUR	
NEPHROLEPIS Schott **Davalliaceae**					
auriculata (L.) Trimen			TAI		
biserrata (Sw.) Schott		CEY GHA MAL		CNK JAM	
exaltata (L.) Schott	FIJ		HAW	JAM NGI	
hirsutula (Forst.) Presl			IDO		
rivularis (Vahl) Mett. (Syn. *N. undulata* (Afez. & Sw.) J. Sm.)					
undulata (Afez. & Sw.) J. Sm. *(See* **N. rivularis** (Vahl) Mett.)		GHA			
NEPSERA Naud. **Melastomataceae**					
aquatica (Aubl.) Naud.			TRI		
NEPTUNIA Lour. **Mimosaceae** *(Leguminosae)*					
natans (L. *f.*) Druce (Syn. *N. oleracea* Lour.)					
oleracea Lour. *(See* **N. natans** (L. *f.*) Druce)			CAB IND THI	GHA MAD SAF	
NERIUM L. **Apocynaceae**					
oleander L.			AUS	BRA IRA NZ TUR	ISR
NESLIA Desv. **Brassicaceae** *(Cruciferae)*					
paniculata (L.) Desv.			ARG CAN LEB POR SOV SPA	GER USA	AFG ALK AUS POL
NEUROLAENA R. Br. **Asteraceae** *(Compositae)*					
lobata (Sw.) R. Br.			PR	DR JAM	
NICANDRA Adans. **Solanaceae**					
physalodes (L.) Gaertn.	RHO TNZ ZAM	ANG AUS KEN PER SAF	ARG HAW HON IDO POL	BOT CNK ETH IRA JPN MOZ NEP NZ SAL THI	

NICANDRA Adans. Solanaceae *(Cont'd)*	S	P	C	X	F
physalodes (L.) Gaertn. *(Cont'd)*				USA	
NICOTIANA L. Solanaceae					
alata Link & Otto				BRA	
glauca R. Grah.			AUS HAW RHO SAF	ARG CHL NZ URU USA VEN	ISR
glutinosa L.				PER	
longiflora Cav.			ARG		
paniculata L.				PER	
suaveolens Lehm.			AUS		
tabacum L.			HAW		
trigonophylla Dun.				USA	
NIEREMBERGIA Ruiz & Pav. Solanaceae					
hippomanica Miers			ARG		
NIGELLA L. Ranunculaceae					
arvensis L.			POR SPA	MOR	POL
damascena L.			POR	MOR	POL
hispanica L.			MOR POR		
NITELLA (Ag.) Leonhardi Characeae					
batrachosperma Ag.				SUD	
flexilis (L.) Ag.				ENG	
hyalina (Dc.) Ag.		THI			
NOLINA Michx. Liliaceae					
microcarpa S. Wats.				USA	
texana S. Wats.				USA	
NONNEA Medic. Boraginaceae					
alba Dc.				SPA	
lutea Dc.				TUR	
nigricans Dc.				PK	POR
picta Sweet			IRA		AFG
pulla Dc.			SOV	PK TUR	AFG POL

NOTHOSCORDUM Kunth Liliaceae	S	P	C	X	F
fragrans (Vent.) Kunth *(See* **N. inodorum** (W. Ait.) Aschers. & Graebn.)			AUS JPN	ARG BRA MAU	PER
inodorum (W. Ait.) Aschers. & Graebn. (Syn. *N. fragrans* (Vent.) Kunth)			AUS SAF	ARG CHL MAU NZ URU	IDO
NOTOBASIS Cass. **Asteraceae** *(Compositae)*					
syriaca (L.) Cass. (Syn. *Cirsium syriacum* (L.) Gaertn.)	ISR JOR		LEB		AUS
NUPHAR J. E. Sm. **Nymphaeaceae**					
advena (Ait.) Ait. *f.*			USA		
japonicum Dc.			CHN		
lutea (L.) J. E. Sm.			USA	BEL ENG FRA GER	
NYMPHAEA L. **Nymphaeaceae** *(Syn. CASTALIA* Salisb.)					
alba L. (Syn. *Castalia alba* (L.) Wood)		GER NOR SWE	POR	AST BEL BOR COL ENG FRA IRA IRQ ISR ITA JAM JOR NEP PK POL TUN TUR	
amazonum Mart. & Zucc.		SUR			
caerulea Savign.		RHO		BOT CNK EGY HAW ISR JOR KEN SAF SUD TNZ UGA ZAM	
lotus L.		PK		BOT CAB CNK DAH ECU EGY GHA	

NYMPHAEA L. Nymphaeaceae *(Cont'd)*	S	P	C	X	F
lotus L. *(Cont'd)*				GUA HAW IDO IND MEL MEX NEP RHO SEG SUD THI	
maculata Schum. & Thonn.				GHA	
mexicana Zucc.				USA	
nouchali Burm. *f.* (Syn. *N. stellata* Willd.)					
odorata Ait.			USA	CAN IND	
pubescens Willd.				IND	
stellata Willd. *(See* **N. nouchali** Burm. *f.)*		IND		BND CAB IDO JOR MAL PHI PK THI VIE	PLW
tuberosa Paine				USA	
NYMPHOIDES Hill **Gentianaceae** *(Syn.LIMNANTHEMUM* S. G. Gmel.)					
aquaticum Fern				USA	
cristata (Roxb.) O. Ktze. (Syn. *Limnanthemum cristatum* (Roxb.) Griseb.)					
humboldtianum (H. B. K.) O. Ktze.		SUR			
indica (L.) O. Ktze. (Syn. *Limnanthemum indicum* (L.) Griseb.)			IDO	THI	AUS FIJ SAF
peltata (Gmel.) Britten & Rendle				BEL	
NYPA Steck **Arecaceae** *(Palmae)*					
fruticans Steck				VIE	
NYSSA L. **Nyssaceae**					
aquatica L.				USA	
sylvatica Marsh.				USA	
OCIMUM L. **Lamiaceae** *(Labiatae)*					
adscendens Willd.				IND	
americanum L. (Syn. *O. canum* Sims)	MOZ	IND		GHA	

OCIMUM L. Lamiaceae *(Labiatae) (Cont'd)*	S	P	C	X	F
basilicum L.		SUD	HAW PR	FIJ JAM PLW	AFG FIN IDO
canum Sims *(See* O. americanum L.)				IND	
gratissimum L.		PLW		GHA MEL	CAB
micranthum Willd.				JAM	
sanctum L.			PR	IND PLW	IDO
viride Willd.				ANG	
OCTODON Thonn. **Rubiaceae**					
setosum Hiern.				ANG GHA	
ODONTITES Hall. **Scrophulariaceae**					
rubra Pers.				SOV	
verna Dum.			ENG		
ODONTOSORIA Fee **Lindsaeaceae**					
aculeata (L.) J. Sm.			PR		
OENANTHE L. **Apiaceae** *(Umbelliferae)*					
aquatica (L.) Poir.				BEL GER	
fistulosa L.				BEL	
javanica (Bl.) Dc. (Syn. *O. stolonifera* (Roxb.) Dc.)	JPN	TAI			
pimpinelloides L.				TUR	
stolonifera (Roxb.) Dc. *(See* O. javanica (Bl.) Dc.)				KOR VIE	
OENOTHERA L. **Onagraceae**					
albicaulis Pursh				USA	
biennis L.			CAN SAF	JPN USA	AUS
erythrosepala Borb.			JPN	NZ	
grandiflora Soland.			SAF		
indecora Cambess.			SAF	ARG BRA	
jamesii Torr. & Gray			SAF		
laciniata Hill			JPN SAF	USA	
lamarkiana Ser.			JPN		AFG
linifolia Nutt.				USA	

OENOTHERA L. Onagraceae *(Cont'd)*	S	P	C	X	F
molissima L.				CHL	
nuttallii Sweet			CAN	USA	
odorata Jacq.			AUS JPN		
pallida Lindl.				USA	
parodiana Munz				URU	
parviflora L.			JPN	CAN	
perennis L.			CAN	USA	
prostrata Ruiz & Pav.				ECU	
rosea Sol.			RHO SAF	MAU	ARG AUS
speciosa Nutt.				USA	
striata Ledeb.			HAW SAF	CHL NZ	
tetragona Roth			COL		
tetraptera Cav.			SAF	AUS COL	
OLDENLANDIA L. Rubiaceae					
aspera Dc.				IND	
auricularia (L.) F. V. M. *(See* **Hedyotis auricularia** L.)				VIE	
biflora L. *(See* **Hedyotis biflora** (L.) Lam.)				IND PHI THI	IDO
capensis L. *f.*	NIG				ISR
corymbosa L. *(See* **Hedyotis corymbosa** (L.) Lam.)		GHA IND MAL NIG PHI	IDO TAI	BND CNK	HK
decumbens Hiern				IVO	
diffusa (Willd.) Roxb. *(See* **Hedyotis diffusa** Willd.)			JPN MAL TAI	IND PHI	IDO
herbacea (L.) Roxb. *(See* **Hedyotis herbacea** L.)	IND		RHO TRI	GHA IVO PHI	IDO
lancifolia Dc.	NIG			SUR	
pinifolia Wall. *ex* G. Don *(See* **Hedyotis pinifolia** (Wall. *ex* G. Don) K. Schum.)				VIE	
praecox Pierre *ex* Pitard				VIE	
prostrata Bl.				IDO	

OLDENLANDIA L Rubiaceae *(Cont'd)*	S	P	C	X	F
pterita (Bl.) Miq. *(See* **Hedyotis pterita** Bl.)				THI	
strumosa Hiern				SUD	
thesiifolia K. Schum.				BRA	
umbellata L. *(See* **Hedyotis umbellata** (L.) Lam.)				IND	
OLEARIA Moench Asteraceae *(Compositae)*					
elliptica Dc.				AUS	
OLYRA L. Poaceae *(Gramineae)*					
latifolia L.				PER	
ONCIDIUM Sw. Orchidaceae					
variegatum Sw.		PR		DR	
ONOCLEA L. Aspidiaceae					
sensibilis L.				USA	
ONONIS L. Papilionaceae *(Leguminosae)*					
biflora Desf.			MOR		
procurrens Wallr.			POR		
spinosa L.		CZE	LEB		AUS
ONOPORDUM L. Asteraceae *(Compositae)*					
acanthium L.	AUS		ARG ENG SOV SPA	NZ TUR USA	AFG POL
acaulon L.	AUS				
anisacanthum Boiss.			LEB		ISR
heteracanthum C. A. Mey.			IRA LEB		
illyricum L.	AUS				
leptolepis Dc.			IRA		AFG
macrocanthum Schousb.			MOR		
OPERCULINA Manso Convolvulaceae					
turpethum (L.) Manso	AUS			FIJ PLW	
OPLISMENUS Beauv. Poaceae *(Gramineae)*					
burmanni (Retz.) Beauv.				CR GHA IDO NIG VIE	
compositus (L.) Beauv.			BOR MAL	PHI PLW	

OPLISMENUS Beauv. Poaceae *(Gramineae)* *(Cont'd)*	S	P	C	X	F
compositus (L.) Beauv. *(Cont'd)*			TAI	SAF	
hirtellus (L.) Beauv.			CR GHA		
setarius (Lam.) Roem. & Schult.			DR	MEX	
undulatifolius (Ard.) Beauv.			JPN	PLW	
OPUNTIA Mill. **Cactaceae**					
aurantiaca Lindl.	SAF		AUS		
borinquensis Britton & Rose			PR		
dillenii Haw.			KEN PR SAF	IND	
elatior Mill.				IND	
engelmanni Salm-Dyck				USA	
ficus-indica Mill.			AUS KEN PR		ISR
fulgida Engelm.				USA	
humifusa Rafin.				USA	
imbricata (Haw.) Dc.			SAF	AUS USA	CHL
inermis (Dc.) Dc.			AUS		
leptocaulis Dc.				.USA	
lindheimeri Engelm.				USA	
megacantha Salm-Dyck		HAW	RHO SAF	USA	AUS
monocantha Haw. *(See* **O. vulgaris** Mill.*)*			AUS		
polyacantha Haw.				USA	
schumannii Speg.			SAF		
spinosior (Engelm. & Bigel.) Toumey				USA	
spinulifera Salm-Dyck			SAF		
streptacantha Lem.				AUS	
stricta (Haw.) Haw.				AUS	
tardospina Griffiths			SAF		
tomentosa Salm-Dyck			AUS		
tuna Mill.				JAM	
versicolor Engelm.				USA	
vulgaris Mill. (Syn. *O. monocantha* Haw.)	PLW		AUS FIJ	BRA	ARG ISR

OPUNTIA Mill. Cactaceae *(Cont'd)*	S	P	C	X	F
vulpina A. Web.				ARG	
ORCHIS L. Orchidaceae					
laxiflora Lam.				TUR	
palustris Jacq.				TUR	
ORIGANUM L. Lamiaceae *(Labiatae)*					
bilgeri P. H. Davis				TUR	
vulgare L.				USA	
ORLAYA Hoffm. Apiaceae *(Umbelliferae)*					
kochii Heywood				MOR	
platycarpos (L.) Koch				ISR	POR
ORMENIS Cass. Asteraceae *(Compositae)*					
mixta Dum.			ISR MOR		
paecox (Link) Briq. & Cavill.				MOR	
ORMOCARPUM Beauv. Papilionaceae *(Leguminosae)*					
trichocarpum Burtt-Davy		RHO			
ORNITHOGALUM L. Liliaceae					
narbonense L.			POR	GRE TUR	ISR
nutans L.				TUR	
pyramidale L.			MOR		
pyrenaicum L.		TUN			
umbellatum L.			SPA	USA	
ORNITHOPUS L. Papilionaceae *(Leguminosae)*					
compressus L.			POR	TUR	ISR
pinnatus (Mill.) Druce			POR	NZ	
OROBANCHE L. Orobanchaceae					
aegyptiaca Pers.	AFG ARB IRA JOR	ITA	IRQ PK TUR	EGY ENG IND LEB SOV	HUN ISR
australiana F. Muell.		AUS			
bicolor C. A. Mey. (*See* **O. cernua** Loefl.)				SOV	
boninsimae (Maxim.) Tuyama				JPN	
brassicae Novopokr.				HUN	
cernua Loefl. (Syn. *O. bicolor* C. A. Mey.)	ARB EGY	IND ITA	AFG TUR	ENG IRQ	

OROBANCHE L. Orobanchaceae *(Cont'd)*	S	P	C	X	F
cernua Loefl. *(Cont'd)*	IRA NEP	PK		JOR LEB SOV SUD USA	
cooperi G. Beck				HUN	
crenata Forsk.	EGY JOR TUN	ITA LEB	MOR	COL GRE IND IRQ ISR POR TUR USA	PK
cumana Wallr.			SOV	BUL COL HUN YUG	GRE ITA
gracilis Sm.				GER	
hederae Duby				FRA ITA	
indica Buch.-Ham.				PK	
loricata Reichb.				CZE FRA	
ludoviciana Nutt.				USA	
lutea Baumg.				GER	
minor Sm.	EGY PK UGA	ITA	HUN KEN MAU POL SAF USA	AST AUS CHL CZE DEN ENG FRA GER GRE JOR LEB NET NZ RHO SUD SWE SWT TNZ TUR	ISR
muteli F. Schultz				BUL SOV	
nicotianae Wight				IND	PK
pallidiflora Wimm. & Grab.				SOV	
picridis F. Schultz				ENG FRA GER	
ramosa L.	EGY HUN	CUB ITA	AFG POL	AST BUL	

OROBANCHE L. Orobanchaceae *(Cont'd)*	S	P	C	X	F
ramosa L. *(Cont'd)*	JOR LEB NEP	TUR	SAF SOV YUG	ENG FRA GER GRE IND ISR ROM SUD USA	
rapum-genistae Thuill.				FRA	
rubens Wallr.				TUR	
speciosa A. Dietr.				ITA	
ORONTIUM L. **Araceae**					
aquaticum L.				USA	
ORTHOCLADA Beauv. **Poaceae** *(Gramineae)*					
laxa (L. Rich.) Beauv.				PER	
ORTHOSIPHON Benth. **Lamiaceae** *(Labiatae)*					
pallidus Royle *ex* Benth.				IND	
ORTHROSANTHUS Sweet **Iridaceae**					
chimboracensis (H. B. K.) Baker				VEN	
ORYCHOPHRAGMUS Bunge **Brassicaceae** *(Cruciferae)*					
violaceus O.E. Schulz				CHN	
ORYCTANTHUS Eichl. **Loranthaceae**					
occidentalis Eichl.				JAM	
ORYZA L. **Poaceae** *(Gramineae)*					
alta Swallen				COL	
barthii A. Chev.		NIG			KEN
latifolia Back. *(See* **O. minuta** Presl)				NIC THI	IND
minuta Presl (Syn. *O. latifolia* Back.)		IDO			
perennis Moench		MAD		CEY	SUD THI
punctata Kotschy *ex* Steud.	SWZ			NIG	
rufipogon Griff.				BND	
sativa L.	ARG CR KOR PER SEG THI USA	CAB COL	PLW TAI	BND BRA ECU EGY HON IDO IND IRA IRQ JAM	AFG AUS CHN HAW HK

ORYZA L. Poaceae *(Gramineae) (Cont'd)*	S	P	C	X	F
sativa L. *(Cont'd)*				JPN PR RHO SUR TNZ TUR YUG	
ORYZOPSIS Michx. **Poaceae** *(Gramineae)*					
miliacea (L.) Benth. & Hook. *f.*			LEB	CHL NZ	HAW ISR SAF
nymenoides (Roem. & Schult.) Ricker				USA	
OSMUNDA L. **Osmundaceae**					
cinnamomea L.				USA	
japonica Thunb.			JPN		
OSTRYA Scop. **Betulaceae**					
virginiana (Mill.) C. Koch				USA	
OTHONNOPSIS Jaub. & Spach. **Asteraceae** *(Compositae)*					
intermedia Boiss.				PK	AFG
OTOSPERMUM Willk. **Asteraceae** *(Compositae)*					
glabrum Willk.			MOR		
OTTELIA Pers. **Hydrocharitaceae**					
alismoides (L.) Pers.	THI	IDO JPN TAI		BND CAB CAN CEY CHN HON IND KOR MAL PHI SUD VIE	HK
ulvaefolia Walp.			NZ	AUS SUD	ETH MAD MOZ RHO
OTTOCHLOA Dandy **Poaceae** *(Gramineae)*					
nodosa (Kunth) Dandy (Syn. *Panicum nodosum* Kunth)		IDO MAL	BOR	CEY	BUR IND PHI
OXALIS L. **Oxalidaceae** *(Syn. XANTHOXALIS Small)*					
acetosella L.			AUS SPA	COL IND	
anthelmintica A. Rich.			KEN		
articulata Savign.			ARG	BRA	

OXALIS L. Oxalidaceae *(Cont'd)*	S	P	C	X	F
articulata Savign. *(Cont'd)*				NZ	
bahiensis Prog.				PLW	
barrelieri L.			TRI	ANT IDO	
bowiei Lindl.		AUS			
cernua Thunb.			SPA	CHL ISR	AUS SAF TUN
compressa L. *f.*		AUS			
cordobensis Kunth				ARG	
corniculata L. (Syn. *Xanthoxalis corniculata* (L.) Small)	MEX TNZ	ETH IND JPN TAI VEN	ARG AUS CHN COL EGY ENG HAW IDO IRA IRQ ISR KEN MOR PHI PR RHO SAF SAL SPA	BOR BRA CAM CEY CHL CR CUB FIJ GHA GUA GUI HUN IVO JAM KOR LIB MAL MAU NIG NZ PER PK SOV TRI USA	AFG HK ITA MOZ
corymbosa Dc.	CEY		AUS IDO IND MAL	BRA FIJ GHA PLW TRI USA	
debilis H. B. K.		MAU			
dillenii Jacq.				USA	
dombeyi St. Hil.				PER	
europaea Jord.			ENG	BND USA	
flava L.		AUS			
florida Salisb.				USA	
latifolia H. B. K.	CEY IND KEN TNZ UGA	HON IDO NZ SPA	AUS GUA SAF SAL	ARB CNK ECU ETH FRA IRA	

OXALIS L. Oxalidaceae *(Cont'd)*	S	P	C	X	F
latifolia H. B. K. *(Cont'd)*				MAU MEL MOZ NEP NGI PER RHO TAI URU ZAM	
mallobolba Cav.			ARG		
martiana Zucc. (Syn. *Ionoxalis martiana* (Zucc.) Small)			AUS CHN HAW IND JPN SPA TAI	BOL BRA USA	PR
micrantha Bert. *ex* Colla				CHL	
neaei Dc.			SAL	MEX	
obliquifolia Steud. *ex* A. Rich.			KEN		
pes-caprae L.		AUS	IND MOR POR SAF	NZ USA	
purpurata Jacq.	KEN				JPN
purpurea L.		AUS		NZ	
repens Thunb.		MAU	KEN	IND PHI THI VIE	RHO SAF
rubra St. Hil.				USA	
semiloba Sond.	RHO	KEN		MOZ SAF	
sepium St. Hil				IDO	
stricta L.		MEX	AUS USA	NZ	
violacea L.			JPN	BRA	
OXYDENDRUM Dc. Ericaceae					
arboreum (L.) Dc.				USA	
OXYGONUM Burch. Polygonaceae					
atriplicifolium (Meissn.) Mart.			KEN	SUD	
sinuatum (Meissn.) Dammer	TNZ		KEN		
OXYPETALUM R. Br. Asclepiadaceae					
solanoides Hook. & Arn.				ARG	BRA

	S	P	C	X	F
OXYSTELMA R. Br. **Asclepiadaceae**					
esculentum R. Br.			EGY		
OXYTENIA Nutt. **Asteraceae** *(Compositae)*					
acerosa Nutt.				USA	
OXYTROPIS Dc. **Papilionaceae** *(Leguminosae)*					
lambertii Pursh			USA		
macounii (Greene) Rydb.				USA	
saximontana Nels.				USA	
splendens Dougl.				USA	
PAEDERIA L. **Rubiaceae**					
chinensis Hance			TAI	JPN	
foetida L.	NGI	MAU	HAW		BRA CHN IDO
scandens (Lour.) Merr.				JPN	
PAGESIA Rafin. **Scrophulariaceae**					
dianthera (Sw.) Pennell				ANT TRI	
PALICOUREA Aubl. **Rubiaceae**					
macrobotrys (Ruiz & Pav.) Dc.				PER	
marcgravii St. Hil.				BRA	
paraensis (Muell.-Arg.) Standl.				PER	
triphylla Dc.				PER	
PALLENIS Cass. **Asteraceae** *(Compositae)*					
spinosa Cass.				JOR	AUS ISR POR
PANICUM L. **Poaceae** *(Gramineae)*					
adspersum Trin.		PR		COL DR USA	
agrostoides Spreng.				MEX	
ambigium Trin.				IDO	
amplexicaule Rudge		MAL			
arizonicum Scribn. & Merr.				USA	
attenuatum Willd.				PHI	
auritum Presl				BOR	CEY IND
austroasiaticum Ohwi	IND				

PANICUM L. Poaceae *(Gramineae) (Cont'd)*	S	P	C	X	F
barbatum Lam. *(See* **Setaria barbata** (Lam.) Kunth)		IDO			HAW NGI
barbinode Trin. *(See* **Brachiaria mutica** (Forsk.) Stapf)		COL		PER	CEY HAW
bergii Arechav.			ARG	URU	
bisulcatum Thunb. (Syn. *P. coloratum* L.)			JPN		CHN IND
brevifolium L.			IDO TAI	CNK GHA	CEY IND
capillare L.	IND		CAN	CHL NZ USA	AUS HAW
caudiglume Hack.			IDO		
chloroticum Nees		BRA		SUR	
ciliatum Ell.		MEX			
colona L. *(See* **Echinochloa colona** (L.) Link)	IDO SPA	IND		PK	AUS HAW
coloratum L. *(See* **P. bisulcatum** Thunb.)			EGY		
crus-galli L. *(See* **Echinochloa crus-galli** (L.) Beauv.)		IND		SOV TUR	AFG AUS HAW IDO
dactylon L. *(See* **Cynodon dactylon** (L.) Pers.)				SOV	
dichotomiflorum Michx.		CAN	ARG USA	BND NZ	BRA HAW
effusum R. Br.			AUS		
elephantipes Nees			ARG		
erectum Pollacci	ITA				
eruciforme J. E. Sm. *(See* **Brachiaria eruciformis** (J. E. Sm.) Griseb.)			IRA LEB		AFG IDO
excurrens Trin. *(See* **Setaria plicata** (Lam.) T. Cooke)			TAI		
fasciculatum Sw.	CUB HON JAM MEX SUR TRI	COL ECU VEN	PER	ANT PAN PR SAL USA	
fimbriatum (Link) Kunth *(See* **Digitaria adscendens** (H. B. K.) Henr.)				MEX	PR
flavescens Sw. *(See* **Setaria barbata** (Lam.) Kunth)				IND	SAF
flavidum Retz. *(See* **Paspalidium flavidum** (Retz.) A. Camus)				BND PHI THI	

PANICUM L. Poaceae *(Gramineae) (Cont'd)*	S	P	C	X	F
gattingeri Nash				USA	
glabrescens Steud.				MAU	SAF
gouini Fourn.		BRA			
hallii Vasey				MEX	
hemitomun Schult.				USA	
hygrocharis Steud.		SUD			
indicum Hack. *(See* **Sacciolepis interrupta** (Willd.) Stapf)					JPN
isachne Roth *(See* **Brachiaria eruciformis** (J. E. Sm.) Griseb.)				IND	
kerstingii Mez				IVO	
laevifolium Hack.	SAF				AUS HAW
lancearum Trin.				DR	
laxum Sw.		GHA		BRA PER SUR TRI	CR
luzonense Presl	THI		IDO	CAB	
maximum Jacq.	AUS COL CR CUB GHA HAW JAM MOZ SAF UGA	BER ECU KEN MEX PR SWZ TNZ TRI VEN	ISR MAL PHI	BOL BOR BRA CAM DAH DR FIJ GUI HON IVO MAU NIG PER PLW RHO SAL SEG TAI THI URU USA VIE	AFG ARB BUR CEY HK IND JPN
meyerianum Nees				SUD	
miliaceum L.		JPN		ITA NZ TUR USA	AFG AUS CEY HAW IND SAF
millegrana Poir.				DR	
monostachyum H. B. K.				BRA	
montanum Roxb.			IDO	VIE	

PANICUM L. Poaceae *(Gramineae)* *(Cont'd)*	S	P	C	X	F
muticum Forsk. *(See* **Brachiaria mutica** (Forsk.) Stapf)			IDO		HAW
myuros H. B. K. *non* Lam. *(See* **Sacciolepis indica** (L.) A. Chase)				THI	
nodosum Kunth *(See* **Ottochloa nodosa** (Kunth) Dandy)	MAL			PHI	HAW
obtusum H. B. K.				USA	
orizicola Vas. *(See* **Echinochloa orizicola** Vas.)				SOV	
palmifolium Willd. *(See* **Setaria palmifolia** (Willd.) Stapf)			IDO		
paludosum Roxb.	IND				
phyllopogon Stapf	ITA				
pilipes Nees & Arn.			BOR	MAL	IDO
polygonatum Schrad.				TRI	
prionitis Nees				URU	
psilopodium Trin.		IND		BND	
purpurascens Raddi *(See* **Brachiaria mutica** (Forsk.) Stapf)	FIJ HAW	AUS BRA CUB PER PHI PR TRI VEN		ANT ARG COL DR HON JAM MEX USA	CEY PK SUR
queenslandicum Domin				AUS	
ramosum L. *(See* **Brachiaria ramosa** (L.) Stapf)					AFG IDO
repens L.	CEY FRA GUI HAW MAL PHI TAI THI	IDO IND SUD	AUS IRQ MOR POR USA	BND CAB CHN CUB DR HK PR RHO SAF VIE	ARB ARG ISR
reptans L. *(See* **Brachiaria reptans** (L.) Gard. & C. E. Hubb.)	IDO	THI		PHI	AFG HAW
sabulorum Lam.		BRA			
sanguinale L. *(See* **Digitaria sanguinalis** (L.) Scop.)			IRA	GER TUR	HAW IND
sarmentosum Roxb.				IDO MAL	BUR IND
spectabile Nees		BRA			AUS HAW

PANICUM L. Poaceae *(Gramineae) (Cont'd)*	S	P	C	X	F
stagninum Retz. *(See* **Echinochloa stagnina** (Retz.) Beauv.)				IND	
texanum Buckl.			USA		AUS
torridum Gaudich.			HAW		
trichocladum Hack. *ex* Schum.	TNZ		KEN		
trichoides Sw.			PR	DR NIC	
trigonum Retz. *(See* **Cyrtococcum trigonum** (Retz.) A. Camus)			IDO	MAL	
trypheron Schult.			IDO		IND
umbellatum Trin.				MAU	
urvilleanum Kunth.			ARG	CHL	BRA
virgatum L.				ARG USA	HAW IND TRI
zizanioides H. B. K. *(See* **Acroceras zizanioides** (H. B. K.) Dandy)		TRI		ANT	SUR
PAPAVER L. Papaveraceae					
aculeatum Thunb.			SAF		AUS
argemone L.			ENG GER POR SPA	TUR	ISR POL
dubium L.			ENG GER MOR POR	NZ PK USA	AFG AUS BND ISR POL
hybridum L.		AUS	GER MOR POR SPA	NZ TUR	ISR
rhoeas L.	GRE HUN ITA POL SPA	IRA IRE POR SWE TUN	AFG COL ENG GER ISR MOR SOV	BEL CAN CHL CHN CZE DEN EGY FRA ICE IRQ JOR JPN KOR LEB NZ TUR USA YUG	AUS
somniferum L.		AUS	MOR	BRA CHL	AFG CAB

PAPAVER L. Papaveraceae *(Cont'd)*	S	P	C	X	F
somniferum L. *(Cont'd)*				NZ TUR	POL
PAPPEA Ecke. & Zeyh. **Sapindaceae**					
capensis Eckl. & Zeyh.				RHO	
PARACARYUM Boiss. **Boraginaceae**					
strictum Boiss.			IRA		
PARAPHOLIS C. E. Hubb. **Poaceae** *(Gramineae)*					
incurva (L.) C. E. Hubb.			POR	BND NZ	AUS SAF
PARIETARIA L. **Urticaceae**					
debilis Forst. *f.*				BRA	
floridana Nutt.				USA	
judaica L.				GRE NZ	ARG ISR
officinalis L.			SPA		
pensylvanica Muhl.				USA	
PARKINSONIA L. **Caesalpiniaceae** *(Leguminosae)*					
aculeata L.				ARG AUS USA	
PARONYCHIA L. **Caryophyllaceae**					
brasiliana Dc.				AUS SAF	BRA
PARTHENIUM L. **Asteraceae** *(Compositae)*					
hysterophorus L.	MEX	CUB	ARG IND PR	DR JAM MAU TRI USA VEN	
PARTHENOCISSUS Planch. **Vitaceae**					
quinquefolia (L.) Planch.				CAN USA	
PASPALIDIUM Stapf **Poaceae** *(Gramineae)*					
flavidum (Retz.) A. Camus (Syn. *Panicum flavidum* Retz.)			PHI	CAB MAU VIE	BUR IND
geminatum (Forsk.) Stapf		MAU	EGY	ANG	ISR
obtusifolium (Del.) Simps.				MOR	
PASPALUM L. **Poaceae** *(Gramineae)*					
ciliatifolium Trin. (*See* **P. conjugatum** Berg.)				USA	AUS

PASPALUM L. Poaceae *(Gramineae)* *(Cont'd)*	S	P	C	X	F
commersonii Lam.		MAL NIG SWZ	IDO TAI	ANG BND GHA MAU	CEY HAW IND SAF
conjugatum Berg. (Syn. *P. ciliatifolium* Trin.)	AUS BOR CEY CNK CR HAW IDO IVO MAL NGI PHI TRI	COL FIJ GHA GUI IND NIG PER PLW TAI	THI	BRA CAB CUB DAH DR JAM LIB MAU MEX SUR USA VEN VIE	BUR PR
dasypleurum Kunze & Desv.				CHL	
dilatatum Poir.	AUS PHI	BRA COL SOV TAI	HAW JPN USA	ARG CHL GUY IND MAU MEL NZ PLW SAF URU VIE	BND CEY ITA KEN MAL PK PR
distachyon Salzm. *ex* Doell				URU	
distichum L.	AUS CHL GUY HON IRQ ISR PK POR SOV SPA	COL FRA IRA JPN NZ SWZ	ARG EGY MOR PER SAF TAI USA	BND BOR BRA CHN ECU IND JAM JOR MAL MAU MEL MEX MIC NEP PHI PLW PR SAL SUR URU VEN	CEY ENG FIJ HAW ITA
fasciculatum Willd. *ex* Fluegge	TRI	VEN		ANT COL DR HON SAL	
fimbriatum H. B. K.			HAW PR	DR TRI USA	
fluitans (Ell.) Kunth				USA	

PASPALUM L. Poaceae *(Gramineae) (Cont'd)*	S	P	C	X	F
geminatum Forsk.				IDO	
haenkeanum Presl				PER	
laeve Michx.			MEX	USA	
lividum Trin.				PER	HAW
maculosum Trin.			BRA		
maritinum Trin.				BRA SUR	
melanospernum Desv.				SUR	
millegrama Schrad.			PR		
nicorae Parodi				ARG	BRA
notatum Fluegge	BRA	CUB	ARG HAW	COL SAL URU USA VEN	AUS IND SAF
nutans Lam.				MAU	
orbiculare Forst.	GHA		HAW TAI	FIJ NIG PLW TRI USA	AUS IND
paniculatum L.	AUS	MAU	TRI	BRA CR FIJ PER PLW SAL	
paspaloides Scribn.			AUS		
plicatulum Michx.		ARG BRA CUB		MEX TRI	HAW IND
polystachyum (Humb. & Bonpl.) Raspail				GHA	
proliferum Arech.			AFG		
pumilum Nees			BRA		
quadrifarium Lam.				URU	
racemosa Lam.				COL PER	
sanguinale (L.) Lam. *(See* **Digitaria sanguinalis** (L.) Scop.)		IND			HAW IDO
scrobiculatum L.	IND PHI SEG TAI THI	IVO KOR MAU	BOR CHN IDO JPN	BRA CAB CNK DAH FIJ GHA MAL MEL NIG	AUS HK

PASPALUM L. Poaceae *(Gramineae) (Cont'd)*	S	P	C	X	F
scrobiculatum L. *(Cont'd)*				NZ PK SUD VIE	
thunbergii Kunth *ex* Steud.			JPN TAI		
urvillei Steud.	BRA		ARG HAW	AUS· CHL NZ USA VIE	IND SAF
vaginatum Sw.			ARG HAW POR SAF TAI	CHL MAU NZ PHI SPA VIE	AUS IDO IND
virgatum L.		PR TRI VEN		BRA CNK NIC PER	AUS HAW
PASSIFLORA L. Passifloraceae					
alba Link & Otto				AUS	
biflora Lam.				HON PAN SAL	
coccinea Aubl.				PER	
edulis Sims				BRA	ISR
foetida L.	BOR MAL NGI	AUS IDO PER	FIJ HAW MIC PHI PR SUR THI TRI	CEY CHN CNK COL· DAH DR GHA HK HON IND JAM MAU MEL NIC NIG PLW SAL SEG USA VIE	
incarnata L.				USA	
pulchella H. B. K.			HAW	USA	
rubra Lam.				JAM TRI	
sexflora A. Juss.				JAM	
suberosa L.	MEL		HAW	AUS	

PASSIFLORA L. Passifloraceae *(Cont'd)*	S	P	C	X	F
suberosa L. *(Cont'd)*			TAI	CEY FIJ MAU	
subpeltata Orteg.			AUS		
PASTINACA L. **Apiaceae** *(Umbelliferae)*					
sativa L.			ARG CAN SOV	CHL NZ PER USA	AFG AUS POL
PAULLINIA L. **Sapindaceae**					
densiflora Sm.				COL	
pinnata L.		PR		ANG GHA	
PAULOWNIA Sieb. & Zucc. **Scrophulariaceae**					
tomentosa (Thunb.) Steud.				USA	
PAVONIA Cav. **Malvaceae**					
cancellata (Cav.) Diss. *(See* **P. procumbens** Boiss.*)*				BRA	
coxii Tad. & Jacob.				IND	
procumbens Boiss. (Syn. *P. cancellata* (Cav.) Diss.)				IND	
spinifex Cav.		PR		DR	
urens Cav.		KEN			
zeylanica Cav.				IND	
PECTIS L. **Asteraceae** *(Compositae)*					
ciliaris L.		PR		JAM	
PEDALIUM L. **Pedaliaceae**					
murex L.				IND	
PEDICELLARIA Schrank **Capparidaceae** *(See* **GYNANDROPSIS** Dc.*)*					
pentaphylla Schrank *(See* **Cleome gynandra** L.*)*		TAI			
PEDICULARIS L. **Scrophulariaceae**					
comosa L.				TUR	
PEGANUM L. **Rutaceae**					
harmala L.				PK TUR USA	AFG AUS ISR SOV
PELTANDRA Rafin. **Araceae**					
virginica (L.) Schott & Endl.				USA	

PELTOPHORUM (T. Vogel) Benth. Caesalpiniaceae *(Leguminosae)*	S	P	C	X	F
africanum Sond.		RHO			
PENNISETUM L. C. Rich. **Poaceae** *(Gramineae)*					
alopecuroides (L.) Spreng.			CHN JPN	AUS NZ	
chilense (Desv.) Jacks.				CHL	ARG
clandestinum Hochst. *(See* **Dicanthelium clandestinum** (L.) Gould)	CR KEN PER UGA URU	AUS BRA COL HAW IND NZ TNZ		CAM CEY ECU GUA USA	ANG CAB CNK EGY ISR JAM MAD MAU MOR NGI NIG PAN PAR RHO SAF SWZ TUN
fructescens Leeke				ARG	
hordeoides (Lam.) Steud.				IVO	
japonicum Trin.			JPN		
macrourum Trin.			AUS	NZ	
pedicellatum Trin.	NIG THI	AUS		ETH MAL	FIJ IND
polystachyon (L.) Schult.	THI	IND		ANG AUS CEY FIJ IVO MEL NIG PHI	
purpureum Schumach.	COL GHA MOZ TRI		HAW PHI PR	AUS BOR BRA CUB HON KEN MAL NGI PLW SAL THI UGA USA VEN	ARG BUR IND PER SAF
ruppellii Steud.				USA	HAW
setaceum (Forsk.) Chiov.			HAW	USA	
setosum L. C. Rich.			HAW	PHI USA VEN	

PENNISETUM L. C. Rich. Poaceae *(Gramineae) (Cont'd)*	S	P	C	X	F
villosum R. Br.			AUS	CHL NZ	
PENSTEMON Mitch. *ex* Schmid. **Scrophulariaceae**					
digitalis Nutt.				USA	
gracilis Nutt.				USA	
rydbergii Nutt.				USA	
PENTAPETES L. **Sterculiaceae**					
phoenicea L.		THI		CAB	IDO
PENTATROPIS R. Br. **Asclepiadaceae**					
microphylla Wight & Arn.				IND	
PEPEROMIA Ruiz & Pav. **Piperaceae**					
pellucida (L.) H. B. K.			PHI PR TRI	BOR DR FIJ GHA IND JAM NIG PAN PLW THI	IDO
PERILLA L. **Lamiaceae** *(Labiatae)*					
frutescens (L.) Britt.			JPN		
PERISTROPHE Nees **Acanthaceae**					
bicalyculata Nees				IND	
PEROTIS W. Ait. **Poaceae** *(Gramineae)*					
indica (L.) O. Ktze.				VIE	
PERSEA Mill. **Lauraceae**					
borbonia (L.) Spreng.				USA	
PERSICARIA L. **Polygonaceae**					
nodosum Pers. *(See* **Polygonum lapathifolium** L.*)*					KOR
PERYMENIUM Schrad. **Asteraceae** *(Compositae)*					
subsquarrosum Rob. & Greenm.		MEX			
PETASITES Mill. **Asteraceae** *(Compositae)*					
hybridus (L.) G. M. & Sch.				BEL	
japonicus (Sieb. & Zucc.) Maxim.				JPN	
officinalis Moench			GER		
PETIVERIA L. **Phytolaccaceae**					
alliacea L.			PR	BRA COL	

PETIVERIA L. Phytolaccaceae *(Cont'd)*	S	P	C	X	F
alliacea L. *(Cont'd)*				DR HON JAM PAN TRI	
PETROSELINUM Hill **Apiaceae** *(Umbelliferae)*					
segetum (L.) Koch			POR		
PETUNIA Juss. **Solanaceae**					
axillaris (Lam.) B. S. P.			ARG		
PEUCEDANUM L. **Apiaceae** *(Umbelliferae)*					
palustre Moench			FIN		
PFAFFIA Mart. **Amaranthaceae**					
iresinoides Spreng.				TRI	
sericea (Spreng.) Mart.			ARG		BRA
stenophylla (Spreng.) Stuchlik.			ARG		
PHACELIA Juss. **Hydrophyllaceae**					
purshii Buckl.				USA	
tanacetifolia Benth.				USA	
PHALARIS L. **Poaceae** *(Gramineae)*					
angusta Nees *ex* Trin.				ARG	BRA SAF
aquatica L.			POR		
arundinacea L.	AFG HUN JPN	IDO KOR MAU NZ POL	ITA POR USA	ARG BEL CAN CHN COL CZE ENG FIN GER PR SWE TUR	AUS CEY HAW IND SAF
brachystachys Link		ISR SPA	LEB MOR POR	JOR	
canariensis L.		JOR	ARG	COL MOR NZ PER TUR USA	AUS BRA ISR SAF
minor Retz.	AFG IND	ARB COL IRA IRQ MEX	ARG EGY ITA LEB MOR NEP	BRA ECU GRE ISR JOR PK	AUS HAW SAF

PHALARIS L. Poaceae *(Gramineae) (Cont'd)*	S	P	C	X	F
minor Retz. *(Cont'd)*			NZ POR	RHO SAL TAS TUR URU USA	
paradoxa L.	AFG ETH ISR TUN	AUS BEL ENG IRA IRQ	GRE LEB MOR POR SPA TUR	EGY FRA GER NZ SWE URU USA	HAW IND SAF
tuberosa L.		AUS	LEB POR	NZ	HAW IND RHO SAF
PHASEOLUS L. Papilionaceae *(Leguminosae)*					
aconitifolius Jacq.				IND	
adenanthus G. F. W. Mey.			PR	HON SAL TRI	
angularis W. F. Wight *(See* **Azukia angularis** (Willd.) Ohwi)			JPN		
atropurpurens Moq. & Sesse *ex* Dc.				PER SAL	
aureus Roxb.				PHI	
lathyroides L.		AUS	HAW PHI	ANT COL DR FIJ JAM PER PLW THI	IDO PR
linearis H. B. K.				TRI	
lunatus L.				USA	IDO
trilobus (L.) W. T. Ait.		IND		PK	IDO
PHAYLOPSIS Willd. Acanthaceae					
angolana S. Moore				CNK	
PHELYPAEA L. Orobanchaceae					
aegyptiace Walp.				SOV	
longifolia C. A. Mey.				SOV	
ramosa C. A. Mey.				SOV TUR	
PHENAX Wedd. Urticaceae					
sonneratii Wedd.				TRI	

PHILOXERUS R. Br. **Amaranthaceae**	S	P	C	X	F
vermicularis (L.) Nutt.			PR		
PHILYDRUM Banks & Sol. *ex* Gaertn. **Philydraceae**					
lanuginosum Banks & Sol. *ex* Gaertn.				VIE	
PHLEUM L. **Poaceae** *(Gramineae)*					
nodosum L. *(See* **P. pratense** L.)				TUR	
paniculatum Huds.				CHN	
pratense L. (Syn. *P. nodosum* L.)			COL FIN SOV SPA	NZ PER USA	AUS HAW IND
PHLOMIS L. **Lamiaceae** *(Labiatae)*					
herba-venti L.				MOR	POR
kurdica K. H. Rechinger				LEB	
orientalis Mill.				LEB	ISR
PHOTINIA Lindl. **Rosaceae**					
arbutifolia (Ait.) Lindl.				USA	
PHRAGMITES Trin. **Poaceae** *(Gramineae)*					
australis (Cav.) Trin. (Syn. *P. communis* Trin.) (Syn. *P. vulgaris* (Lam.) Crep.)	IRA	AUS GER ITA	PHI SOV TAI USA	ARB ARG BND BUL CAN CEY CHL CZE EGY ENG FIN FRA GRE IRQ ISR JPN MAD MAL NET NZ PER POR PR RHO ROM SAF SUD TAS THI TUR YUG	
communis Trin. *(See* **P. australis** (Cav.) Trin.)		AUS GER GRE ITA NZ	ARG FIN IRA IRQ ISR	CHL EGY ENG FRA MAL	AFG CHN IND KOR

PHRAGMITES Trin. Poaceae *(Gramineae)* *(Cont'd)*	S	P	C	X	F
communis Trin. *(Cont'd)*			JPN POR SOV SPA TAI USA	NET PER SAF THI TUR	
karka (Retz.) Trin.		IND PK		AUS IDO PHI VIE	AFG BUR CEY GHA KOR
mauritiana Kunth				SUD	EGY MAD RHO SAF
vulgaris (Lam.) Crep. *(See* **P. australis** (Cav.) Trin.*)*				PHI	
PHTHIRUSA Mart. **Loranthaceae**					
bicolor Engl.			PR		
pauciflora Eichl.			JAM		
purpurea (L.) Engl.			PR		
PHYLA Lour. **Verbenaceae**					
betulaefolia Greene			TRI		
nodiflora (L.) Greene (Syn. *Lippia nodiflora* L.)		IND	ARG EGY	AUS	AFG
PHYLLANTHUS L. **Euphorbiaceae**					
amarus Schum. & Thonn.	HON JAM	HAW IVO NIG SUD	GHA IDO IND MIC PHI SUR TRI	ANT AUS CAB CNK DAH MAL PER PLW SEG THI VIE	
corcovadensis Muell.-Arg.		BRA			
diffusus Kl.		COL			PR
fraternus Webster			IND		
galeottianus Baill.			MEX		
graminicola Britton			ANT TRI		
grandifolius L.			PER		
lathyroides H. B. K.			VEN		CR PR
maderaspatensis L.		AUS SUD	RHO	GHA IND MOZ	IDO

PHYLLANTHUS L. Euphorbiaceae *(Cont'd)*	S	P	C	X	F
matsumurae Hayata			JPN		
minutiflorus F. Muell. *ex* Muell.-Arg.				AUS	
niruri L.	GHA HAW HON JAM MEX SUD	ANG COL IDO IND MAL PLW	BOR ETH GUA MIC NEP PER PHI PR SAL SUR TAI THI	AUS BOT BRA CAB CEY CHN CNK DAH DR ECU FIJ JPN MAU MEL NGI NIC PK RHO SEG TRI USA VIE	
nummulariaefolius Poir.				MOZ	
odontadenius Muell.-Arg.				GHA NIG	
pseudo-conami Muell.-Arg.				PER	
simplex Retz. *(See* **P. virgatus** Forst. *f.)*			PHI	FIJ IND PLW	
stipulatus (Rafin.) Webs.				MEX	
sublanatus Schum. & Thonn.				GHA	
tenellus Roxb.				AUS MAU	
urinaria L.	SUR	KOR MAL PER	BOR IDO JPN PHI TAI TRI	ANT BRA CAB CEY CHN CNK FIJ FRA GHA HAW HK JAM MAU NGI PLW THI USA ZAM	
virgatus Forst. *f.)* (Syn. *P. simplex* Retz.)				IDO	

PHYLLITIS Hill **Aspleniaceae**	S	P	C	X	F
scolopendrium (L.) Newman				TUR	
PHYLLOSTACHYS Sieb. & Zucc. **Poaceae** *(Gramineae)*					
mitis A. & C. Riviere		NZ			
PHYSALIS L. **Solanaceae**					
alkekengi L.				BOL BRA CHN USA	AFG
angulata L.	GHA SAF TAI	FIJ IDO RHO SWZ VEN	AUS IRQ JPN KEN KOR MIC NIG PR SEG SUD THI TRI	ANG ANT BOR BOT CAB CNK COL DAH DR ECU EGY GUA HON IVO JAM MEL MEX MOZ NGI NIC PER PHI PLW SAL SUR TNZ USA VIE	CHN GUI HK
divaricata D. Don			KEN		
heterophylla Nees			COL USA	AUS BND BRA	
ixocarpa Brot. *ex* Dc.	KEN MEX			USA	
lagascae Roem. & Schult.			TRI	HON PAN SAL	
lanceolata Michx.			AUS		
lobata Torr.				USA	
longifolia Nutt.				USA	
macrophysa Rydb.			AUS		
mendocina Phil.				CHL	
micrantha Link			KEN	NIG	
minima L.		IND PHI	AUS BOR	IVO SUD	AFG

PHYSALIS L. Solanaceae *(Cont'd)*	S	P	C	X	F
minima L. *(Cont'd)*			CHN IDO KEN THI		
nicandroides Schlecht.			SAL		
peruviana L.			HAW IDO KEN RHO	AUS FIJ IND NZ PER PLW USA	
pubescens L.			PR TRI	ANT CHL CR HON SAL USA	
solanacae Mert. *ex* Roth			COL		
subglabrata Mack. & Bush.			USA		
turbinata Medic.			PR TRI		
virginiana Mill.			USA		
viscosa L.			ARG SAF	BRA CHL URU	AUS
wrightii Gray			USA		
PHYSOSTEGIA Benth. **Lamiaceae** *(Labiatae)*					
virginiana (L.) Benth.			USA		
PHYSOSTIGMA Balf. **Papilionaceae** *(Leguminosae)*					
mesoponticum Taub.				ANG	
virginiana (L.) Benth.					
PHYTOLACCA L. **Phytolaccaceae**					
americana L. (Syn. *P. decandra* L.)			AUS JPN USA	ARG NZ	
decandra L. (See **P. americana** L.)			JPN USA		AUS ISR
dodecandra L'herit.			KEN		
octandra L.		AUS NZ	HAW	USA	
rigida Small				USA	
rivinoides Kunth & Bouche			PR	PER	
PICEA A. Dietr. **Pinaceae**					
glauca (Moench) Voss				USA	

PICEA A. Dietr. Pinaceae *(Cont'd)*	S	P	C	X	F
mariana (Mill.) B. S. P.				USA	
pungens Engelm.				USA	
rubens Sarg.				USA	
sitchensis (Bong.) Carr.				USA	
PICRAENA Lindl. **Simaroubaceae**					
excelsa Lindl.				JAM	
PICRAMNIA Sw. **Simaroubaceae**					
antidesma Sw.				JAM	
PICRIDIUM Desf. **Asteraceae** *(Compositae)*					
vulgare Desf.				USA	
PICRIS L. **Asteraceae** *(Compositae)*					
echioides L. (Syn. *Helminthia echioides* (L.) Gaertn.)			ENG LEB MOR SAF	ARG CHL NZ POR URU USA	AUS
hieracioides L.			HAW JPN SOV	USA	AFG AUS
PILEA Lindl. **Urticaceae**					
microphylla (L.) Liebm.			HAW	ANT JAM PLW SAL TRI USA	
nummulariaefolia (Sw.) Wedd.			TRI		
PILIOSTIGMA Hochst. **Caesalpiniaceae** *(Leguminosae)*					
thonningii (Schumach.) Milne-Redhead				ANG	
PIMELEA Banks & Soland. **Thymelaeaceae**					
linifolia Sm.			AUS		
PIMPINELLA L. **Apiaceae** *(Umbelliferae)*					
saxifraga L.			FIN SOV		
PINELLIA Tenore **Araceae**					
ternata (Thunb.) Brietend.		JPN			
PINUS L. **Pinaceae**					
attenuata Lemm.				USA	
banksiana Lamb.				USA	
clausa (Chapm.) Vasey				USA	

PINUS L. Pinaceae *(Cont'd)*	S	P	C	X	F
contorta Dougl.				USA	
echinata Mill.				USA	
edulis Engelm.				USA	
elliotii Engelm.				USA	
jeffreyi A. Murr.				USA	
lambertiana Dougl.				USA	
monticola Dougl.				USA	
palustris Mill.				USA	
ponderosa Dougl.				USA	
resinosa Ait.				USA	
rigida Mill.				USA	
sabiniana Dougl.				USA	
serotina Michx.				USA	
strobus L.				USA	
sylvestris L.				USA	
taeda L.				USA	
virginiana Mill.				USA	
PIPER L. Piperaceae					
aduncum L.		PR TRI		DR FIJ IDO USA	
amalago L.				JAM	
betle L.				IDO	
caninum Bl.				PHI	
guianense C. Dc.				CNK	
hispidum Sw.				PER	
jamaicense C. Dc.				JAM	
nigrinodum C. Dc.				JAM	
tuberculatum Jacq.		TRI		PAN	
umbellatum L.				GHA JAM	
PIPTOCARPHA R. Br. Asteraceae *(Compositae)*					
poeppigiana (Dc.) Baker				PER	
PISCIDIA L. Papilionaceae *(Leguminosae)*					
piscipula (L.) Sarg.				JAM	

PISTIA L. Araceae	S	P	C	X	F
stratiotes L.	CEY GHA IDO THI	ANG BND BOR ETH HAW IND MAD MOZ NIG SEG TNZ	ARG KEN MAL USA	BUR CAB CHN CNK COL DAH EGY GUI HON IVO MLI NGR NIC PHI PK PR RHO SAF SAL SUD URU VEN VIE VOL ZAM	HK MRE
PISUM L. Papilionaceae (*Leguminosae*)					
elatius Bieb.		LEB POR			ISR
PITHECELLOBIUM Mart. Mimosaceae (*Leguminosae*)					
dulce (Roxb.) Benth.			HAW		
filamentosum Benth.				BRA	
lanceolatum Benth.				COL	
ligustrinum Klotzsch *ex* Benth.				VEN	
pachypus Pittier				SAL	
zanzibaricum S. Moore (*See* **Acacia zanzibarica** (S. Moore) Taub.)				KEN	
PITTOSPORUM Banks & Soland. Pittosporaceae					
undulatum Vent.			HAW	USA	
PITYROGRAMMA Link Sinopteridaceae					
calomelanos (L.) Link			HAW PR	BOR GHA MAL PHI	HON
PLANERA J. F. Gmel. Ulmaceae					
aquatica (Walt.) J. F. Gmel.				USA	
PLANTAGO L. Plantaginaceae					
aristata Michx.				NZ USA	ARG
asiatica L.			JPN		
bicallosa Decne.				BRA	

PLANTAGO L. Plantaginaceae *(Cont'd)*	S	P	C	X	F
camtschatica Cham.			JPN		
coronopus L.			SPA	GRE NZ	AFG AUS ISR POR
hirtella H. B. K.				NZ URU	
indica L.			POR	USA	POL
lagopus L.			EGY LEB MOR	GRE TUR	ISR POR
lanceolata L.	ITA	CAN ECU IRA MAU NZ	ARB ARG AUS DEN ENG ETH FIN FRA GER HAW IRQ JOR JPN LEB NET POR PR RHO SAF SOV SPA USA YUG	AST CHL CHN GRE HUN ICE TAI TUR URU	AFG ISR KOR POL
major L.		ECU SWE	ALK ARG AUS CAN CHN COL EGY ENG FIN GER HAW IND LEB POL PR SAF SOV SPA TAI USA	ARB BEL BOL CHL CR DEN DR FIJ FRA HK HUN ICE IDO JAM JPN MAL MAU NET NZ PER PLW TUR VEN VIE YUG	AFG ISR ITA
media L.			IRA SOV	GER USA	POL

PLANTAGO L. Plantaginaceae *(Cont'd)*	S	P	C	X	F
patagonica Jacq.			ARG		AUS
psyllium L.			MOR		AUS ISR POR
purshii Roem. & Schult.				USA	
pusilla Nutt.				USA	
rugelii Decne.			CAN USA		
tomentosa Lam.				BRA	
varia R. Br.			AUS		
virginica L.			HAW JPN	USA	
PLATANUS L. Platanaceae					
occidentalis L.				USA	
PLATYCAPNOS Bernh. Fumariaceae					
spicatus (L.) Bernh.			POR		
PLATYSTOMA Benth. & Hook. *f.* Lamiaceae *(Labiatae)*					
africanum Beauv.		GHA		NIG	
PLECTRANTHUS L'herit. Lamiaceae *(Labiatae)*					
flaccidus Guerke				TNZ	
PLECTRONIA L. Rubiaceae					
parviflora Bedd.				IND	
PLEIOBLASTUS Nakai Poaceae *(Gramineae)*					
variegata (Sieb.) Makino			JPN		
PLUCHEA Cass. Asteraceae *(Compositae)*					
indica (L.) Less.	HAW			VIE	IDO
lanceolata Oliv. & Hiern	IND				
odorata (L.) Cass.	HAW		PR	DR HON JAM SAL	
purpurascens (Sw.) Dc.			PR	JAM	
sagittalis (Lam.) Cabr.		BRA			
sericea (Nutt.) Cov.				USA	
PLUMBAGO L. Plumbaginaceae					
scandens L.				BRA	
zeylanica L.			HAW	GHA IDO	

POA L. Poaceae *(Gramineae)*	S	P	C	X	F
annua L.	AFG ARG ENG GER HON IRE IRQ ITA JPN NEP NOR POL SPA	ALK AUS BEL COL KOR NZ PLW POR SWE	AST BOR CAN CHN EGY FIN HAW HUN IDO JOR MOR NET PER RHO SAF SOV TAI TUN	BRA CHL CR DEN ECU FRA ICE IND IRA KEN MAL MAU MEX MOZ PHI PK SAL TNZ TUR UGA URU USA	CEY ISR TAS YUG
aquatica L.				AUS	
bulbosa L.			IRA	ITA SPA USA	AUS HAW IND ISR
compressa L.			SOV	NZ USA	AUS HAW
pratensis L.		JPN	FIN SOV SPA	ITA NZ USA	AUS HAW IND SAF
sphondylodes Trin.		JPN			
trivialis L.			FIN JPN SOV	CHL ITA NZ TUR	AFG AUS HAW IND ISR
POGONATHERUM Beauv. Poaceae *(Gramineae)*					
crinitum (Thunb.) Kunth			TAI	IND	
POINSETTIA R. Grah. Euphorbiaceae					
heterophylla (L.) Small *(See* **Euphorbia heterophylla** L.)			PR	DR MEX	
POLANISIA Raf. Capparidaceae					
chelidonii Dc. *(See* **Cleome chelidonii** L. *f.)*					IDO
graveolens Rafin.				USA	
icosandra (L.) Wight & Arn. *(See* **Cleome viscosa** L.)				PHI VIE	
trachysperma Torr. & Gray				USA	
viscosa Dc. *(See* **Cleome viscosa** L.)				IDO	

POLEMONIUM L. Polemoniaceae	S	P	C	X	F
micranthum Benth.				USA	
POLYCARPAEA Lam. Caryophyllaceae					
corymbosa Lam.				IND	
POLYCARPON Loefl. *ex* L. Caryophyllaceae					
indicum (Retz.) Merr.		IND			
prostratum (Forsk.) Pax				GHA NIG	
tetraphyllum (L.) L.	SAF			AUS CHL	IND POR
POLYCEPHALIUM Engl. Icacinaceae					
poggei Engl.				CNK	
POLYGALA L. Polygalaceae					
acuminata Willd.				PER	
adenophylla St. Hil.				BRA	
arenaria Willd.				GHA	
chinensis L.				IND	
erioptera Dc.				IND SUD	
glochidiata H. B. K.				MEX	
glomerata Lour.				VIE	
guineensis Willd.				GHA	
japonica Houtt.			JPN		
luteo-alba Gagnep.				VIE	
paniculata L.			IDO	FIJ NGI PLW	PR
POLYGONUM L. Polygonaceae					
achoreum Blake			CAN		
acre H. B. K.				BRA URU USA	
acuminatum H. B. K.			ARG	TRI	ISR
alatum D. Don (Syn. *P. capitatum* Meissn.)			IDO		
amphibium L.			FIN GER KEN POR	BEL ENG NET TUR USA	
aquisetiferme Sibth. & Sm.			EGY		
argyrocoleon Steud.				USA	AFG

POLYGONUM L. Polygonaceae *(Cont'd)*	S	P	C	X	F
aviculare L. (Syn. *P. heterophyllum* Lindm.)	ALK ENG HUN IRQ NZ POL SAF SPA	ARG AST AUS BEL BRA CHL FIN FRA GER IRA ITA JPN KEN NOR ROM SWE TUN	CAN COL ETH HON ICE ISR NET PER SOV USA YUG	AFG CHN EGY GRE IRE LEB MEX MOR NEP PK POR RHO SAL TUR URU	KOR
barbatum L.			IDO IND PHI TAI	AUS PK THI VIE	CAB
bellardii All.		AUS			
bistorta Garcke			GER	PK	AFG USA
bistortoides Pursh				USA	
blumei Meissn. *(See* **P. longisetum** De Bruyn)	PHI	JPN	TAI		
caespitosum Bl.			CHN JPN TAI		
capitatum Meissn. *(See* **P. alatum** D. Don)				NEP	
chinense L.		IND	TAI	JPN	IDO
cilinode Michx.			CAN	USA	
coccineum Muhl.			USA		
conspicuum (Nakai) Nakai			JPN		
convolvulus L.	ALK ARG ENG GER KEN SAF SOV SPA USA	AUS BRA BUL CAN CHL CZE FIN ITA NZ TNZ TUN	GRE IRA JPN MOR PHI POR ROM	CHN FRA GHA ICE IND IRE NOR POL SWE TUR YUG	AFG BEL NET PER
cuspidatum Sieb. & Zucc.				CHN NZ USA	JPN
densiflorum Bl.			HAW		
dumetorum L.			ARG SOV		POL

POLYGONUM L. Polygonaceae *(Cont'd)*	S	P	C	X	F
erecto-minus Makino			JPN		
erectum L.				USA	
flaccidum Meissn.			JPN		AFG
fugax Small			JPN		
glabrum Willd.			IND	FIJ PK SUD	
hastato-sagittatum Makino			JPN		
heterophyllum Lindm. *(See* **P. aviculare** L.*)*			JPN SOV		
higegaweri Steud.		JPN			
hydropiper L.	BRA IRQ POL TAI	FIN HUN JPN NZ SOV	AST AUS ENG GER HON IND MAL NET POR USA	AFG BEL BND BOR CAN CHN COL GRE IDO IRA IRE ITA JOR KOR MEX NEP PK SAL SWE TUR YUG	
hydropiperoides Michx.			COL	AUS BEL BND FIN USA	CHN HK
japonicum Meissn.			JPN TAI		
lapathifolium L. (Syn. *P. nodosum* Pers.) (Syn. *Persicaria nodosum* Pers.)	AFG CAN GER HUN KOR POL POR SOV YUG	AST ENG NOR NZ SWE TAI TUR	AUS CHN FIN IRQ JPN LEB NET SPA TUN	ARG BEL CHL CZE FRA GRE ICE IDO IRA ITA JOR PK THI USA	HK
linicola Sutulow			SOV		
longisetum De Bruyn (Syn. *P. blumei* Meissn.)		JPN		PHI TAI	

POLYGONUM L. Polygonaceae *(Cont'd)*	S	P	C	X	F
longisetum De Bruyn *(Cont'd)*				USA	
maackianum Regel			JPN		
maritima Vell.				TUR	
minus Huds.			AUS GER		POL
minutulum Makino (*See* **P. taquettii** Lev.)				JPN	
mite Schrank				GER	
neglectum Bess.			SOV		
nepalense Meissn.	ETH	CEY IND	JPN KEN TAI		
nipponense Makino			JPN		
nodosum Pers. (*See* **P. lapathifolium** L.)			SOV	CHN JPN	
orientale L.			JPN	AUS USA	IDO
pennsylvanicum L.	ALK			USA	
perfoliatum L.			JPN		
persicaria L.	AFG GER IRE ITA PHI	AST BEL CAN CHL ENG HUN JPN NOR NZ POL SWE	ARG AUS IRQ KEN POR SOV SPA TUN USA	BND CHN COL DEN EGY FRA ICE IRA MEX NEP NET PK TUR UGA YUG	
plebeium R. Br.		IND	EGY TAI	NEP PK	AUS VIE
pubescens Bl.			JPN		
punctatum Ell.		BRA	ARG	COL IND USA VEN	
sachalinense F. Schmidt			JPN	NZ USA	
sagittatum L. (*See* **P. sieboldii** Meissn.)			JPN	USA	
salicifolium Brouss.			EGY		
scabrum Moench (Syn. *P. tomentosum* Schrank *non* Willd.)		CAN	SOV	USA	ISR

POLYGONUM L. Polygonaceae *(Cont'd)*	S	P	C	X	F
scandens L.				CAN USA	
segetum H. B. K.		COL			
senegalense Meissn.				EGY	GHA ISR
senticosum (Meissn.) Fr. & Sav.		JPN			
serrulatum Lag.				EGY	POR
setaceum Baldw.				USA	
sieboldii Meissn. (Syn. *P. sagittatum* L.)					
taquettii Lev. (Syn. *P. minutulum* Makino)					
tataricum L. *(See* **Fagopyrum tataricum** (L.) Gaertn.)				SOV	
thunbergii Sieb. & Zucc.		JPN			
tomentosum Schrank *non* Willd. *(See* **P. scabrum** Moench)			THI		POL
viscosum Ham.			TAI		
POLYMERIA R. Br. **Convolvulaceae**					
longifolia Lindl.				AUS	
POLYPODIUM L. **Polypodiaceae**					
exiguum Hew. *(See* **Microsorium heterophyllum** (L.) A. D. Hawkes)				JAM	
phyllitidis L. *(See* **Campyloneurum phyllitidis** (L.) Presl)				JAM	
vulgare L.			SPA		
POLYPOGON Desf. **Poaceae** *(Gramineae)*					
australis Brongn.				CHL	
interruptus H. B. K.			HAW	USA	
monspeliensis (L.) Desf. (Syn. *Alopecurus monspeliensis* L.)	IND		EGY HAW IRQ JPN MOR POR SPA	AUS CHL IRA MAU NZ USA	BRA ISR SAF
POLYPREMUM L. **Loganiaceae**					
procumbens L.				USA	
POLYSTICHUM Roth **Dryopteridaceae**					
munitum (Kaulf.) Presl				USA	
POLYTRIAS Hack. **Poaceae** *(Gramineae)*					
amaura (Buse) O. Ktze.		IDO		THI	

POLYTRIAS Hack. Poaceae *(Gramineae)* *(Cont'd)*	S	P	C	X	F
amaura (Buse) O. Ktze. *(Cont'd)*				VIE	
POLYZYGUS Dalz. Apiaceae *(Umbelliferae)*					
tuberosus Dalz.				IND	
PONTEDERIA L. Pontederiaceae					
cordata L.				USA	
POPULUS L. Salicaceae					
alba L.				USA	
balsamifera L.				USA	
deltoides Marsh.				USA	
fremonti S. Wats.				USA	
grandidentata Michx.				USA	
heterophylla L.				USA	
nigra L.				USA	
sargentii Dode				USA	
tremuloides Michx.				USA	
trichocarpa Torr. & Gray				USA	
PORLIERIA Ruiz & Pav. Zygophyllaceae					
angustifolia (Engelm.) Gray				USA	
POROPHYLLUM Guett. Asteraceae *(Compositae)*					
ellipticum Cass. *(See* **P. ruderale** (Jacq.) Cass.*)*				PER TRI	
lanceolatum Dc.			ARG		
panctatum Blake				HON	
ruderale (Jacq.) Cass. (Syn. *P. ellipticum* Cass.)					
PORTULACA L. Portulacaceae					
cryptopetala Speg.				BRA	
formosana Poelln.		TAI			
grandiflora Hook.			ARG	BRA GHA	
lanceolata Engelm.			ARG		
oleracea L.	ARG AUS BRA COL EGY HAW ISR MEX NGI PHI	ANG BOR BUL CAN CNK GRE GUA IDO IND IRA	CHN IRQ MOR NIG NZ PLW	AFG ANT ARB BND CAB CHL DR ETH FIJ FRA	CR POL

PORTULACA L. Portulacaceae *(Cont'd)*	S	P	C	X	F
oleracea L. *(Cont'd)*	SAF SPA TAI THI TNZ TUR ZAM	JAM JPN KEN KOR LEB MOZ NIC PAR PER POR PR RHO SAL SOV SUD TRI UGA USA VEN YUG		GER GHA HK HON HUN ITA IVO JOR MAL MAU MIC PK SUR URU VIE	
pilosa L.		TAI THI		ARG DR TRI	PR VEN
quadrifida L.	GHA		PR TAI	FIJ IND IVO KEN MOZ THI TRI VEN	IDO SUD
POTAMOGETON L. Potamogetonaceae					
amplifolius Tuckerm.				USA	
crispus L.		AUS EGY IRQ NZ PK SUD	IND IRA POR USA	AST BEL BOT ENG GER GRE HUN ISR JOR JPN KOR NET POL RHO SWE THI	
cristatus Regel & Maack.			JPN		
distintus A. Benn.	JPN	KOR			
diversifolius Rafin.		USA			
epihydrum Rafin.				USA	
filiformis Pers.				ENG USA	
fluitans Roth				DR	
foliosus Raf.		USA			

POTAMOGETON L. Potamogetonaceae *(Cont'd)*	S	P	C	X	F
illinoensis Morong			USA		
indicus Roxb.				IND	
lucens L.		SPA		ENG NET	ISR
natans L.		EGY GER HUN IRQ NOR SPA SWE TUR	POR TAI	AFG AST BEL CAN CHN CZE ENG FIN GRE IND ITA JOR JPN KOR LEB MAU MEL MEX NET PK POL USA	
nodosus Poir.		EGY SUD USA		AFG AST CAN ENG FRA HAW IND IRA ISR ITA JAM JOR PK POL	
ochreatus Raoul		IND USA	SPA	AUS BEL ENG GER NZ SUD TUR	ISR KOR
pectinatus L.	IND USA	EGY ENG TUR		AFG AST AUS BEL CAN CNK GRE HON HUN IRA ISR ITA JOR JPN KEN MAU	

POTAMOGETON L. Potamogetonaceae *(Cont'd)*	S	P	C	X	F
pectinatus L. *(Cont'd)*				NET NOR PER PK POL SAF SPA SUD SWE TNZ UGA	
perfoliatus L.	IND	AUS NOR SUD SWE		AFG BEL CAN EGY GER GRE HUN IRA ISR JOR JPN NET PK POL TUR	KOR USA
praelongus Wulf.				USA	
pusillus L.			ARG USA		
richardsonii (A. Benn.) Rydb.			USA		
robinsii Oakes				USA	
schweinfurthii Arth.		TNZ			
striatus Ruiz & Pav.			ARG		
strictifolius A. Benn.				USA	
tricarinatus F. Muell. & A. Benn.		AUS			
vaginatus Turc.				USA	
zosteriformis Fern.				USA	
POTENTILLA L. Rosaceae					
anserina L.			FIN GER SOV	CHL USA	AUS POL
argentea L.			CAN SOV	NZ USA	
arguta Pursh				USA	
bifurca L.				SOV	
canadensis L.			SOV USA		
canescens Bess.				TUR	
chinensis Ser.			JPN		

POTENTILLA L. Rosaceae *(Cont'd)*	S	P	C	X	F
diversifolia Lehm.				USA	
fruticosa L.				USA	
intermedia L.			SOV	USA	
kleiniana Wight & Arn.			CHN JPN		
norvegica L.			CAN FIN SOV USA	NZ	
recta L.			CAN USA	NZ	
reptans L.			IRA JPN LEB SPA	CHL ITA NZ TUR USA	AFG AUS
simplex Michx.				USA	
supina L.			EGY SOV	IND PK	
tormentilla Neck.				ITA	
POTERIUM L. Rosaceae					
magnolii Spach			POR		
polygamum Waldst.				AUS	
sanguisorba L.			SOV	ITA NZ	
spinosum L.			LEB		ISR PK
villosum Sibth. & Sm.			IRA		
POTHOMORPHE Miq. Piperaceae					
dombeyana Miq.				PER	
peltata (L.) Miq.		PR		COL DR PER	
POUZOLZIA Gaudich. Urticaceae					
guineensis Benth.				GHA NIG	
indica (L.) Gaudich. *(See* **P. zeylanica** (L.) Benn.)				IND	
zeylanica (L.) Benn. (Syn. *P. indica* (L.) Gaudich.)				PHI	
PRATIA Gaudich. Campanulaceae					
concolor (R. Br.) Druce			AUS		
nummularia (Lam.) A. Br. & Aschers. *(See* **Lobelia angulata** Forst. *f.)*			IDO TAI		

PRESTONIA R. Br. **Apocynaceae**	S	P	C	X	F
acutifolia (Benth.) Schum.				PER	
quinquangularis (Jacq.) Spreng.			TRI	ANT	
PRIONITIS Adans. **Apiaceae** *(Umbelliferae)*					
falcaria Dum. *(See* **Falcaria vulgaris** Bernh.*)*				SOV	
PRIVA Adans. **Verbenaceae**					
cordifolia (L. *f.*) Druce				IND	
laevis Juss.				CHL	
lappulacea (L.) Pers.			COL TRI	HON JAM PAN SAL	
leptostachya Juss.				IND	
PROBOSCIDEA Schmid. **Martyniaceae**					
jussieui Steud.			AUS		
louisianica (Mill.) Thell.			AUS	NZ USA	MEX
lutea Stapf				ARG	
PROSOPIS L. **Mimosaceae** *(Leguminosae)*					
campestris Griseb.		PAR			
chilensis (Molina) Stuntz			HAW		
farcta Macbride			ISR		
glandulosa Torr.				AUS PER	
humilis Gill. *ex* Hook.				ARG	
juliflora (Sw.) Dc.		MEX	USA	AUS DR IND IRQ VEN	
pallida H. B. K.				AUS	
pubescens Benth.				USA	
ruscifolia Griseb.		PAR	ARG		
spicigera L.				PK	
stephaniana Kunth *ex* Spreng.		TUR	IRA	SOV	
PRUNELLA L. **Lamiaceae** *(Labiatae)*					
vulgaris L.			CAN CHN ENG FIN GER HAW JPN	AUS CHL ITA NZ	

PRUNELLA L. Lamiaceae *(Labiatae) (Cont'd)*	S	P	C	X	F
vulgaris L. *(Cont'd)*			SOV USA		
PRUNUS L. Rosaceae					
americana Marsh.			USA		
angustifolia Marsh.			USA		
avium (L.) L.			USA		
cerasus L.			USA		
emarginata Dougl.			USA		
nigra Ait.			USA		
pennsylvanica L. *f.*			USA	CHL HAW	
serotina Ehrh.			USA		
umbellata Ell.			USA		
virginiana L.			USA		
PSEUDECHINOLAENA Stapf Poaceae *(Gramineae)*					
polystachya (H.B.K.) Stapf (Syn. *Echinochloa polystachya* (H. B. K.) Hitchc.)	MEX			CNK GHA	CEY IND SAF
PSEUDELEPHANTOPUS Rohr Asteraceae *(Compositae)*					
funckii (Turcz.) Cabr.				VEN	
spicatus (Juss. *ex* Aubl.) C. F. Baker (Syn. *Elephantopus spicatus* Aubl.)	FIJ PLW	COL HON IDO	GUA HAW PHI PR	CAB DR GHA JAM MEL NIC PAN PER SAL TRI USA	CR
PSEUDOTSUGA Carr. Pinaceae					
menziesii (Mirb.) Franco				USA	
PSIADIA Jacq. Asteraceae *(Compositae)*					
arabica Jaub. & Spach			KEN		
PSIDIUM L. Myrtaceae					
cattleianum Sabine			HAW	USA	
guajava L.	FIJ HAW MEL	AUS MEX PLW TNZ	JAM PR TRI	ANT CHN CNK CR DAH DR HON IND JOR	CAB HK JPN USA

PSIDIUM L. Myrtaceae *(Cont'd)*	S	P	C	X	F
guajava L. *(Cont'd)*				NEP	
				NGI	
				NIC	
				PER	
				RHO	
				SEG	
				SUD	
				THI	
				VEN	
PSORALEA L. Papilionaceae *(Leguminosae)*					
americana L.			MOR		
bituminosa L.			SPA		
corylifolia L.		IND			
lanceolata Pursh				USA	
tenuiflora Pursh				USA	
PSYCHINE Desf. Brassicaceae *(Cruciferae)*					
stylosa Desf.				MOR	
PSYCHOTRIA L. Rubiaceae *(Syn. CEPHAELIS Sw.)*					
alba Ruiz & Pav.				PER	
ruelliaefolia Muell.-Arg. (Syn. *Cephaelis ruelliaefolia* Cham. & Schlecht.)					
PTELEA L. Rutaceae					
trifoliata L.				USA	
PTERIDIUM Scop. Dennstaedtiaceae					
aquilinum (L.) Kuhn (Syn. *P. lanuginosum* Clute) (Syn. *P. latiusculum* (Desv.) Maxon) (Syn. *Pteris aquilina* L.)	ENG JPN NZ PHI TAI	AUS BRA CAN COL IND MAD SWE UGA	AFG AST CR FIN GRE HAW HON HUN IDO IRQ ITA KEN NOR POL SPA USA	BEL BUL CEY CHN CNK ETH FIJ FRA GER GHA HK IRA ISR MAU NET PER PK PLW PR RHO SAF SAL SOV THI TUR YUG ZAM	
caudatum (L.) Maxon			PR		

PTERIDIUM Scop. Dennstaedtiaceae *(Cont'd)*	S	P	C	X	F
esculentum (Forst.) Nakai		FIJ NZ	AUS		
lanuginosum Clute *(See* **P. aquilinum** (L.) Kuhn)				SOV	
latiusculum (Desv.) Maxon *(See* **P. aquilinum** (L.) Kuhn)				USA	
revolutum (Bl.) Nakai				AUS	
PTERIS L. **Pteridaceae**					
aquilina L. *(See* **Pteridium aquilinum** (L.) Kuhn)				ITA	
ensiformis Burm.			IDO TAI		
vittata L.				GHA MAL	IDO
PTEROCAULON Ell. **Asteraceae** *(Compositae)*					
rugosum (Vahl) Malme		BRA			
sphacelatum (Labill.) B. & H. *ex* Bailey			AUS		
virgatum (L.) Dc.			PR		
PTEROGYNE Tul. **Caesalpiniaceae** *(Leguminosae)*					
nitens Tul.				BRA	
PTEROLEPIS Miq. **Melastomataceae**					
glomeratum Miq.			TRI		
PUCCINELLIA Parl. **Poaceae** *(Gramineae)*					
airoides (Nutt.) Wats. & Coult.				USA	
distans (L.) Parl.				USA	
lemmoni (Vasey) Scribn.				USA	
stricta Keng				CHN	
PUERARIA Dc. **Papilionaceae** *(Leguminosae)*					
javanica Benth. *(See* **P. phaseoloides** (Roxb.) Benth.)				PHI	
lobata (Willd.) Ohwi *(See* **P. triloba** (Lour.) Makino)			JPN USA	PLW	
phaseoloides (Roxb.) Benth. (Syn. *P. javanica* Benth.)		MAL	IDO PR		AUS IND
thomsoni Benth.				VIE	
thunbergiana (Sieb. & Zucc.) Benth. *(See* **P. triloba** (Lour.) Makino)			JPN		AUS CHL
triloba (Lour.) Makino (Syn. *P. lobata* (Willd.) Ohwi) (Syn. *P. thunbergiana* (Sieb. & Zucc.) Benth.)					

PULICARIA Gaertn. Asteraceae *(Compositae)*	S	P	C	X	F
arabica (L.) Cass.			EGY		
crispa Sch.-Bip.		SUD		GHA IND	ISR
dysenterica (L.) Gaertn.				TUR	
prostrata (Gilib.) Aschers.			SOV		
PUPALIA Juss. **Amaranthaceae**					
atropurpurea Moq.				IND	
lappacea (L.) Juss.		GHA	IND SAF		
PURSHIA Dc. **Rosaceae**					
tridentata (Pursh) Dc.		GHA		USA	
PYCREUS Beauv. **Cyperaceae**					
albo-marginatus Nees (Syn. *Cyperus albo-marginatus* Mart. & Schrad. *ex* Nees)					
eragrostis (Vahl) Palla *(See* **Cyperus sanguinolentus** Vahl)			JPN PHI		
ferrugineus C. B. Clarke				MAU	
flavescens Beauv. *ex* Reichb. (Syn. *Cyperus flavescens* L.)					
lanceolatus C. B. Clarke	NIG				
mundtii Nees		MAD			CUB RHO TNZ
nitens (Retz.) Nees			PHI		IND
odoratus Urb. *(See* **Cyperus polystachyos** Rottb.)			MAU		
polystachyos (Rottb.) Beauv. *(See* **Cyperus polystachyos** Rottb.)	TAI		JPN	FIJ	IND
propinquus Nees (Syn. *Cyperus fasciculatus* Ell.)					
sanguinolentus (Vahl) Nees *(See* **Cyperus sanguinolentus** Vahl)		IND	TAI	VIE	
tremulus (Poir.) C. B. Clarke		MAD			MAU
PYROSTEGIA Presl **Bignoniaceae**					
ignea Presl				BRA	
venusta (Ker-Gawl.) Miers				PER	
PYRRHOPAPPUS Dc. **Asteraceae** *(Compositae)*					
carolinianus (Walt.) Dc.				USA	
PYRUS L. **Rosaceae**					
angustifolia Ait.				USA	

PYRUS L. Rosaceae *(Cont'd)*	S	P	C	X	F
arbutifolia (L.) L. *f.*				USA	
coronaria L.				USA	
loensis (Wood) Bailey				USA	
malus L.				USA	
melanocarpa (Michx.) Willd.				USA	
QUAMOCLIT Moench **Convolvulaceae**					
coccinea (L.) Moench				AUS BRA CR FIJ HON	IDO
pinnata (Desv.) Boj. *(See* **Ipomoea quamoclit** L.*)*		PR		AUS FIJ IND	IDO
QUERCUS L. **Fagaceae**					
agrifolia Nee				USA	
alba L.				USA	
arizonica Sarg.				USA	
bicolor Willd.				USA	
cerris L.				TUR	
chrysolepis Liebm.				USA	
coccinea Muenchh.				USA	
douglasii Hook. & Arn.				USA	
dumosa Nutt.				USA	
durandii Buckl.				USA	
durata Jeps.				USA	
ellipsoidalis E. J. Hill				USA	
emoryi Torr.				USA	
falcata Michx.				USA	
gambelii Nutt.				USA	
garryana Dougl.				USA	
havardi Rydb.				USA	
hypoleucoides A. Camus				USA	
ilicifolia Wangenh.				USA	
imbricaria Michx.				USA	
incana Bartr.				USA	
kelloggii Newb.				USA	
laevis Walt.				USA	

QUERCUS L. Fagaceae *(Cont'd)*	S	P	C	X	F
laurifolia Michx.				USA	
lobata Nee				USA	
lyrata Walt.				USA	
macrocarpa Michx.				USA	
marylandica Muenchh.				USA	
michauxii Nutt.				USA	
muehlenbergii Engelm.				USA	
myrtifolia Willd.				USA	
nigra L.				USA	
oblongifolia Torr.				USA	
palustris Muenchh.				USA	
phellos L.				USA	
prinoides Willd.				USA	
prinus L.				USA	
pubescens Willd.				TUR	
pungens Liebm.				USA	
rubra L.				USA	
sessiliflora Salisb.				TUR	
shumardii Buckl.				USA	
stellata Wangenh.				USA	
turbinella Greene				USA	
vaccinifolia Kellogg				USA	
velutina Lam.				USA	
virginiana Mill.				USA	
wislizenii A. Dc.				USA	
RAMPHICARPA Reichb. **Scrophulariaceae**					
fistulosa Benth. *(See R. longiflora Benth.)*				NIG	
longiflora Benth. (Syn. *R. fistulosa* Benth.)					
RANDIA L. **Rubiaceae**					
mitis L.			PR		
RANUNCULUS L. **Ranunculaceae** *(Syn.CERATOCEPHALUS Moench)*					
abortivus L.				USA	
acer L.		BEL	GER SOV	ITA	POL

RANUNCULUS L. Ranunculaceae *(Cont'd)*	S	P	C	X	F
acris L.		NOR NZ SWE	CAN CHN ENG FIN SPA USA	IRA MEX	
alismaefolius Geyer *ex* Benth.				USA	
aquatilis L.				BEL ENG GER TUR	
arvensis L.	AFG POL	KOR SWE	ENG GER HUN IRA LEB MOR POR TUN	AST BEL ECU EGY FRA GRE IRQ ISR ITA JOR JPN KEN NEP PK SOV SPA TUR USA	AUS
auricomus L.			FIN		
bulbosus L.			ENG GER SPA	ITA NZ USA	POL
calthaefolius (Guss.) Jord.			JOR		
cantoniensis Dc.			JPN TAI		
circinatus Sibth.				BEL GER	
falcatus L. (Syn. *Ceratocephalus falcatus* Pers.)					
ficaria L.			GER	BEL ITA NZ	
flabellaris Rafin.				USA	
flammula L.			GER		
fluitans Lam.				BEL	
japonicus Thunb.			CHN JPN TAI		
lomatocarpus Fisch. & Mey.			LEB		AFG ISR
muricatus L.			AUS JPN	CHL PK	AFG ISR

RANUNCULUS L. Ranunculaceae *(Cont'd)*	S	P	C	X	F
muricatus L. *(Cont'd)*			POR	USA	
ophioglossifolius Vill.			POR		AUS ISR
quelpaertensis (Lev.) Nakai			JPN		
repens L.	FIN ITA POL	BEL ENG NOR SWE	BOR CAN GER HUN IRA NET POR SOV SPA TUN USA	ARG AST BRA CHL CHN FRA GRE ICE JPN NEP NZ TUR	AUS
sardous Crantz			POR	NZ	POL
sceleratus L.			AUS GER JPN SOV SPA	BEL CHN IND NZ TUR USA	ISR POR
ternatus Thunb.			CHN JPN		
testiculatus Crantz				USA	
trichophyllus Chaix				ENG USA	
RAPHANUS L. **Brassicaceae** *(Cruciferae)*					
arvense Wallr. *(See* **R. raphanistrum** L.)				SOV	
microcarpus Willk.			POR		ARG
raphanistrum L. (Syn. *R. arvense* Wallr.) (Syn. *R. sylvestris* Aschers.)	AFG AUS ENG HUN ISR MEX MOZ POL SAF	ARG BEL BRA COL FIN GER ITA KEN PER POR SOV SPA SWE TUN	BOR CAN HON IRA JOR LEB MOR USA	BUL CHL CZE DEN EGY ETH FRA GRE ICE IRQ NET NOR NZ RHO TAS TUR URU YUG	CHN
sativus L.		ARG	CAN FIN HAW SPA	BRA CHL NZ USA	AFG AUS ISR

	S	P	C	X	F
RAPHANUS L. **Brassicaceae** *(Cruciferae)* *(Cont'd)*					
sylvestris Aschers. (See **R. raphanistrum** L.)			SOV		
RAPISTRUM Crantz **Brassicaceae** *(Cruciferae)*					
hispanicum (L.) Crantz				MOR	
orientale Dc.				GRE MOR	
rugosum (L.) All.	ISR POL SOV	ARG AUS BRA IRA SPA	ENG LEB MOR POR TUN	CHL EGY IRQ ITA JOR NZ RHO TUR URU ZAM	
RATIBIDA Rafin. **Asteraceae** *(Compositae)*					
columnifera (Nutt.) Woot. & Standl.				USA	
pinnata (Vent.) Barnh.				USA	
RAULINOREITZIA R. M. King & H. Robinson **Asteraceae** *(Compositae)*					
tremula (Hook. & Arn.) R. M. King & H. Robinson (Syn. *Eupatorium tremulum* Hook. & Arn.)					
REGNELLIDIUM Lindm. **Marsileaceae**					
diphyllum Lindm.		BRA			
REICHARDIA Roth **Asteraceae** *(Compositae)*					
intermedia (Fan. *ex* Dc.) Dinsm.			POR		ISR
picroides (L.) Roth			HAW		
RENEALMIA L. *f.* **Zingiberaceae**					
antillarum Gagnep.				DR	PR
RESEDA L. **Resedaceae**					
alba L.		JOR		GRE MOR NZ USA	AUS ISR
lutea L.		AUS SAF	IRA LEB POR SOV SPA	CHL MOR NZ TUR USA	AFG ISR
luteola L.				MOR SPA	
orientalis Boiss.				TUR	
phyteuma L.				SPA	
RHAGADIOLUS Scop. **Asteraceae** *(Compositae)*					
cathartica L.			CAN		

RHAGADIOLUS Scop. Asteraceae *(Compositae)* *(Cont'd)*	S	P	C	X	F
edulis Gaertn.				MOR	
stellatus Dc.				LEB MOR SPA	ISR POR
RHAMNUS L. **Rhamnaceae**					
californica Eschsch.				USA	
caroliniana Walt.				USA	
cathartica L.				CAN USA	
crocea Nutt.				USA	
frangula L.				CAN USA	
purshiana Dc.				USA	CHL
RHINANTHUS L. **Scrophulariaceae**					
apterus Ostenf. (Syn. *Alectorolophus apterus* (Fr.) Ostenf.)			SOV		
crista-galli L.			SOV	USA	
glaber Lam.				BEL	
major Ehrh. (Syn. *Alectorolophus major* Reichb.)				SOV	
serotinus Oborny.				GER	
RHIZOPHORA L. **Rhizophoraceae**					
mangle L.			HAW	NIC USA	
RHODODENDRON L. **Ericaceae**					
canadense (L.) Torr.				USA	
canescens (Michx.) Sweet				USA	
macrophyllum D. Don				USA	
maximum L.				USA	
occidentale (Torr. & Gray) Gray				USA	
ponticum L.				TUR	
RHODOMYRTUS Reichb. **Myrtaceae**					
macrocarpa Benth.			AUS		
tomentosa (W. Ait.) Hassk.			HAW MAL	TAI	USA
RHODOSTACHYS Phil. **Bromeliaceae**					
bicolor Benth. & Hook. (Syn. *Bromelia bicolor* Ruiz & Pav.)					

RHUS L. Anacardiaceae	S	P	C	X	F
aromatica Ait.				USA	
copallina L.				USA	
diversiloba Torr. & Gray			USA	CAN	
glabra L.			USA		
lanceolata (A. Gray) Britt.				USA	
laurina Nutt.				USA	
microphylla Engelm.				USA	
ovata S. Wats.				USA	
pyroides Burch.				RHO	
radicans L.			CAN USA		CHL
toxicodendron L.			USA	CAN	AUS CHL
trilobata Nutt.				USA	
typhina L.				CAN USA	
virens Lindl.				USA	
RHYNCHANTHERA Dc. Melastomataceae					
grandiflora (Aubl.) Dc.				BRA	
RHYNCHELYTRUM Nees Poaceae *(Gramineae)*					
repens (Willd.) C. E. Hubb. (Syn. *R. roseum* (Nees) Stapf & C. E. Hubb.) (Syn. *Tricholaena repens* (Willd.) Hitchc.) (Syn. *T. rosea* Nees)	AUS BRA GHA	MAL ZAM	ARG CNK COL HAW KEN PER PHI SUD TNZ	ANG BOT CAM DAH FIJ GUI HON IDO IVO JAM MAU MEX MOZ NIG PLW RHO SEG THI UGA USA VOL	IND SAF
roseum (Nees) Stapf & C. E. Hubb. (*See* **R. repens** (Willd.) C. E. Hubb.)		BRA MEX	IDO	IND NGI PER	ARG HAW PHI SAF
RHYNCHOCORYS Griseb. Scrophulariaceae					
elephas (L.) Griseb.			IRA		

RHYNCHOSIA Lour. Papilionaceae *(Leguminosae)* (Syn. *DOLICHOLUS* Medic.)	S	P	C	X	F
memnonia (Del.) Dc.			SUD	ANG	
minima (L.) Dc. (Syn. *Dolicholus minimus* (L.) Medic.)			HAW IND TAI	ANT PK SAL	IDO ISR
sublobata (Schumach.) Meikle				GHA	
RHYNCHOSPORA Vahl **Cyperaceae**					
aurea Vahl (*See* **R. corymbosa** (L.) Britt.)			BOR	VIE	
corymbosa (L.) Britt. (Syn. *R. aurea* Vahl)					
rubra (Lour.) Makino				THI	
tenuis Link		ARG BRA			
RIBES L. **Saxifragaceae**					
americanum Mill.				USA	
binominatum Heller				USA	
bracteosum Dougl.				USA	
californicum Hook. & Arn.				USA	
cereum Dougl.				USA	
cynosbati L.				USA	
glandulosum Grauer				USA	
glutinosum Benth.				USA	
hirtellum Michx.				USA	
inerme Rydb.				USA	
lacustre Poir.				USA	
laxiflorum Pursh				USA	
lobbii A. Gray				USA	
marshallii Greene				USA	
menziesii Pursh				USA	
missouriense Nutt.				USA	
montigenum Mcclatchie				USA	
nevadense Kellogg				USA	
oxyacanthoides L.				USA	
petiolare Dougl.				USA	
roezli Regel				USA	
sanguineum Pursh				USA	CHL
speciosum Pursh				USA	

RIBES L. Saxifragaceae *(Cont'd)*	S	P	C	X	F
triste Pall.				USA	
tularense Standl.				USA	
velutinum Greene				USA	
viscosissimum Pursh				USA	
RICCIA Mich. **Ricciaceae** *(See* **RICCIOCARPUS** Corda*)*					
natans L. *(See* **Ricciocarpus natans** (L.) Corda*)*					AUS
RICCIOCARPUS Corda **Ricciaceae** *(Syn. RICCIA* Mich.*)*					
natans (L.) Corda (Syn. *Riccia natans* L.)		JPN		IDO	
RICHARDIA L. **Rubiaceae**					
brasiliensis Gomez		BRA RHO SWZ	ARG HAW IDO SAF	USA	KEN
scabra L.	HAW		SAL USA	AUS HON MEX	
RICINUS L. **Euphorbiaceae**					
communis L.		TNZ	AFG AUS BRA CNK HAW JAM KEN RHO SAF	ARG BOT CAB CHL CHN COL CR DR ECU FIJ FRA GHA HON IDO IRA JOR JPN KOR MEL MIC MOZ NEP NZ PER PLW POL PR SAL SEG THI UGA URU USA VEN YUG ZAM	HK ISR

RIDOLFIA Moris Apiaceae *(Umbelliferae)*	S	P	C	X	F
segetum (L.) Moris	TUN	LEB	ISR MOR POR		
RIVINIA L. **Phytolaccaceae**					
humilis L.	MEL		AUS	COL FIJ HON JAM NIC	BRA
ROBINIA L. **Papilionaceae** *(Leguminosae)*					
pseudoacacial				NZ USA	AFG AUS CHL ISR
ROCHEFORTIA Sw. **Boraginaceae**					
spinosa Urb.				VEN	
ROCHELIA Reichb. **Boraginaceae**					
disperma Hochr.				IRA	ISR
ROEMERIA Medic. **Papaveraceae**					
hybrida Dc.				SPA	MOR TUR
refracta Dc.				USA	AFG
rhoeadiflora Bois.				IRA	
ROLANDRA Rottb. **Asteraceae** *(Compositae)*					
fruticosa (L.) O. Ktze.				TRI	
ROMULEA Maratti **Iridaceae**					
rosea (L.) Eckl.			AUS	NZ	
RORIPPA Scop. **Brassicaceae** *(Cruciferae)*					
amphibia (L.) Besser				BEL GER	
aquatica (Eaton) Palmer & Steyermark				PER	
atrovirens (Hornem.) Ohwi & Hara *(See* **R. indica** (L.) Hiern)		JPN TAI			
austriaca (Crantz) Bess.				USA	
cantoniensis (Lour.) Ohwi			JPN		
dubia (Pers.) Hara (Syn. *Nasturtium sublyratum* (Miq.) Fr. & Sav.)			JPN		
hilariana (Walpers) Cabrera				ARG	
indica (L.) Hiern (Syn. *R. atrovirens* (Hornem.) Ohwi & Hara) (Syn. *Nasturtium indicum* (L.) Dc.)			JPN		
islandica (Oeder) Borb. (Syn. *R. palustris* (L.) Bess.)			FIN JPN	CAN NZ	ARG AUS
(Syn. *Nasturtium palustre* (Pollich) Dc.)				USA	

RORIPPA Scop. Brassicaceae *(Cruciferae) (Cont'd)*	S	P	C	X	F
microsperma (Dc.) L. H. Bailey				CHN	
montana Small				CHN	
palustris (L.) Bess. *(See* R. islandica (Oeder) Borb.*)*			SOV	JPN	
sylvestris (L.) Bess.				CAN USA	
ROSA L. **Rosaceae**					
arkansana Porter				USA	
bracteata Wendl.				USA	
californica Cham. & Schlecht.				USA	
canina L.				TUR	
eglanteria L.				USA	
gymnocarpa Nutt.				USA	
laevigata Michx.				USA	
multiflora Thunb. *ex* Murr.			USA	NZ	
nutkana Presl				USA	
rubiginosa L.		AUS NZ		TUR	
sulphurea Ait.				TUR	
tomentosa Sm.				TUR	
woodsii Lindl.				USA	
ROTALA L. **Lythraceae**					
densiflora (Roth) Koehne				IND	
indica (Willd.) Koehne		AFG JPN KOR PHI TAI	VIE	CAB CEY CHN IDO ITA PK THI	
leptopetala (Bl.) Koehne			IDO TAI	CEY	AFG
mexicana Cham. & Schlecht.			JPN		
ramosior (L.) Koehne				TRI USA	
rotundifolia Koehne			TAI		
uliginosa Miq.			KOR		
ROTHIA Pers. **Papilionaceae** *(Leguminosae)*					
trifoliatia (Roxb.) Pers.				IND	

ROTTBOELLIA L. *f.* Poaceae *(Gramineae)*	S	P	C	X	F
compressa L. *f.* (*See* **Hemarthria compressa** (L. F.) R. Br.)			TAI		AUS
exaltata L. *f.*	GHA JAM MOZ PHI RHO TNZ TRI ZAM	CUB KEN MAD SUD USA VEN	IND	AUS BOL BRA CHN COL DR ETH GUI IDO IVO MEL NIG PAN PER SEG THI UGA	BUR HK MAL SAF
ROUREA Aubl. **Connaraceae**					
surinamensis Miq.		PR			
ROYENA L. **Ebenaceae**					
sericea Bernh.		RHO			
RUBIA L. **Rubiaceae**					
cordifolia L.			IDO	JPN	AFG
RUBUS L. **Rosaceae**					
alceaefolius Poir.		AUS			
allegheniensis Porter			USA		
annamensis Cardot			AUS		
arcticus L.			FIN		
brasiliensis Mart.				BRA	
caesius L.		YUG	ARG	NZ	AFG
chevalieri Toussaint				VIE	
cuneifolius Pursh		SAF			
flagellaris Willd.				USA	
fruticosus L.		AUS NZ		TUR USA	AFG
idaeus L.			FIN	TUR USA	AUS
laciniatus Willd.				AUS NZ USA	
lasiocarpus Sm.				VIE	
leucodermis Dougl.				USA	
moluccanus L.		HAW MAD	AUS	FIJ IND	

RUBUS L. Rosaceae *(Cont'd)*	S	P	C	X	F
moluccanus L. *(Cont'd)*		MAU			
occidentalis L.				USA	
palmatus Thunb.				JPN	
parviflorus L.			AUS	USA	
penetrans L. H. Bailey		HAW		USA	
procerus P. J. Muell.				USA	
roridus Lindl.				MAU	
rosaefolius Sm.			AUS HAW PR	BRA MAU	
spectabilis Pursh				USA	
taiwanianus Matsum.			TAI		
trivialis Michx.				USA	
ulmifolius Schott	CHL				AFG
villosus Ait.			AUS		
vitifolius Cham. & Schlecht.				USA	
vulgaris Weihe & Nees			AUS		
RUDBECKIA L. Asteraceae *(Compositae)*					
amplexicaulis Vahl				USA	
hirta L.				USA	
laciniata L.				JPN	
occidentalis Nutt.				USA	
serotina Nutt.				USA	
triloba L.				USA	
RUELLIA L. Acanthaceae					
coccinea (L.) Vahl			PR		
graecizans Back. (*See* **Styphanophysum longifolium** Poir.)			HAW		
patula Jacq.				IND	
prostrata Poir.				IND PLW	IDO
tuberosa L.	THI		IND JAM MAU TRI	PLW VEN VIE	CAB IDO PR
tweediana Griseb.			PR		
RUGELIA Shuttlew. *ex* Chapm. **Asteraceae** *(Compositae)*					
repens Nees		IND			

RUMEX L. Polygonaceae	S	P	C	X	F
abyssinicus Jacq.			ETH	KEN	
acetosa L.		FIN		AUS CHN ENG JPN SPA TAI	CAN GER ICE NZ USA	AFG IND POL SOV
acetosella L.		AFG BRA HUN ITA PHI POL	AUS BEL CAN CHL COL ECU FIN IDO KOR NOR NZ SOV SWE USA	AST ENG GER HAW ICE IND JPN KEN MEX PER SPA TUN	ARG CHN CR DEN GRE IRA JOR MEL NET PK RHO SAF SAL TUR VEN YUG	
acutus L.			IND		PK	
alpinus L.					GER	
altissimus Wood					USA	
angiocarpus Murb.		SAF		KEN POR RHO		
bequaerti De Wild.			ETH	KEN		
brownii Campd.				AUS		NZ
bucephalophorus L.		POR	ISR		MOR	
conglomeratus Murr.				ARG AUS ENG JPN POR	GER PER TUR USA	AFG ISR NZ.
crispus L.		MEX	ARG BEL BRA COL ECU FRA SOV TUR USA	AUS CAN DEN ENG FIN GER IRA IRQ JPN KEN NOR POR SPA SWE	CHL CHN CR CZE GRE GUA HUN ICE IRE ITA MAU NZ PER TAI TUN URU VEN	AFG KOR MIC POL YUG
cuneifolius Campd.				ARG		SAF
dentatus L.				EGY	IND	AFG

RUMEX L. Polygonaceae *(Cont'd)*	S	P	C	X	F
dentatus L. *(Cont'd)*			IRQ	PK	ISR
domesticus Hartm.		NOR		CAN FIN ICE	
hastatulus Baldw.				USA	
hastatus D. Don				PK	AFG USA
hydrolapathum Huds.				BEL ENG	
hymenosepalus Torr.				USA	
japonicus Meissn.		JPN			
longifolius Dc.			CAN		
mexicanus Meissn.				USA	
obtusifolius L.		AUS BEL NZ	ARG CAN ENG JPN SPA	BND BRA CHL FIN GER ICE USA VEN	COL POL SOV
occidentalis S. Wats.				USA	
paraguayensis Parodi			ARG		
patientia L.			HAW	CAN DR USA VEN	AFG
pseudonatronatus Borb.			CAN		
pulcher L.	MEX	TUN	ARG MOR POR SPA	CHL GRE NZ USA	AUS ISR
rugosus Campd.				CAN	
salicifolius Weinm.				USA	
sanguineus L.				GRE	
stenophyllus Ledeb.			CAN		
thyrsiflorus Fingerh.				CAN	
RUMFORDIA Dc. Asteraceae *(Compositae)*					
media Blake				MEX	
RUNGIA Nees Acanthaceae					
repens Nees		IND			
RUPPIA L. Ruppiaceae					
maritima L.				USA	

RUSCUS L. Liliaceae	S	P	C	X	F
aculeatus L.				TUR	
RYTIDOPHYLLUM Mart. **Gesneriaceae**					
tomentosum Mart.				JAM	
SABAL Adans. **Arecaceae** *(Palmae)*					
mexicanum Mart.			MEX		
minor (Jacq.) Pers.				USA	
palmetto Lodd. *ex* Schult. *f.*				USA	
SACCHARUM L. **Poaceae** *(Gramineae)*					
arundinaceum Retz.				IND VIE	
benghalense Retz.		BND			HAW IND
narenga (Nees) Hack.				VIE	
officinarum L.			TAI		
spontaneum L.	IDO IND THI	PHI PR		ANT ARB AUS BND EGY GHA HAW IRA JOR JPN MAL MAU MEL MIC NEP NGI PK SOV SUD VIE	AFG BUR SAF
SACCIOLEPIS Nash **Poaceae** *(Gramineae)*					
angusta (Trin.) Stapf				THI	
indica (L.) A. Chase (Syn. *Hymenachne indica* Buese) (Syn. *Panicum myuros* H. B. K. *non* Lam.)			HAW TAI	FIJ IND PLW THI VIE	
insulicola (Steud.) Ohwi			IDO	PHI	
interrupta (Willd.) Stapf (Syn. *Panicum indicum* Hack.)	NIG		IDO	BND IND VIE	
myosuroides (R. Br.) A. Camus			MAL		
myurus (Lam.) A. Chase (Syn. *Hymenachne myurus* (Lam.) Beauv.)				VIE	
polymorpha A. Chase *ex* E. G. & A. Camus				VIE	

SAGINA L. Caryophyllaceae	S	P	C	X	F
apetala Ard.				USA	
japonica (Sw.) Ohwi		JPN			
nodosa (L.) Fenzl			SOV		FIN POL
procumbens L.			AUS ENG FIN	NZ USA	AFG POL
SAGITTARIA L. Alismataceae					
aginashi Makino		JPN			
calycina Engelm.				USA	
chilensis Cham. & Schlecht.		KOR		CHL	
cuneata Sheldon				CAN GRE USA	
falcata Pursh				USA	
guayanensis H. B. K.			MAL	AFG BND BOR CAB CHN GHA IND MAD MLI NIG PK SEG SUD SUR TAI THI VIE	
latifolia Willd.			USA	CAN ECU GUA HON IDO IND	
montevidensis Cham. & Schlecht.		ARG		AUS BRA CAN IDO TNZ USA	
platyphylla (Engelm.) J. G. Smith				USA	
pygmaea Miq.		JPN	TAI	CHN	
sagittifolia L.		GER HAW ITA SWE TAI	IRA PHI POR	ARG AUS BEL BND CAB CHN ENG FRA	AFG HK

SAGITTARIA L. Alismataceae *(Cont'd)*	S	P	C	X	F
sagittifolia L. *(Cont'd)*				GRE IDO IND JPN KOR MAL MEX NEP NET NOR PK POL THI VIE	
subulata (L.) Buchenau				USA	
trifolia L.		JPN KOR	TAI	AFG CHN	
SALICORNIA L. Chenopodiaceae					
quinqueflora Bunge *ex* Ung.			AUS		
SALIX L. Salicaceae					
alba L.				NZ USA	AFG ISR
amygdaloides Anderss.				USA	
babylonica L.				NZ	AFG
bebbiana Sarg.				USA	
caprea L.		NZ	FIN		
caroliniana Michx.				USA	
caudata (Nutt.) Heller				USA	
cinerea L.				NZ	
discolor Muhl.				NZ	
exigua Nutt.				USA	
fragilis L.		NZ			
interior Rowles				USA	
laevigata Bebb				USA	
lasiandra Benth.				USA	
lutea Nutt.				USA	
nigra Marsh.				USA	
petiolaris J. E. Sm.				USA	
phylicifolia L.			FIN		
SALPICHROA Miers Solanaceae					
origanifolia (Lam.) Baill.			ARG	BRA NZ URU	

SALSOLA L. Chenopodiaceae	S	P	C	X	F
collina Pall				SOV	
foetida Delile				PK	
kali L.	AFG ARG	CAN HUN	IRQ ITA MOR SAF USA	AUS CHL CHN EGY GRE HAW IDO IRA JOR LEB MEX NOR NZ PK POL SOV TUR	ISR
pestifer A. Nelson			CAN HAW USA		
ruthenica Iljin			IRA	SOV	
soda L.			ARG	CHN	
SALVIA L. Lamiaceae (Labiatae)					
acetabulosa L.			LEB		
aethiopis L.				USA	
algeriensis Desf.				MOR	
apiana Jeps.				USA	
argentea L.				MOR	
barrelieri Etling.				MOR	
brachiata Roxb.			TAI		
coccinea Juss. ex Murr.			AUS HAW TRI	FIJ MEX USA	IDO
dasycalyx Fern.				MEX	
glutinosa L.				TUR	
horminum L.				MOR	
leucophylla Greene				USA	
mellifera Greene				USA	
mexicana L.				MEX	
moureti Battand & Pittard			MOR		
nemorosa Crantz			IRA		SOV
occidentalis Sw.			HAW TRI	CR IDO JAM	

SALVIA L. Lamiaceae *(Labiatae) (Cont'd)*	S	P	C	X	F
occidentalis Sw. *(Cont'd)*				PAN PLW SAL	
plebeia R. Br.			IND	AUS PK	AFG
pratensis L.				GER	
procurrens Benth.				URU	
reflexa Hornem.			AUS	NZ	
sclarea L.				TUR	
serotina L.			PR	DR JAM	
sonomensis Greene				USA	
spinosa L.			LEB		AFG
syriaca L.			LEB	TUR	ISR
verbenaca L.			AUS	NZ	ISR
verticillata L.				SOV TUR	
SALVINIA Adans. **Salviniaceae**					
auriculata Auctt. *non* Aubl. *(See* **S. molesta** D. S. Mitchell)	CEY GUY IDO ZAM	BRA CNK IND KEN RHO	AUS SAF	ARG COL FIJ NZ	BOT CUB MAL NIG TRI
cucullata Roxb.		IDO THI	LAO MAL	BND VIE	
hastata Desv.				MAD MOZ	
molesta D. S. Mitchell (Syn. *S. auriculata* Auctt. *non* Aubl.)					
natans (L.) All.		ITA JPN KOR MAU NZ	IDO IRQ SPA TAI	ARG BND CHN HON HUN IND POL THI	HK
nymphellula Desv.		GHA			
rotundifolia Willd.				USA	CNK
SAMBUCUS L. **Caprifoliaceae**					
callicarpa Greene				USA	
canadensis L.				USA	
chinensis Lindl. (Syn. *S. formosana* Nakai)					

SAMBUCUS L. Caprifoliaceae *(Cont'd)*	S	P	C	X	F
formosana Nakai *(See* **S. chinensis** Lindl.)			TAI		
glauca Nutt.				USA	CHL
javanica Reinw. *ex* Bl.				IDO	
racemosa L.				USA	
simpsonii Rehder				JAM	
SANGUISORBA L. Rosaceae					
canadensis L.				USA	
minor Scop.				TUR USA	
officinalis L.				TUR USA	
SANSEVIERIA Thunb. Liliaceae					
guineensis (Jacq.) Willd.		PR		DR	
SANTALUM L. Santalaceae					
album L.				IND	
SANTOLINA L. Asteraceae *(Compositae)*					
chamaecyparissus L.				TUR	
SAPIUM P. Br. Euphorbiaceae					
grahami (Stapf) Prain				GHA	
jamaicense Sw.				DR	
sebiferum (L.) Roxb.			TAI	USA	
SAPONARIA L. Caryophyllaceae					
officinalis L.			AUS CAN SPA	CHL TUR USA	POL
vaccaria L. (Syn. *Vaccaria vulgaris* Host)	LEB		CAN SAF	GRE IND PK USA	AFG AUS
SAPOTA Mill. Sapotaceae					
achras Mill. *(See* **Manilkara zapota** (L.) Van Royen)				JAM	
SARCOBATUS Nees Chenopodiaceae					
vermiculatus (Hook.) Torr.				USA	
SARCOSTEMMA R. Br. Asclepiadaceae					
australe R. Br.				AUS	
cynanchoides Decne.				USA	

SASSAFRAS Nees Lauraceae	S	P	C	X	F
albidum (Nutt.) Nees				USA	
SATUREJA L. Lamiaceae *(Labiatae)*					
pseudosimensis Brenan		ETH			
vulgare (L.) Fritsch				USA	
SAUROPUS Bl. Euphorbiaceae					
androgynus (L.) Merr.				IDO	
SAURURUS L. Saururaceae					
cernuus L.				USA	
chinensis (Lour.) Baill.			JPN		
SAUSSUREA Dc. Asteraceae *(Compositae)*					
affinis Spreng. *ex* Dc. (See **Hemistepta lyrata** Bunge)			CHN TAI		
candicans C. B. Clarke				PK	AFG
SAUVAGESIA L. Ochnaceae					
brownei Planch.				JAM	
erecta L.				PER	
SCABIOSA L. Dipsacaceae					
atropurpurea L.				CHL MOR	
maritima L.				SPA	
palaestina L.				IRA TUR	
prolifera L.				JOR	
rotata Bieb.				TUR	
semipapposa Salzm. *ex* Dc.				MOR	
stellata L.				SPA	
veronica L.				TUR	
SCANDIX L. Apiaceae *(Umbelliferae)*					
australis L.				MOR	
iberica Bieb.			LEB		ISR
pecten-veneris L.	GRE TUN	ENG IRA ISR LEB MOR POR SPA	CHL NZ TUR USA		AFG AUS
SCHEDONNARDUS Steud. Poaceae *(Gramineae)*					
paniculatus (Nutt.) Trel.				USA	

SCHINUS L. Anacardiaceae	S	P	C	X	F
molle L.				USA	AFG AUS ISR
terebinthifolius Raddi			HAW	USA	AUS ISR
SCHISMUS Beauv. **Poaceae** *(Gramineae)*					
barbatus (L.) Thell.				USA	
SCHIZACHYRIUM Nees **Poaceae** *(Gramineae)*					
brevifolium (Sw.) Nees *ex* Buse (Syn. *Andropogon brevifolius* Sw.)	NIG				IND SAF
paniculatum (Kunth) Herter		BRA			
scoparium (Michx.) Nash (Syn. *Andropogon scoparius* Michx.)					
SCHKUHRIA Roth **Asteraceae** *(Compositae)*					
isopappa Benth.				KEN	
pinnata (Lam.) O. Ktze.	SAF	ARG AUS KEN RHO			
virgata Dc.				MEX	
SCHLERANTHUS L. **Caryophyllaceae**					
annuus L.			FIN SAF USA	CHL GER ITA	
SCHOENOPLECTUS (Reichb.) Palla **Cyperaceae**					
lacustris (L.) Palla *(See* **Scirpus lacustris** L.)				ENG	
SCHOENUS L. **Cyperaceae**					
apogon Roem. & Schult.			JPN		AUS
SCHRANKIA Willd. **Mimosaceae** *(Leguminosae)*					
leptocarpa Dc.				BRA GHA	
microphylla (Dryander) Macbr.				USA	
nuttallii (Dc.) Standl.				USA	
SCHULTESIA Mart. **Gentianaceae**					
guyanensis Malme				TRI	
SCHWENKIA L. **Solanaceae**					
americana L.		GHA		IVO	
SCILLA L. **Liliaceae**					
maritima L.				GRE	

SCILLA L. Liliaceae *(Cont'd)*	S	P	C	X	F
scilloides (Lindl.) Druce (Syn. *S. thunbergii* Miyabe & Kudo)			JPN		
thunbergii Miyabe & Kudo *(See* **S. scilloides** (Lindl.) Druce)			TAI		
SCIRPUS L. **Cyperaceae**					
acutus Muhl.			USA	BND	
americanus Pers.				USA	
articulatus L.	IND	THI			IDO
atrovirens Willd.				USA	
australis Murr.				TUR	
californicus (C. A. Mey.) Steud.				ARG CHL USA	
cubensis Poepp. & Kunth		GHA			CNK MAD RHO
cyperinus (L.) Kunth				USA	
erectus Poir.		IND	PHI	CHN KOR	IDO
fluviatilis (Torr.) A. Gray			JPN	AUS USA	
grossus L. *f.*		CAB IDO THI VIE	MAL PHI	BOR PK TUR	
holoschoenus L.			LEB POR		AFG ISR
hotarui Ohwi *(See* **S. juncoides** Roxb. *var.* hotarui (Ohwi) Ohwi)		JPN			
juncoides Roxb.	AFG	BOR MAD	IDO JPN PHI TAI	CAB CHN ECU HAW PK POR THI VIE	LEB MAL
juncoides Roxb. *var.* hotarui (Ohwi) Ohwi (Syn. *S. hotarui* Ohwi)					
lacustris L. (Syn. *Schoenoplectus lacustris* (L.) Palla)				BEL FRA	
lateriflorus J. F. Gmel.		IDO	MAL		
litoralis Schrad.			EGY IRQ	IRA	AFG IDO IND
maritimus L. (Syn. *Bolboschoenus maritimus* Asch. & Godr.)	AFG ITA PHI	CAB IND IRQ	JPN MOR POR	BEL CAN FRA	ISR

SCIRPUS L. Cyperaceae *(Cont'd)*	S	P	C	X	F
maritimus L. *(Cont'd)*	SPA	NZ PER ROM SEG		GER GRE HUN IRA JOR KEN KOR NOR PK RHO SOV SUD SUR SWE TNZ TUR VIE ZAM	
mucronatus L.	ITA SPA	BND IDO MAL POR	PHI	CAB EGY FIJ GER HUN IRA JPN KOR MEL POL THI TNZ USA VIE ZAM	CHN
palustris L.				TUR	
praelongatus Poir.	NIG				
smithii A. Gray			JPN		
supinus L.	MAL	IND	PHI SPA	CAB EGY GER HUN IDO IRA ISR JOR POL RHO SEG SUD TNZ VIE ZAM	HK
sylvaticus L.			FIN		
triangulatus Roxb.			JPN		
triqueter L.			JPN		AFG
tuberosus Desf.			EGY	IND	
validus Benth.			HAW	AUS USA	
wallichii Nees		TAI			

SCLERIA Berg. Cyperaceae	S	P	C	X	F
bancana Miq.				VIE	
barteri Boeck.		GHA			
bracteata Cav.				BRA NIC	
canescens Boeck.			PR	MEX	
hebecarpa Nees				MAL	
levis Retz.				MAL	
lithosperma (L.) Swartz				PLW	
melaleuca Riechb. *ex* Schlect. & Cham.		DR	TRI		
multifoliata Boeck.			BOR	CAB CEY MAL	
myriocarpa Kunth				BRA	
oryzoides Presl *(See* S. poaeformis Retz.*)*				THI VIE	
poaeformis Retz. (Syn. *S. oryzoides* Presl)		AUS	THI	VIE	
polycarpa Boeck.				PLW	
pterota Presl				DR HON	CR
scrobiculata Ness & Mey. *ex* Ness				FIJ PHI PLW	
sumatrensis Retz.	BOR		MAL		
SCLEROBLITUM Ulbr. **Chenopodiaceae**					
atriplicinum (F. Muell.) Ulbr.			AUS		
SCLEROCARPUS Jacq. **Asteraceae** *(Compositae)*					
africanus Jacq.				GHA SUD	
coffeaecola Klatt		VEN			
SCLEROCARYA Hochst. **Anacardiaceae**					
birroea Hochst.				RHO	
SCLEROCHLOA Beauv. **Poaceae** *(Gramineae)*					
dura (L.) Beauv.				USA	
SCLEROPOA Griseb. **Poaceae** *(Gramineae)*					
rigida (L.) Griseb.				USA	AUS ISR
SCOLYMUS L. **Asteraceae** *(Compositae)*					
hispanicus L.		TUN	LEB MOR SPA	ARG AUS	ISR POR

SCOLYMUS L. Asteraceae *(Compositae) (Cont'd)*	S	P	C	X	F
maculatus L.		ISR	AUS IRQ LEB MOR		POR

SCOPARIA L. Scrophulariaceae					
dulcis L.	GHA HON	MAL	BOR CHN HK IDO PER TAI TRI	AUS BRA CAB CEY CNK COL CR DAH FIJ GRE IND JAM MEX NEP NIG PHI PR SAL SEG SUD SUR THI VIE	
montevidensis R. E. Fries				ARG	

SCORPIURUS L. Papilionaceae *(Leguminosae)*					
muricata L.		ISR	EGY POR		
sulcata L.			EGY MOR POR		
vermiculata L.			MOR POR	SPA	

SCORZONERA L. Asteraceae *(Compositae)*					
lanata Bieb.				TUR	
purpurea L.				TUR	

SCROPHULARIA L. Scrophulariaceae					
lanceolata Pursh				USA	
marilandica L.				USA	
nodosa L.				BEL TUR	

SCUTELLARIA L. Lamiaceae *(Labiatae)*					
dependens Maxim.				JPN	
galericulata L.			FIN POR		
strigillosa Hemsl.				JPN	

SCUTIA (Dc.) Brongn. **Rhamnaceae**	S	P	C	X	F
myrtina Kurz.				KEN	
SEBASTIANIA Spreng. **Euphorbiaceae**					
chamaelea (L.) Muell.-Arg.				GHA	
corniculata (Vahl) Muell.-Arg.			TRI	SUR	
SECALE L. **Poaceae** (*Gramineae*)					
cereale L.			ARG FIN IRA	TUR USA	AFG AUS HAW IND
montanum Guss.				TUR	
SECURIDACA L. **Polygalaceae**					
virgata Sw.			PR	DR	
SECURIGERA Dc. **Papilionaceae** (*Leguminosae*)					
securidaca (L.) Dalla Torre & Sarnth.			ISR		
virosa Thunb.		TNZ		GHA	
SEDUM L. **Crassulaceae**					
acre L.				USA	
bulbiferum Makino			JPN		
formosanum N. E. Br.			TAI		
lineare Thunb.			JPN		
purpureum (L.) Link				USA	
SEHIMA Forsk. **Poaceae** (*Gramineae*)					
ischaemoides Forsk.				SUD	
SELAGINELLA Beauv. **Selaginellaceae**					
belluta Cesati				PHI	
opaca Warb.				IDO	
plana Hieron.			IDO		
SENEBIERA Dc. **Brassicaceae** (*Cruciferae*)					
coronopus Poir. (*See* **Coronopus squamatus** (Forsk.) Asch.)			SPA		
didyma Pers. (Syn. *S. pinnatifida* Dc.)				MAU	AUS
pinnatifida Dc. (*See* **S. didyma** Pers.)				ARG	
SENECIO L. **Asteraceae** (*Compositae*)					
abyssinicus Sch.-Bip. *ex* Hochst.				KEN	
aegyptius L.			EGY		
aquaticus Hill				BEL	

SENECIO L. Asteraceae *(Compositae) (Cont'd)*	S	P	C	X	F
argentinus Baker			ARG		
aureus L.				USA	
bonariensis Hook. & Arn.			ARG		
brasiliensis (Spr.) Less.			ARG	BRA URU	
burchellii Dc.		ARG	SAF		
campestris Dc.				CHN	
compactus Rydb.				CHN	
consanguineus Dc.			SAF		
coronopipolius Desf.			IRQ	IRA	
daltonii F. Muell.				AUS	
desfontainei Druce			EGY		
discifolius Oliver				KEN	
discolor (Sw.) Dc.				JAM	
erraticus Bertol.				CHL	
erucifolius Ledeb.			ENG		
formosus H. B. K.				VEN	
glabellus Poir.				USA	
grisebachii Baker			ARG		
heterotrichius Dc.				BRA	
ilicifolius L.			SAF		
incognitus Cabrera				ARG	
integrifolius Clairv.			JPN		
jacobaea L.	NZ		AUS CAN ENG USA	ARG AST BEL BRA CZE DEN FRA GER HUN IRE ITA NET NOR POL ROM SAF SOV SPA SWE SWT TAS URU YUG	TRI

SENECIO L. Asteraceae *(Compositae) (Cont'd)*	S	P	C	X	F
latifolius Dc.			RHO		
lautus (Forst. *f.*) Willd.			AUS	NZ	
leucanthemifolius Poir.		TUN			
longilobus Benth.				USA	
mikanioides Otto			HAW	NZ USA	AUS
moorei R. E. Fries		KEN			
oligophyllus Baker		BRA			
oxyriaefolius Dc.		SAF			
pinnatifidus Less.				ARG	
platylepis Dc.				AUS	
quadridentatus Labill.			AUS		
riddellii Torr. & Gray				USA	
rupestris Waldst. & Kit.				TUR	
ruwenzoriensis S. Moore				KEN	
sonchifolius (L.) Moench *(See* **Emilia sonchifolia** (L.) Dc. *ex* Wight*)*			IDO		
sylvaticus L.		NZ	HAW	USA	
triangularis Hook.				USA	
vernalis Waldst. & Kit.			LEB	GER SOV TUR	ISR POL
viscosus L.			ENG	USA	
vulgaris L.	AST ENG NET NZ POL SWE	BEL CAN GER GRE HUN JOR KOR NOR SPA USA	ARG COL CZE EGY FIN IRA ISR KEN TUN	AFG CHL CHN ECU FRA HK ICE IRE IRQ ITA JPN LEB MOZ POR SOV TUR YUG	AUS
SENECIOIDES L. Asteraceae *(Compositae)*					
cinerea (L.) O. Ktze. *(See* **Vernonia cinerea** (L.) Less.*)*			PR	DR	
SENEGALIA Rafin. Mimosaceae *(Leguminosae)*					
westiana (Dc.) Britton & Rose (Syn. *Acacia westiana* Dc.)			PR		

	S	P	C	X	F
SEQUOIA Endl. **Pinaceae**					
sempervirens (D. Don) Endl.				USA	
SERENOA Hook. *f.* **Arecaceae** *(Palmae)*					
repens (Bartr.) Small				USA	
SERIOLA L. **Asteraceae** *(Compositae)*					
aethnensis L.				GRE	ISR
SERJANIA Plum. **Sapindaceae**					
acoma Radlk.				BRA	
polyphylla Poir. *ex* Steud.			PR		
rubicaulis (Ruiz & Pav.) Benth.				PER	
SERRATULA L. **Asteraceae** *(Compositae)*					
cerinthifolia Sibth. & Sm.			LEB		ISR
SESAMUM L. **Pedaliaceae**					
alatum Thonn.				GHA	
indicum L. (*See* **S. orientale** L.)			TAI	NIG	
orientale L. (Syn. *S. indicum* L.)					
radiatum Schumach. & Thonn.				GHA	
SESBANIA Scop. **Papilionaceae** *(Leguminosae)*					
aculeata Poir. (*See* **S. cannabina** (Retz.) Pers.)		FIJ		BND IND PK	AUS
benthamiana Domin	AUS				
bispinosa Steud. *ex* Wight (*See* **S. cannabina** (Retz.) Pers.)	IND	AUS	DR		
cannabina (Retz.) Pers.) (Syn. *S. aculeata* Poir.) (Syn. *S. bispinosa* Steud. *ex* Wight)				AUS	
drummondii (Rydb.) Cory				USA	
exaltata (Rafin.) Cory				USA	
exasperata H. B. K.				SUR	
grandiflora (L.) Pers.			IDO		CAB
javanica Miq. (Syn. *S. roxburgii* Merr.)					
punicea (Cav.) Benth.				ARG USA	BRA
roxburgii Merr. (*See* **S. javanica** Miq.)			TAI	THI	
sesban (L.) Merr.	SEG		IDO PHI TAI	ANG	

SESBANIA Scop. Papilionaceae *(Leguminosae) (Cont'd)*	S	P	C	X	F
vesicaria (Jacq.) Ell.				USA	
SESSEA Ruiz & Pav. **Solanaceae**					
brasiliensis Toledo		BRA			
SESUVIUM L. **Aizoaceae**					
portulacastrum (L.) L.				FIJ VIE	IDO PR
SETARIA Beauv. **Poaceae** *(Gramineae)*					
acromelaena (Hochst.) Dur. & Schinz	ETH			KEN	
aequalis Stapf		ANG			
aurea Hochst.				VIE	
barbata (Lam.) Kunth (Syn. *Panicum barbatum* Lam.) (Syn. *P. flavescens* Sw.)		IDO MAU	PR TRI	FIJ GHA JAM NIG VIE	BUR CEY HAW IND NGI
conspersum (L.) Beauv.				BRA	
faberii Herrm.			JPN USA	CAN	
geniculata (Lam.) Beauv.	ARG	BRA ECU HON PER USA VEN	HAW JAM PR TAI	ANT BER BOR CAB CHL CHN COL CR DR GRE IDO JPN MAL MEX NZ PHI PLW SAL SUR THI TRI URU	AUS BUR IND SAF
gigantea (Fr. & Sav.) Makino			JPN		
glauca (L.) Beauv. (Syn. *S. lutescens* (Weig.) F. T. Hubb.)	AFG AST HUN IND ISR KOR TUN USA	AUS ECU SOV YUG	CAN EGY HAW IDO JPN PHI PLW POR ROM SPA	BND BUL CHL CHN COL CR CZE ENG FIJ FRA GER GRE IRA IRQ	HK

SETARIA Beauv. Poaceae *(Gramineae) (Cont'd)*	S	P	C	X	F
glauca (L.) Beauv. *(Cont'd)*			.	ITA	
				JAM	
				JOR	
				LEB	
				MEX	
				MIC	
				MOZ	
				NEP	
				NZ	
				PER	
				PK	
				POL	
				VEN	
grisebachii Fourn.				MEX	
homonyma (Steud.) Chiov.	TNZ	KEN			IND
					SAF
incrassata (Hochst.) Hochst. *ex* Hack.				SUD	
intermedia (Roth) Roem. & Schult.				IND	
				MAU	
italica (L.) Beauv.			JPN	GRE	AUS
				PHI	BUR
				USA	HAW
					IND
longiseta P. Beauv.				NIG	
lutescens (Weig.) F. T. Hubb. *(See* **S. glauca** (L.) Beauv.*)*	MEX	POR	CAN	CHL	JPN
			COL	FIJ	
			HAW	PER	
			USA	VEN	
pallide-fusca (Schumach.) Stapf & Hubb.	SEG	FIJ		AUS	SAF
	SUD	IND		CNK	
	UGA	KEN		ETH	
	ZAM	MAU		IDO	
		RHO		IVO	
				JPN	
				MOZ	
				NIG	
				PHI	
				PK	
				PLW	
				THI	
				TNZ	
				TUN	
palmifolia (Willd.) Stapf *(Syn. Panicum palmifolium* Willd.*)*	IDO		HAW	AUS	FIJ
	IND		TAI	JAM	
				PLW	
				VIE	
plicata (Lam.) T. Cooke *(Syn. Panicum excurrens* Trin.*)*					
poiretiana (Schult.) Kunth *(Syn. S. sulcata* (Bertol.) Raddi*)*		TRI			IND
pumila (Poir.) Roem. & Schult.				GER	
sphacelata (Schumach.) Stapf & C. E. Hubb.	AUS			GHA	
sulcata (Bertol.) Raddi *(See* **S. poiretiana** (Schult.) Kunth*)*	BRA				TRI

SETARIA Beauv. Poaceae *(Gramineae) (Cont'd)*	S	P	C	X	F
verticillata (L.) Beauv.	HAW ISR PER SPA TNZ TUR	KEN LEB SAF TUN ZAM	ARG CAN EGY IRQ MOR RHO USA	ANG ARB AUS CHL ETH FRA GER GRE HUN IND ITA MAU NIG NZ SOV SUD THI UGA	CEY
viridis (L.) Beauv.	IRA SPA USA	CAN JPN SOV YUG	ARG CHN EGY IND IRQ LEB MOR ROM TAI	AFG AUS BND BUL CHL CZE ETH GER HUN IDO ITA KEN NET NZ PHI PK RHO TUN	ENG FRA ISR KOR POL SAF
welwitschii Rendle				ANG	
SHEPHERDIA Nutt. **Elaeagnaceae**					
argentea Nutt.				USA	
canadensis (L.) Nutt.				USA	
SHERARDIA L. **Rubiaceae**					
arvensis L.			ENG MOR SPA	CHL GER GRE MEX POR USA	AUS ISR POL
SIBARA Greene **Brassicaceae** *(Cruciferae)*					
virginica (L.) Rollins				USA	
SICYOS L. **Cucurbitaceae**					
angulata L.	MEX			USA	
SIDA L. **Malvaceae**					
acuta Burm. *f.*	AUS BRA MEL MEX	FIJ MAL SAL	KEN NIG TAI TRI	CEY CHN CNK COL	CAB IDO

SIDA L. Malvaceae *(Cont'd)*	S	P	C	X	F
acuta Burm. *f. (Cont'd)*	PHI PLW THI			CUB DR GHA HK HON IND IVO JAM MAU NGI PAN PER SUR USA VIE	
alba L.			EGY KEN RHO SUD	IVO TNZ ZAM	
angustifolia L.		MEX			
carpinifolia L. *f.*			PR	BRA COL DR TRI VEN	FIJ KEN
cordifolia L.		AUS BRA UGA	HAW IND KEN RHO SAF TAI	GHA HON IVO SAL USA	
corrugata Lindl.				AUS	
corymbosa R. E. Fries				NIG	
cuneifolia Gray			KEN		
fallax Walp.			HAW		
glomerata Cav.			DR	VEN	TRI
glutinosa Cav.				IND	
hederacea (Dougl.) Torr.				AUS USA	
jamaicensis Vell.				SUR	
javensis Cav. (Syn. *S. veronicaefolia* Lam.)					
linifolia Cav.			TRI	BRA GHA IVO	FIJ
micrantha St. Hil.				BRA	
mysorensis Wight & Arn.			TAI		
ovata Forsk.			KEN		
paniculata L.				PER	
physocalyx A. Gray				USA	

SIDA L. Malvaceae *(Cont'd)*	S	P	C	X	F
platycalyx F. Muell. *ex Benth.*				AUS	
potentilloides St. Hil.			ARG	BRA	
retusa L. *(See* **S. rhombifolia** L. *ssp.* retusa (L.) Borss.*)*	NGI			PHI	AUS IDO
rhombifolia L.	BRA GUA HON JAM MAU MEX	COL ECU MEL NGI PLW SWZ	ARG AUS CNK HAW IDO KEN PHI SAF TAI TNZ TRI UGA	ANT BOR BOT CAM CHN CR EGY FIJ GHA GUI IND IRA IVO LIB MAL MIC NGR NIG NZ PER PR RHO SAL SEG SUD SUR THI URU USA VEN ZAM	
rhombifolia L. *ssp.* retusa (L.) Borss. (Syn. *S. retusa* L.)					
samoensis Rech.				PLW	
spinosa L.	AUS	MEX USA	ARG HAW IND	CHL PER URU	
stipulata Cav.			NIG	GHA	
subspicata F. Muell. *ex Benth.*				AUS	
urens L.				GHA IVO JAM NIG PER TRI	
veronicaefolia Lam. *(See* **S. javensis** Cav.*)*			IND	BND NIG	IDO
SIDERITIS L. Lamiaceae *(Labiatae)*					
montana L.			IRA	TUR	
SIEGESBECKIA L. Asteraceae *(Compositae)*					
agrestis Poepp. & Endl.				ECU	

SIEGESBECKIA L. Asteraceae *(Compositae) (Cont'd)*	S	P	C	X	F
cordifolia H. B. K.				COL	SAF
glabrescens (Makino) Makino			JPN		
orientalis L.		MAU SAF	HAW IDO TAI	AUS CHL FIJ PLW UGA USA	
pubescens (Makino) Makino			JPN		CHL COL FIJ HAW IDO MAU SAF TAI
SILENE L. **Caryophyllaceae**					
alba Muhl. *ex* Rohrb.			ENG	CAN	
anglica L.			AUS KEN		
antirrhina L.			USA		
cerastoides All.			ECU		
colorata Poir.			POR		ISR
conica L.			SPA	USA	AUS
conoidea L.			IRA LEB	CHN PK	IND ISR
crassipes Fenzl			LEB		ISR
cserei Baumg.			CAN	USA	
cucubalus Wibel			CAN FIN LEB POR SPA	CHL USA	AUS
dichotoma Ehrh.				SOV USA	AUS POL
dioica Fisch. & Mey.			ENG		
fuscata Link *ex* Brot.			POR		
gallica L.	BRA ITA KEN POR	TNZ	ARG COL GUA HAW ISR MOR PER POL SAF SPA TUN	AUS CAN CHL ECU EGY HON HUN IRA JOR JPN MAU NZ RHO TUR	

SILENE L. Caryophyllaceae *(Cont'd)*	S	P	C	X	F
gallica L. *(Cont'd)*				UGA URU USA	
gonocalyx Boiss.			LEB		
inflata Sm.		TAS	CAN	ITA SOV SPA TUR	AUS ISR POL
noctiflora L.			CAN ENG SPA USA		AUS
rubella L.		TUN	EGY POR	MOR	ISR
venosa Aschers.			LEB	GRE SOV	AUS
vulgaris (Moen.) Garcke			MOR	SOV	
SILPHIUM L. Asteraceae *(Compositae)*					
perfoliatum L.				USA	
SILYBUM Adans. Asteraceae *(Compositae)*					
marianum (L.) Gaertn. (Syn. *Carduus marianus* L.)	AFG AUS BRA TUN	ARG COL ISR ITA JOR	EGY IRA IRQ PER POL SAF SPA TUR	CAN CHL ETH GRE KEN MOR NZ PK TAS URU USA	POR SOV
SINAPIS L. Brassicaceae *(Cruciferae)*					
alba L.			CAN ENG POR SPA	ARG ITA MOR NZ	AUS ISR POL SOV
arvensis L. *(See* **Brassica kaber** (Dc.) L. C. Wheeler *var.* pinnatifida (Stokes) L. C. Wheeler)	AFG AST GRE ISR LEB POL SOV SPA	AUS BEL BUL CAN CZE ENG GER HUN JOR ROM SWE TUN TUR USA YUG	ARG BOR EGY GUA IRA IRQ MOR NET POR	ALK DEN FIN FRA ICE IRE ITA MEL NOR NZ TAS	
juncea L.	EGY				

SISYMBRIUM L. Brassicaceae (Cruciferae)	S	P	C	X	F
altissimum L.			ARG CAN USA	AUS NZ PER TUR	AFG SOV
bolgense Bieb.			SOV		
columnae Jacq.			SPA		
erysimoides Desf.			MOR		
exacoides Dc.			LEB		
irio L.		AUS	ARG EGY IRQ	GRE IND PK USA	AFG ISR
loeselii L.			CAN IND IRA	SOV USA	AFG POL
officinale (L.) Scop.		AUS HUN IRA JOR	ARG AST CAN ENG HAW ITA JPN POL POR SPA	ARB BRA CHL GRE IRQ ISR KEN LEB MEL MEX NOR NZ SOV TUR USA	
orientale L.		AUS	JPN	NZ	AFG ISR
runcinatum Lag. ex Dc.			MOR		
sophia L.			CAN IND	SOV	AFG AUS POL
thellungii A. E. Schulze			SAF	AUS	
SISYRINCHIUM L. Iridaceae					
atlanticum Bick.			JPN TAI		
chilense Hook.				MAU NZ	
iridifolium H. B. K.			JPN		
SITANION Rafin. Poaceae (Gramineae)					
hystrix (Nutt.) J. G. Smith				USA	
SIUM L. Apiaceae (Umbelliferae)					
erectum Huds.				BEL	
latifolium L.				AUS	BEL

SIUM L. Apiaceae *(Umbelliferae) (Cont'd)*	S	P	C	X	F
suave Walt.			CAN	USA	
SMILAX L. Smilacaceae					
balbisiana Kunth				JAM	
bona-nox L.				USA	
coriacea Spreng.				BRA DR	PR
cumanensis Humb. & Bonpl. *ex* Willd.				TRI	
glauca Walt.				USA	
laurifolia L.				USA	
ornata Lem.				JAM	
rotundifolia L.				USA	
walteri Pursh				USA	
SNOWDENIA C. E. Hubb. Poaceae *(Gramineae)*					
polystachya (Fresen.) Pilger	ETH		KEN		
SOLANUM L. Solanaceae					
aculeastrum Dun.			RHO		
aculeatissimum Jacq.			HAW	AUS BRA JAM USA	IDO
alatum Moench.		LEB			
atropurpureum Schrank				BRA	
auriculatum Ait.		AUS		MAU	SAF
balbisii Dun.				BRA	
biflorum Lour.			IDO TAI		
bonariense L.			ARG	BRA URU	
capsicastrum Link				URU	
caribaeum Dun.			PR	MEX	
carolinense L.	USA		CAN	BND NEP NZ	AUS JPN
chacoense Bitter				ARG	
chenopodioides Lam.				CHL	
ciliatum Lam.			PR		
cinereum R. Br.			AUS		
commersonii Dun. *ex* Poir.				ARG BRA URU	

SOLANUM L. Solanaceae *(Cont'd)*	S	P	C	X	F
diflorum Vell.			ARG		
dubium Fresen.		SUD			
dulcamara L.			CAN	BEL NZ TUR USA	AFG
elaeagnifolium Cav.	IND	AUS	ARG SAF	CHL MEX USA	
ellipticum R. Br.			AUS		
esuriale Lindl.			AUS		
flagellare Sendt.			BRA		
glaucum Dun.		ARG			
gracile Otto *ex* W. Baxt.		BRA	ARG SAF		
grossedentatum A. Rich.		KEN			
hamulosum C. T. White			AUS		
hirtum Vahl			COL SAL		
hispidum Pers.			AUS		
hystrix R. Br.		AUS			
incanum L.			ETH KEN RHO	ANG	ISR
indicum L.			TAI	IND MAU	
jamaicense Mill.			TRI		IDO
laciniatum Ruiz & Pav.			AUS		
lepidotum Humb. & Bonpl. *ex* Dun.			COL		
lycopersicum L. *(See* **Lycopersicon lycopersicum** (L.) Karsten)			HAW		
mammosum L.			JAM		
marginatum L. *f.*			COL	CHL NZ	AUS
mauritianum Scop.			AUS SAF	NZ	
melongena L.			IND		
miniatum Bernh. *ex* Willd.			LEB		
nigrum L.	AUS HAW IRA ISR NEP NZ	BEL BRA CAN FRA GRE HUN	ARG CHN COL EGY ENG ETH	BND BOR CAB CAM CHL CNK	AFG CR KOR POL UGA

SOLANUM L. Solanaceae *(Cont'd)*	S	P	C	X	F
nigrum L. *(Cont'd)*	SOV TAI	ITA KEN MAU PK POR SPA TUN	HK IDO IRQ JPN MOR RHO SAF USA YUG	CUB FIJ FIN GER GHA GUI HON ICE IND IVO JAM MAD MAL MEL MEX MLI NIG PER PLW SAL SEG THI TNZ TRI TUR VEN	
nodiflorum Jacq.	HAW		SAF	NZ	
orthocarpum Pichi-sermolli				PK	
panduriforme Drege *ex* Dun.			KEN		RHO
paniculatum L.				BRA	
photeinocarpum Nakam. & Odash.			JPN		
poeppigianum Sendt.				PER	
pseudocapsicum L.			ARG AUS HAW	BRA NZ	
pygmaeum Cav.			ARG		
retroflexum Dun.			SAF		
rostratum Dun.	MEX		AUS SAF	BND NZ USA	
sarrachoides Sendt.			ARG	NZ USA	
seaforthianum Andr.				AUS SAL	
semiarmatum F. Muell.				AUS	
sisymbrifolium Lam.		ARG BRA	SAF	URU	
sodomeum L.		AUS	HAW MOR	NZ USA	
stramoniifolium Jacq.			TRI	ANT	
sturtianum F. Muell.			AUS		

SOLANUM L. Solanaceae *(Cont'd)*	S	P	C	X	F
subinerme Jacq.				JAM	
surattense Burm. *f.*				IND	
torvum Sw.	MEL NGI	AUS CEY FIJ GHA PLW	IDO MAU PR	ANT BND BOR CAM CHN CNK CR DAH GUI HON IVO JAM LIB MAL MEX NIG PAN PHI SEG THI USA	HK
triflorum Nutt.		AUS		USA	
tuberosum L.			FIN	IND TUR USA	
verbascifolium L.			IDO	BRA CR HON NIC NIG SAL	AUS
villosum Mill.	ISR		SAF	USA	
welwitschii C. H. Wright				CNK	
xanthocarpum Schrad. & Wendl.			IND	PK	ARG
SOLENOSTEMON Thonn. **Labiatae**					
monostachyus (P. Beauv.) Briq.				NIG	
SOLIDAGO L. Asteraceae *(Compositae)*					
altissima L. *(See* **S. canadensis** L.*)*			JPN	USA	
canadensis L. (Syn. *S. altissima* L.)			CAN USA		AUS
chilensis Meyen			ARG		
gigantea W. Ait.			JPN USA		
graminifolia (L.) Salisb.			USA		
microglossa Dc.			BRA	ARG CHL	
missouriensis Nutt.			USA		

SOLIDAGO L. Asteraceae *(Compositae) (Cont'd)*	**S**	**P**	**C**	**X**	**F**
nemoralis W. Ait.			USA		
occidentalis Nutt.			USA		
rigida L.			USA		
rugosa Mill.			USA		
virgaurea L.			FIN JPN		AFG SOV
SOLIVA Ruiz & Pav. Asteraceae *(Compositae)*					
anthemifolia (Juss.) R. Br. *ex* Less.		TAI		AUS BRA MAU NZ	
pterosperma (Juss.) Less.			ARG AUS	BRA NZ	
sessilis Ruiz & Pav.				BRA CHL NZ	
SONCHUS L. Asteraceae *(Compositae)*					
arvensis L.	AFG HUN NOR POL SOV TNZ	BUL GRE HAW IND JPN KOR SWE TUN TUR	ALK AST AUS CAN ENG FIN GUA IDO NET SPA USA	BEL CHN CZE DEN EGY FRA GER ICE IRA IRE ITA MEL MEX MOZ NEP NZ PK PLW RHO SEG THI YUG	HK
asper (L.) Hill	POL	ARB BEL ENG FRA GER HUN IND MAU NOR SOV SWE TUN	ARG AUS CAN CHN COL EGY IDO IRA ITA JPN KEN MOR SAF SPA TNZ USA	BRA CHL CR ECU FIJ FIN GRE IRQ JOR MEX NET NGI NZ PER PK PLW POR RHO SUD	AFG

SONCHUS L. Asteraceae *(Compositae)* *(Cont'd)*	S	P	C	X	F
asper (L.) Hill *(Cont'd)*				THI TUR UGA URU	
brachyotus Dc.			JPN		
cornutus Hochst. *ex* Steud.		SUD			
exauriculatus (Oliv. & Hiern) O. Hoffm.		KEN TNZ			
glaucescens Jord.			LEB		
oleraceus L.	ENG MEX	AUS BRA ISR LEB PK SPA TUN ZAM	ALK ARG BOR CAN CHN COL EGY FIN HAW HK IDO IND IRQ JPN KEN MOR RHO SAF TAI USA	ANG CAM CHL CR GER GHA GRE GUI IRA IRE ITA JAM KOR MAU MLI MOZ NIG NZ PER POR PR SEG SOV SUD TNZ TUR URU VIE	AFG ARB ETH POL YUG
tartaricus L. *(See* **Lactuca tatarica** (L.) C. A. Mey.)				SOV	
tenerrimus L.			SPA	MOR	ISR
SONNERATIA L. *f.* Sonneratiaceae					
acida L. *f.* *(See* **S. caseolaris** (L.) Engl.)				VIE	
caseolaris (L.) Engl. (Syn. *S. acida* L. *f.*)				SOV	
SOPHORA L. Papilionaceae *(Leguminosae)*					
alopecuroides L.				SOV	
pachycarpa Schrank *ex* C. A. Mey.				SOV	
secundiflora (Ortega) Lag.				USA	
sericea Nutt.				USA	
SORBUS L. Rosaceae					
americana Marsh.				USA	

SORGHUM Moench Poaceae *(Gramineae)*	S	P	C	X	F
affine (Presl) E. G. & A. Camus				VIE	
almum Parodi			ARG SAF	AUS USA	
arundinaceum Roem. & Schult.		VEN		ANG GHA	
bicolor (L.) Moench (Syn. *Andropogon sorghum* (L.) Brot.)				USA	AFG IND SAF
halepense (L.) Pers. (Syn. *Andropogon arundinacea* Scop.) (Syn. *A. halepensis* (L.) Brot.) (Syn. *Holcus halepensis* Pers.)	ARG AUS COL CUB FIJ GRE HAW IND ISR LEB MEL MEX PER PK PLW POL ROM SOV TUR USA VEN YUG	BOR CHL GUA ITA JAM MOZ NZ PAR PHI SAF SPA	BRA IRA IRQ MOR	ARB BOL BUL DR HON IDO MIC NGI NIC PR SAL THI TNZ URU	AFG CEY CHN HK JOR POR TAI
miliaceum Snowden				AUS	IND
plumosum Beauv.				AUS	
propinquum (Kunth) Hitchc.				PHI	
sudanense (Piper) Stapf				ARG SUD USA	AUS HAW SAF
verticilliflorum (Steud.) Stapf	MAU	MOZ SAF UGA	AUS KEN	JAM	IND
virgatum (Hack.) Stapf			EGY		
vulgare Pers.		IND VEN	AUS TAI		HAW
SPANANTHE Jacq. Apiaceae *(Umbelliferae)*					
paniculata Jacq.				COL TRI	
SPARGANIUM L. Sparganiaceae					
americanum Nutt.				USA	
angustifolium Michx.				USA	
chlorocarpum Rydb.				USA	
erectum L.				BEL GER	

	S	P	C	X	F
SPARGANIUM L. **Sparganiaceae** *(Cont'd)*					
erectum L. *(Cont'd)*				NET	
eurycarpum Engelm.				USA	
fluctuans (Morong) Robins.				USA	
ramosum Curt.			IRA POR SPA TUR	ENG	
stoloniferum Buch.-Ham.			JPN		AFG
SPARTINA Schreb. **Poaceae** *(Gramineae)*					
alterniflora Loisel.				USA	
bakeri Merr.				USA	
cynosuroides (L.) Roth				USA	
densiflora Brongn.				CHL	
patens (Ait.) Muhl.				USA	
SPARTIUM L. **Papilionaceae** *(Leguminosae)*					
junceum L.			HAW	USA	AUS ISR
SPATHOGLOTTIS Bl. **Orchidaceae**					
plicata Bl.			HAW	PLW	
SPECULARIA Heist. *ex* Fabric. **Campanulaceae**					
biflora (Ruiz & Pav.) Fisch. & Mey.			ARG		
perfoliata (L.) A. Dc.				USA	
SPERGULA L. **Caryophyllaceae**					
arvensis L. (Syn. *S. maxima* Weihe) (Syn. *S. sativa* Boenn.) (Syn. *S. vulgaris* Boenn.)	ALK FIN IRE KEN PHI TNZ	AUS BRA CHL COL ENG GER IND NOR NZ SWE TAS	ARG CAN HAW ISR JPN POR SAF SPA TAI USA	BEL DEN ECU ETH FRA HUN ICE KOR SOV	JAM PK POL UGA
linicola Bor. *ex* Nym.				SOV	POL
maxima Weihe (*See* **S. arvensis** L.)				SOV	
sativa Boenn. (*See* **S. arvensis** L.)				SOV	
vulgaris Boenn. (*See* **S. arvensis** L.)				SOV	
SPERGULARIA (Pers.) J.S. & K.B. Presl. **Caryophyllaceae**					
marina (L.) Griseb.			EGY		

SPERGULARIA (Pers.) J.S. & K.B. Presl. **Caryophyllaceae** *(Cont'd)*	S	P	C	X	F
media (L.) Presl				CHL	
platensis Fenzl				CHL	
rubra (L.) J.S. & K.B. Presl.				USA	
SPERMACOCE L. **Rubiaceae**					
hispida L.f. *(See* **Borreria articularis** (L.f.) F. N. Williams)				VIE	
laevis Lam. *(See* **Borreria laevis** (Lam.) Griseb.)				USA	
latifolia Aubl. *(See* **Borreria alata** (Aubl.) Dc.)	CEY			MEX	
pilosa Dc.				ANG	
tenuior L.				ANT IDO MEX TRI	
SPHAERANTHUS L. **Asteraceae** *(Compositae)*					
africanus L.		PHI		THI VIE	IDO
bullatus Mattf.		TNZ	KEN		
indicus L.				CEY IND VIE	IDO
senegalensis Dc.				GHA	
suaveolens (Forsk.) Dc.			KEN		
SPHAEROMARISCUS E. G. Camus **Cyperaceae**					
microcephalus J.S. & K.B. Presl. E. G. Camus				VIE	
SPHENOCLEA Gaertn. **Sphenocleaceae**					
zeylanica Gaertn.	GUY NIG PHI SUR TRI	GHA IND MAD SEG THI	TAI	CHN CUB IDO MAL PK USA VIE	HK
SPHENOMERIS Maxon **Lindsaeaceae**					
chinensis (L.) Maxon			HAW		
SPIGELIA L. **Loganiaceae**					
anthelmia L.		IDO	NIG PR TRI	ANT DR GHA JAM PAN SUR	
SPILANTHES Jacq. **Asteraceae** *(Compositae)*					
acmella (L.) Murr.			IDO	IND	

SPILANTHES Jacq. **Asteraceae** *(Compositae)* *(Cont'd)*	S	P	C	X	F
americana Hieron. *ex* Sod.				SAL VEN	
calva Dc.			IDO		
macraei Hook. & Arn.				CHL	
uliginosa Sw.			TRI	IVO	
urens Jacq.				PER	
SPIRAEA L. **Rosaceae**					
alba Du Roi				USA	
douglasii Hook.				USA	
japonica L. *f.*				USA	
latifolia (Ait.) Borkh.				USA	
tomentosa L.				USA	
SPIRODELA Schleid. **Lemnaceae**					
oligorrhiza (Kurz.) Hegelm. (Syn. *Lemna oligorrhiza* (Hegelm.) Kurz.)					
polyrhiza (L.) Schleid. (Syn. *Lemna polyrhiza* L.)	GER GHA ITA JPN	POR		BEL BRA CAB CAN CHN CNK EGY HAW HUN IDO IND ISR JAM JOR KEN KOR NEP NOR RHO SWE TAI THI TNZ UGA USA VIE	
SPONDIAS L. **Anacardiaceae**					
lutea L. *(See* **S. mombin** L.)				BRA	
mombin L. (Syn. *S. lutea* L.)				HON JAM	
SPOROBOLUS R. Br. **Poaceae** *(Gramineae)*					
africanus (Poir.) Robyns & Tourn. (Syn. *S. capensis* (Willd.) Kunth)			HAW	NZ USA	CEY IND
airoides (Torr.) Torr.				USA	HAW

SPOROBOLUS R. Br. Poaceae *(Gramineae) (Cont'd)*	S	P	C	X	F
airoides (Torr.) Torr. *(Cont'd)*					IND
berteroanus (Trin.) Hitchc. & A. Chase *(See* **S. poiretii** (Roem. & Schult.) Hitchc.*)*		IDO		COL	ARG HAW
capensis (Willd.) Kunth *(See* **S. africanus** (Poir.) Robyns & Tourn.*)*			AUS HAW	MAU USA	SAF
coromandelianus (Retz.) Kunth				IND	
cryptandrus (Torr.) A. Gray				USA	
diander (Retz.) Beauv.	IND		HAW MAL TAI	BND FIJ NGI PHI PLW USA VIE	BUR CEY IDO
heterolepis (A. Gray) A. Gray				USA	
indicus (L.) R. Br.			JPN PHI TRI	ANT BND CHL FIJ JAM NIC PER SUR	AUS PR
neglectus Nash				USA	
poiretii (Roem. & Schult.) Hitchc. (Syn. *S. berteroanus* (Trin.) Hitchc. & A. Chase)			ARG COL HAW USA	FIJ MEX	BRA
pyramidalis Beauv.				GHA IVO NIG	ANG
robustus Kunth				ANG	
tenuissimus (Schrank) O. Ktze.				TRI	IND
tremulus (Willd.) Kunth		IND		BND	CEY IDO
vaginiflorus (Torr.) Wood				USA	
virginicus (L.) Kunth			AUS HAW		CEY IDO IND SAF
STACHYS L. Lamiaceae *(Labiatae)*					
agraria Schlecht. & Cham.			SAL	MEX	
annua L.		TUN		SOV USA	POL
arvensis (L.) L.	AST BRA HAW	AUS POL POR SWE	ARG ENG ITA TUR	CAN IRA JAM JOR MAU NOR	AFG CHN ISR

STACHYS L. Lamiaceae *(Labiatae) (Cont'd)*	S	P	C	X	F
arvensis (L.) L. *(Cont'd)*				NZ PER USA	
durandiana Coss.				MOR	
elliptica H. B. K.				ECU	
floridana Shuttlew				USA	
grandidentata Lindl.				CHL	
lanata Jacq.			IRA		
nivea Labill.			LEB		
ocymastrum (L.) Briq.				MOR	
palustris L.		SOV	FIN	BEL GER ICE NZ USA	AUS POL
parviflora Benth.				PK	AFG
petiolosa Briq.				ARG	
pubescens Tenore			IRA		
sieboldii Miq.				USA	
STACHYTARPHETA Vahl Verbenaceae					
angustifolia Vahl	NIG	GHA		IVO	
australis Moldenke			HAW IDO	BRA	
cayennensis (L. C. Rich.) Vahl	ECU HAW NGI PLW		COL HON JAM TAI THI TRI USA	ANT AUS DAH GHA GUI IDO IVO LIB MEL NIG PER SUR	BRA SAL
indica (L.) Vahl		MAL		ANT GHA IND JAM MAU TRI USA	AUS IDO
jamaicensis (L.) Vahl	HAW	GHA MEL	AUS ECU IDO PHI PLW PR SUR TAI TRI	ANT CAB CHN COL CR IND JAM MAL MAU	CEY

STACHYTARPHETA Vahl Verbenaceae *(Cont'd)*	S	P	C	X	F
jamaicensis (L.) Vahl *(Cont'd)*				NGI THI USA VIE	
mutabilis (Jacq.) Vahl			AUS HAW	FIJ	
urticaefolia (Salisb.) Sims	HAW MEL NGI PLW		AUS FIJ	IND MIC SUR USA	
STANLEYA Nutt. **Brassicaceae** *(Cruciferae)*					
pinnata (Pursh) Britt.				USA	
STELLARIA L. **Caryophyllaceae**					
alsine Grimm *(See* **S. uliginosa** Murr.*)*	JPN PHI		TAI	CAN CHN NZ USA	KOR
aquatica (L.) Scop. (Syn. *Malachium aquaticum* (L.) Fries)	TAI		CHN IND JPN	CAN HUN IRA ITA JOR KOR NEP NOR PK SOV SWE THI TUN TUR	HK
cuspidata Willd. *ex* Schlecht.				CHL	
graminea L.			CAN FIN SOV SPA	NZ USA	POL
media (L.) Cyr.	ALK ENG FIN IRE JPN SPA	AUS BEL CHL ETH GRE ISR NGI NOR NZ SOV SWE TAS TNZ TUN USA	ARG BUL CAN CHN COL FRA GER HAW IND IRA IRQ KEN POR SAF YUG	BRA CZE DEN ICE IDO ITA MAU MEX PK TUR URU	AFG HK HUN KOR POL
neglecta Weihe			JPN		
pallida Dum. *ex* Nym.			EGY		
uliginosa Murr. (Syn. *S. alsine* Grimm)	JPN KOR POL		CHN ITA	PHI PK	HK

	S	P	C	X	F
STEMODIA L. **Scrophulariaceae**					
parvifolia W. T. Ait. *(See* **S. verticillata** (Mill.) Bold.)				MAU	HAW
verticillata (Mill.) Bold. (Syn. *S. parvifolia* W. T. Ait.)				IDO	
viscosa Roxb.				IND	
STENOCHLAENA J. Sm. **Blechnaceae**					
palustris (Burm.) Bedd.			MAL		
STENOPHYLLUS Rafin. **Cyperaceae** *(See* **BULBOSTYLIS** Kunth)					
barbatus (Rottb.) Cooke *(See* **Bulbostylis barbata** (Rottb.) C. B. Clarke)				IND	
STENOTAPHRUM Trin. **Poaceae** *(Gramineae)*					
dimidiatum (L.) Brongn. (Syn. *S. glabrum* Trin.)				MAU	
glabrum Trin. *(See* **S. dimidiatum** (L.) Brongn.)				MAU	
secundatum (Walt.) O. Ktze.				HAW PR	FIJ JAM PLW USA	AUS
STEPHANIA Lour. **Menispermaceae**					
cephalantha Hayata			TAI		
elegans Hook. *f.* & Thoms.				IND	
hernandifolia Walp. *(See* **S. japonica** (Murr.) Miers)				IDO	
japonica (Murr.) Miers (Syn. *S. hernandifolia* Walp.)					
STEPHANOMERIA Nutt. **Asteraceae** *(Compositae)*					
tenuifolia Torr.				USA	
STICTOCARDIA Hall. *f.* **Convolvulaceae**					
tiliaefolia (Desr.) Hall. *f.*				PLW	
STIGMAPHYLLON A. Juss. **Malpighiaceae**					
tomentosum A. Juss.			PR		
STILLINGIA L. **Euphorbiaceae**					
sylvatica L.				USA	
STIPA L. **Poaceae** *(Gramineae)*					
arvenacea L.				USA	
brachychaeta Godr.			ARG		
cernua Stebbins & Love			HAW		
comata Trin. & Rupr.				USA	HAW

STIPA L. Poaceae *(Gramineae) (Cont'd)*	S	P	C	X	F
hyalina Nees			ARG		
neesiana Trin. & Rupr.				CHL NZ	
retorta Cav.			POR		
spartea Trin.				USA	
tortilis Desf.			IRA		ISR POR
viridula Trin.				USA	HAW
STIZOLOBIUM P. Br. **Papilionaceae** *(Leguminosae)*					
pruritum (Wight) Piper		MEX PR			
STOEBE L. **Asteraceae** *(Compositae)*					
vulgaris Levyns			RHO		
STRATIOTES L. **Hydrocharitaceae**					
aloides L.				BEL GER	
STRIGA Lour. **Scrophulariaceae**					
angustifolia (Don) Saldanha (Syn. *S. euphrasioides* (Benth.) Benth.)	IND			BND BUR CEY ETH IDO RHO SAF TNZ	
asiatica (L.) O. Ktze. *(See* **S. lutea** Lour.)	PK SAF	RHO UGA ZAM	KEN	CAB IDO IND MAL NIG NZ SOV USA	
aspera Benth.				NIG	
densiflora (Benth.) Benth.	IND	PK		NIG RHO SAF	
elegans Benth.				RHO	
euphrasioides (Benth.) Benth. *(See* **S. angustifolia** (Don) Saldanha)	IND			NIG RHO	IDO
forbesii Benth.			RHO SAF	NIG	KEN
gesnerioides (Willd.) Vatke (Syn. *S. orobanchoides* (R. Br. *ex* Endl.) Benth.)		NIG	RHO SAF	ARB AUS CAB CEY EGY ETH GUI	

STRIGA Lour. Scrophulariaceae *(Cont'd)*	S	P	C	X	F
gesnerioides (Willd.) Vatke *(Cont'd)*				IND JPN KEN MOZ SUD USA	
hermonthica (Del.) Benth.	ARB NIG SUD	UGA		CAM CHA CNK DAH EGY ETH GHA GUI KEN MLI MOZ NGR RHO SAF SEG TNZ VOL	
hirsuta (Benth.) Benth. *(See* **S. lutea** Lour.)				MAU	
lutea Lour. (Syn. *S. asiatica* (L.) O. Ktze.) (Syn. *S. hirsuta* (Benth.) Benth.)	IND SAF USA	MAU PK ZAM		ARB AUS BND BUR CAB CAM CEY CHN CNK EGY GHA GUI IDO IVO JPN KEN LIB MAD MAL MOZ NGI NIG NZ RHO SEG SUD THI TNZ UGA VIE	HK
orobanchoides (R. Br. *ex* Endl.) Benth. *(See* **S. gesnerioides** (Willd.) Vatke)				NIG SUD	
senegalensis Benth.				NGR	
thunbergii Benth.				ZAM	
STROPHIOSTOMA Turcz. **Boraginaceae**					
sparsiflorum Turcz.			SOV		

STRUCHIUM P. Br. Asteraceae *(Compositae)*	S	P	C	X	F
sparganophorum (L.) O. Ktze.			TRI	FIJ PLW	
STRYPHNODENDRON Mart. Mimosaceae *(Leguminosae)*					
barbatimam Mart.				BRA	
obovatum Benth.				BRA	
STYLOSANTHES Sw. Papilionaceae *(Leguminosae)*					
guianensis (Aubl.) Sw.				PER	
hamata Taub.				JAM	
sundaica Taub.				AUS	IDO
viscosa Sw.				JAM	
STYPANDRA R. Br. Liliaceae					
glauca R. Br.			AUS		
STYPHANOPHYSUM Pohl Acanthaceae					
longifolium Poir. (Syn. *Ruellia graecizans* Back.)					
STYPHELIA (Soland *ex* G. Forst) Sm. Epacridaceae					
douglasii (A. Gray) Hochr.			HAW		
tameiameiae (Cham.) F. Muell.			HAW		
STYRAX L. Styracaceae					
camporum Pohl				BRA	
SUAEDA Forsk. Chenopodiaceae					
australis Moq.				VIE	
fruticosa Forsk.				PK USA	
SUCCISA Neck. Dipsacaceae					
australis Reichb.				USA	
SUCKLEYA A. Gray Chenopodiaceae					
suckleyana (Torr.) Rydb.				USA	
SWAINSONA Salisb. Papilionaceae *(Leguminosae)*					
galegifolia R. Br.				AUS	
salsula (Pall.) Taub.				USA	
SWIETENIA Jacq. Meliaceae					
macrophylla King				HON	
SYMPHORICARPOS Duham. Caprifoliaceae					
albus (L.) Blake				USA	
occidentalis Hook.				USA	

SYMPHORICARPOS Duham. Caprifoliaceae *(Cont'd)*	S	P	C	X	F
orbiculatus Moench				USA	
SYMPHYTUM L. **Boraginaceae**					
asperum Lepechin			SOV	NZ USA	
officinale L.			SPA	BEL GER NZ USA	POL
SYMPLOCARPUS Salisb. *ex* Nutt. **Araceae**					
foetidus (L.) Nutt.				USA	
SYNEDRELLA Gaertn. **Asteraceae** *(Compositae)*					
nodiflora (L.) Gaertn.	GHA HON IVO	COL MEL PHI TRI	HAW IDO IND JAM MIC NIG PER PLW PR TAI	ANT CAB CHN CNK DAH DR ECU FIJ NGI SAL SEG SUR THI UGA USA VEN VIE	BOR
vialis A. Gray				AUS	
SYNGONIUM Schott **Araceae**					
podophyllum Schott			SAL	GUA HON PAN	
SYRINGA L. **Oleaceae**					
vulgaris L.				USA	
SYZYGIUM Gaertn. **Myrtaceae**					
cumini (L.) Skeels (Syn. *Eugenia cumini* (L.) Druce) (Syn. *E. jambolana* Lam.)					
jambos (L.) Alst. (Syn. *Eugenia jambos* L.)					
TABEBUIA Gomez **Bignoniaceae**					
haemantha Dc.			PR		
TABERNAEMONTANA L. **Apocynaceae**					
chrysocarpa Blake				CR	
fuchsiaefolia A. Dc.	BRA				

TACCA J. R. & G. Forst. Taccaceae	S	P	C	X	F
leontopetaloides (L.) O. Ktze.				PLW	
TAENIATHERUM Nevski Poaceae *(Gramineae)*					
asperum (Sim.) Nevski (Syn. *Elymus caput-medusae* Boiss.)			USA		
TAENITIS Willd. Sinopteridaceae					
blechnoides (Willd.) Sw.				MAL	
TAGETES L. Asteraceae *(Compositae)*					
minuta L.	BRA GUA SAF TNZ ZAM	ARG GUY KEN RHO SWZ UGA	AUS ETH HAW PER	ANG BOT CHL EGY NZ THI URU USA	
patula L.				COL DR MEX	
tenuifolia Cav.				HON	
TALINUM Adans. Portulacaceae					
paniculatum (Jacq.) Gaertn.			ARG IND SAL	COL FIJ MEX VEN	
triangulare (Jacq.) Willd.	GHA			CNK COL NIG PHI	
TAMARIX L. Tamaricaceae					
aphylla (L.) Karst.				USA	
gallica L.				IRA USA	
pentandra Pall.				USA	
TANACETUM L. Asteraceae *(Compositae)*					
vulgare L.			AUS CAN SPA	NZ USA	SOV
TANAECIUM Sw. Bignoniaceae					
exitiosum Dugand		COL			
TAPIRIRA Aubl. Anacardiaceae					
guianensis Aubl.				BRA	
TARAXACUM Wiggers Asteraceae *(Compositae)*					
erythrospermum Andrz. *ex* Bess. *(See* **T. officinale** Wiggers)				USA	ARG
japonicum Koidz.			JPN		

TARAXACUM Wiggers Asteraceae *(Compositae) (Cont'd)*	S	P	C	X	F
officinale Wiggers (Syn. *T. erythrospermum* Andrz. *ex* Bess.)	AST ITA POL TUR	BEL CAN FIN GER NOR NZ SWE USA	AFG ALK ARG AUS COL CZE ENG FRA GUA HAW HON ICE IDO JPN MAU NET PLW SAF SOV SPA TUN	BRA CHL CHN CR DEN ECU GRE IRQ JOR MEX PER PK RHO SAL URU VIE YUG	HK KOR
platycarpum Dahlst.			JPN		
serotinum Poir.			SAF	TUR	SOV
vulgare Schrank		FIN		ICE SOV	
TARCHONANTHUS L. Asteraceae *(Compositae)*					
camphoratus L.		KEN TNZ		RHO	
TAXODIUM Rich. **Taxodiaceae**					
distichum (L.) Richard				USA	
TAXUS L. **Taxaceae**					
brevifolia Nutt.				USA	
floridana Nutt.				USA	
TECOMA Juss. **Bignoniaceae**					
stans (L.) H. B. K.		ARG		NIC USA	
TELIOSTACHYA Nees **Acanthaceae**					
alopecuroides Nees		TRI		DR	
TELOXYS Moq. **Chenopodiaceae**					
aristata Moq.				SOV	
TEPHROSIA Pers. **Papilionaceae** *(Leguminosae)*					
candida (Roxb.) Dc.		PR		IND	
cathartica Urb.		PR			
cinerea Pers.		PR			MEX
ehrenbergiana Schweinf.				ANG	
elegans Schum.				IVO	

TEPHROSIA Pers. Papilionaceae *(Leguminosae) (Cont'd)*	S	P	C	X	F
procumbens Buch.-Ham.				IND	
purpurea (L.) Pers.		IND		USA	
spinosa (L. *f.*) Pers.				IND	
tinctoria Pers.				IND	
uniflora Pers.		SUD			
TERAMNUS P. Br. Papilionaceae *(Leguminosae)*					
labialis (L. *f.*) Spreng.			IND	ANT DR	
volubilis Sw.				TRI	
TERMINALIA L. Combretaceae					
catappa L.				CR HON PAN	
oblongata F. Muell.		AUS			
sericea Burch. *ex* Dc.				ANG RHO	
TESSARIA Ruiz & Pav. Asteraceae *(Compositae)*					
absinthioides Dc.				ARG CHL	
integrifolia Ruiz & Pav.				PER	
TETRACERA L. Dilleniaceae					
breyniana Schlecht.				BRA	
indica (Christm. & Panz.) Merr.				MAL	
scandens (L.) Merr.				MAL	
TETRADYMIA Dc. Asteraceae *(Compositae)*					
canescens Dc.				USA	
glabrata A. Gray				USA	
TETRAGONIA L. Tetragoniaceae					
expansa Murr. *(See* **T. tetragonioides** (Pallas) O. Ktze.)	NZ	JPN			AUS ISR
tetragonioides (Pallas) O. Ktze. (Syn. *T. expansa* Murr.)				AUS	
TETRAGONOLOBUS Scop. Papilionaceae *(Leguminosae)*					
purpureus Moench		TUN			
TETRASTIGMA Planch. Vitaceae					
umbellata (Hemsl.) Nakai			TAI		
TEUCRIUM L. Lamiaceae *(Labiatae)*					
cubense Jacq.				MEX	

TEUCRIUM L. Lamiaceae *(Labiatae) (Cont'd)*	S	P	C	X	F
integrifolium Benth.				AUS	
resupinatum Desf.				MOR	
scordium L.		NZ	SOV		
spinosum L.			MOR		
TEXIERA Jaub. & Spach **Brassicaceae** *(Cruciferae)*					
glastifolia (Dc.) Jaub. & Spach *(See* **Glastaria deflexa** Boiss.*)*			LEB		ISR
THALIA L. **Marantaceae**					
geniculata L.				SAL SUR VEN	PER
THALICTRUM L. **Ranunculaceae**					
minus L.				TUR	
simplex L.				TUR	
THELYPTERIS Schmidel **Thelypteridaceae** *(Syn. LASTREA* Bory)					
gracilescens (Bl.) Ching (Syn. *Lastrea gracilescens* (Bl.) Moore)					
THEMEDA Forsk. **Poaceae** *(Gramineae)*					
arguens (L.) Hack.		IDO	IND	JAM MAL NGI	HAW
australis (R. Br.) Stapf				NGI	HAW
gigantea (Cav.) Hack.				PHI	HAW IDO
quadrivalvis (L.) O. Ktze.				AUS MAU	
triandra Forsk.				NZ PHI	
villosa (Poir.) A. Camus				VIE	IND
THERMOPSIS R. Br. **Papilionaceae** *(Leguminosae)*					
rhombifolia (Nutt.) Richards				USA	
THESIUM L. **Santalaceae**					
australe R. Br.				AUS	
humile Vahl		TUN	MOR		
THESPESIA Soland. *ex* Corr. **Malvaceae**					
garckeana F. Hoffm. (Syn. *Azanza garckeana* (F. Hoffm.) Exell & Hillc.)					
THLASPI L. **Brassicaceae** *(Cruciferae)*					
arvense L.	ALK CAN	AST BEL	ARG AUS	AFG COL	

THLASPI L. Brassicaceae *(Cruciferae) (Cont'd)*	S	P	C	X	F
arvense L. *(Cont'd)*	GER KOR POL	HUN IRA NZ SWE TUN	CHN ENG FIN JPN NET NOR SOV SPA USA	CZE ICE ITA JOR PK TUR	
montanum L.				SPA	
perfoliatum L.			LEB POR SOV	TUR USA	ISR
THRELKELDIA R. Br. **Chenopodiaceae**					
proceriflora F. Muell.				AUS	
THUAREA Pers. **Poaceae** *(Gramineae)*					
involuta (G. Forst.) R. Br.			TAI		
THUJA L. **Cupressaceae**					
occidentalis L.				USA	
plicata Donn.				USA	
THUNBERGIA Retz. **Acanthaceae**					
alata Boj. *ex* Sims			PR	AUS CR DR FIJ JAM MAU PLW	
annua Hochst. *ex* Nees		SUD			
fragrans Roxb.		HAW	PR	MAU	
THYMELAEA Mill. **Thymelaeaceae**					
passerina Lange				MOR	
THYMUS L. **Lamiaceae** *(Labiatae)*					
serpyllum L.				USA	
vulgaris L.				NZ PER	
THYSANOLAENA Nees **Poaceae** *(Gramineae)*					
maxima (Roxb.) O. Ktze.				VIE	
TIARIDIUM Lehm. **Boraginaceae**					
indicum (L.) Lehm. *(See* **Heliotropium indicum** L.*)*			PR	DR MEX	
TIBOUCHINA Aubl. **Melastomataceae**					
longifolia Baill.				VEN	
semidecandra Cogn.		HAW			

	S	P	C	X	F
TIDESTROMIA Standl. **Amaranthaceae**					
lanuginosa (Nutt.) Standl.				USA	
TILIA L. **Tiliaceae**					
americana L.				USA	
heterophylla Vent.				USA	
TILLANDSIA L. **Bromeliaceae**					
landbecki Phil.				CHL	
TINANTIA Scheidw. **Commelinaceae**					
erecta Scheidw.				SAL	
TINOSPORA Miers **Menispermaceae**					
cordifolia Miers				IND	
TITHONIA Desf. *ex* Juss. **Asteraceae** *(Compositae)*					
diversifolia (Hemsl.) A. Gray			AUS HAW IND	FIJ MAU THI USA	
TOCOCA Aubl. **Melastomataceae**					
quadrialata (Naud.) Macbride				PER	
TOLPIS Adans. **Asteraceae** *(Compositae)*					
barbata (L.) Gaertn.			MOR	NZ POR	AUS
TORDYLIUM L. **Apiaceae** *(Umbelliferae)*					
aegyptiacum (L.) Lam.		JOR	LEB		ISR
apulum L.		JOR		GRE	
officinale L.				TUR	
TORENIA L. **Scrophulariaceae**					
bicolor Dalz.				IND	
concolor Lindl.				PHI	
spicata Engl.				NIG	
thouarsii O. Ktze.				NIG	
TORILIS Adans. **Apiaceae** *(Umbelliferae)*					
anthriscus Gaertn.			CHN SPA		
arvensis (Huds.) Link		ETH	POR		
japonica (Houtt.) Dc.			JPN	USA	
leptophylla (L.) Reichb. *f.*			POR		
neglecta Schult.			LEB		ISR
nodosa (L.) Gaertn.			LEB MOR	AUS BRA	ISR

TORILIS Adans. Apiaceae *(Umbelliferae) (Cont'd)*	S	P	C	X	F
nodosa (L.) Gaertn. *(Cont'd)*			POR	CHL GRE MAU NZ	
scabra (Thunb.) Dc. (Syn. *Caucalis scabra* (Thunb.) Makino)					
syriaca Boiss. & Blanche			LEB		
TORREYA Arn. **Taxaceae**					
californica Torr.				USA	
TORULINIUM Desv. **Cyperaceae**					
ferax (L. C. Rich.) Ham. *(See* **Cyperus odoratus** L.)		SUR		ANT TRI	
TOURNEFORTIA L. **Boraginaceae**					
bicolor Swartz				PER	
cuspidata H. B. K.				PER	
hirsutissima L.			PR TRI	JAM	HON
TRACHYMENE Rudge **Apiaceae** *(Umbeliferae)*					
oleracea (Domin) B. L. Burtt				AUS	
TRADESCANTIA L. **Commelinaceae**					
albiflora Kunth.			AUS		
crassifolia Cav.	MEX				
fluminensis Vell.			TAI	NZ	
gracilis H. B. K.				BRA	
volubilis L.			PR		
TRAGIA L. **Euphorbiaceae**					
benthami Baker				GHA	
cannabina L. *f.*				IND SUD	
involucrata Linn.				IND	
mercurialis L.				ANG	
nepetifolia Cav.				MEX	
spathulata Benth.				GHA	
stylaris Muell.-Arg.				MEX	
volubilis L.				DR	
TRAGOPOGON L. **Asteraceae** *(Compositae)*					
buphtalmoides (Dc.) Boiss.			LEB		ISR
dubius Scop.			CAN	USA	

TRAGOPOGON L. Asteraceae *(Compositae)* *(Cont'd)*	S	P	C	X	F
graminifolius Dc.			IRA		
hybridus L. *(See* **Geropogon glaber** L.)					POR
latifolius Boiss.				TUR	
major Jacq.			SAF	ARG CAN USA	SOV
orientalis L.			SOV		
porrifolius L.			CAN SAF	GRE NZ USA	
pratensis L.			CAN ENG SAF SPA	GER ITA TUR USA	SOV
TRAGUS Hall. **Poaceae** *(Gramineae)*					
berteronianus Schult.			RHO SAF	MOZ	
biflorus Schult.				IND	
racemosus Scop.	SAF		ARG	IND SPA	RHO
TRAPA L. **Trapaceae**					
natans L.			USA	BEL BOT CAB CHN CNK CZE GER GRE GUI HUN IDO IND IRA ITA JPN KEN KOR MOZ NEP POL ROM SAF SEG SUD THI TNZ UGA	
TREMA Lour. **Ulmaceae**					
aspera Bl.		AUS			
guineensis (Schum.) Priemer				GHA NIG	

TREMA Lour. Ulmaceae *(Cont'd)*	S	P	C	X	F
micrantha Bl.				CR HON SAL USA	
orientalis (Bl.) Bl.			HAW TAI		
TRIANTHEMA L. **Aizoaceae**					
australis Melville		AUS			
decandra L.				IND	
galericulata Melville				AUS	
monogyna L.		IND		PK	
pentandra L.	PK	SUD	RHO	ANG	ISR
portulacastrum L.	AUS GHA IND PHI THI	CAB GUA NIC	COL HON NIG PER PR SEG	CR ECU IDO IVO KEN MEX MOZ NGI PK PLW SAL SUD TNZ USA VEN VIE	
triquetra Rottl. *ex* Willd.				IND	
TRIBULUS L. **Zygophyllaceae**					
cistoides L.			AUS HAW MAD PR VEN	DR IDO JAM MAU MEX USA	
terrestris L.	AUS	CAB GRE HAW IND IRA ISR KEN LEB MAD MOZ PK SAF SWZ TUR	ARG ETH IRQ MOR POR RHO SPA TNZ USA	ARB CHN FRA GHA ITA JOR JPN MAU SOV SUD THI TRI VIE YUG ZAM	KOR
zeyheri Sond.				ANG	
TRICHACHNE Nees **Poaceae** *(Gramineae)*					
insularis (L.) Nees	HAW	PAR	BRA	BOL	CEY

TRICHACHNE Nees Poaceae *(Gramineae) (Cont'd)*	S	P	C	X	F
insularis (L.) Nees *(Cont'd)*		VEN	PER PR	COL DR HON MEX PHI SAL USA	FIJ
patens Swallen				MEX	
sacchariflora Nees				BOL PER	
TRICHODESMA R. Br. **Boraginaceae**					
africanum R. Br.				GHA	
amplexicaule Roth				IND	
dekindtianum Guerke				ANG	
indicum R. Br.				IND MAU	
zeylanicum R. Br.		TNZ	KEN RHO	FIJ MAU MOZ ZAM	
TRICHOLAENA Schrad. **Poaceae** *(Gramineae)*					
repens (Willd.) Hitchc. *(See* **Rhynchelytrum repens** (Willd.) C. E. Hubb.)			HAW PR	BRA	MEX
rosea Nees *(See* **Rhynchelytrum repens** (Willd.) C. E. Hubb.)		MAL		DR	AUS HAW PR
TRICHOSTEMA L. **Lamiaceae** *(Labiatae)*					
dichotomum L.				USA	
TRIDAX L. **Asteraceae** *(Compositae)*					
procumbens L.	GHA IVO MOZ NGI THI	GUA HON IDO NIG SAL TNZ	AUS CEY HAW IND KEN MAL PHI PR TAI TRI	ANT CAB CHN CNK COL CR DAH DR ECU FIJ MAU MEL MEX MIC NEP NZ PLW RHO SAF SEG SOV SUD UGA VEN VIE	BRA

TRIFOLIUM L. Papilionaceae *(Leguminosae)*	S	P	C	X	F
agrarium L.				POR USA	
amabile H. B. K.				PER	
arvense L.			POR SOV SPA	NZ USA	AUS ISR POL
campestre Schreb.			IRQ POR	BRA CHL NZ TUR USA	AUS ISR POL SOV
dubium Sibth.			ENG JPN SOV	NZ USA	AUS POL
elegans Savi			SOV		
filiforme L.				COL	ISR
glomeratum L.				NZ POR	
hybridum L.			SOV	COL NZ USA	AUS
incarnatum L.			SPA	USA	
lappaceum L.			POR	TUR	
macraei Hook. & Arn.				CHL	
medium L.			SOV	NZ USA	AUS
montanum L.			SOV		
nigrescens Viv.				POR	
pratense L.			ARG ENG FIN IRA JPN SPA	COL NZ USA	AUS SOV
procumbens L.				IRQ USA	
purpureum Gilib.				TUR	
repens L.		NZ	ARG CHN COL ENG JPN POR SOV SPA	ITA TUR USA VEN	AUS CHL ISR
resupinatum L.			EGY IRQ POR	NZ USA	ISR
scabrum L.				POR	

	S	P	C	X	F
TRIFOLIUM L. Papilionaceae *(Leguminosae) (Cont'd)*					
stellatum L.			SPA		POR
strepens Crantz			SOV		POL
striatum L.				SPA	
subterraneum L.		AUS		NZ	ISR
tomentosum L.			POR	NZ	
TRIGLOCHIN L. **Juncaginaceae**					
maritimum L.	ALK			USA	
palustre L.			FIN	USA	
procerum R. Br.				AUS	
TRIGONELLA L. Papilionaceae *(Leguminosae)*					
foenum-graecum L.			POR	TUR	
hamosa L.			EGY		
incisa L.				PK	
laciniata L.			EGY		
monspeliaca L.			POR		
polycerata L.		PK		IND	
radiata (L.) Boiss.			LEB		
TRIGONOTIS Stev. **Boraginaceae**					
clavata Stev.				CHN	
peduncularis Benth. *ex* S. Moore			CHN JPN		
TRIGUERA Cav. **Solanaceae**					
osbekii (L.) Willk.			MOR		
TRIMEZIA Salisb. *ex* Herb. **Iridaceae**					
martinicensis (L.) Herb.			TRI		
TRIODANIS Rafin. **Campanulaceae**					
biflora Greene		ARG			
TRIPHASIA Lour. **Rutaceae**					
trifolia (Burm. *f.*) Wilson				FIJ	
TRIPLEUROSPERMUM Sch.-Bip. **Asteraceae** *(Compositae)*					
inodorum (L.) Sch.-Bip.			GER	HUN	
maritimum (L.) Koch	FIN	SWE	ENG	CZE SOV	
TRIPLOTAXIS Hutch. **Asteraceae** *(Compositae)*					
stellulifera Hutch.				CNK GHA	

TRIPSACUM L. Poaceae *(Gramineae)*	S	P	C	X	F
laxum Nash			MAL	CEY COL	CR HAW
TRISETOBROMUS Nevski **Poaceae** *(Gramineae)*					
hirtus (Trin.) Nevski				CHL	
TRISETUM Pers. **Poaceae** *(Gramineae)*					
bifidum (Thunb.) Ohwi			JPN		
TRITICUM L. **Poaceae** *(Gramineae)*					
aestivum L.				NEP	
ramosum Trin.				SOV	
repens L. *(See* **Agropyron repens** (L.) Beauv.)				SOV	
TRITONIA Ker-Gawl. **Iridaceae**					
pottsii Benth. & Hook.			HAW		
x **crocosmiflora** (Lemoine) Nich.			HAW	USA	
TRIUMFETTA L. **Tiliaceae**					
acuminata H. B. K.				COL	
althaeoides Lam.			TRI	BRA	
bartramia L.	AUS		HAW IND	PLW USA	FIJ MEL
cordifolia A. Rich. *(See* **T. semitriloba** (L.) Jacq.)				NIG	
flavescens Hochst. *ex* A. Rich.			KEN IDO		
lappula L.				COL HON PAN SAL	
pentandra A. Rich.			NIG		
pilosa Roth			BOR SAF	IND THI	
procumbens Forst. *f.*				FIJ	
rhomboidea Jacq.	FIJ		IDO KEN RHO	ANG MEL NIG SAF	AUS
rotundifolia Lam.				IND	
semitriloba (L.) Jacq. (Syn. *T. cordifolia* A. Rich.)			HAW	BRA DR PER PHI USA	PR
velutina Sieber *ex* Presl				AUS	
welwitschii Mast.				ANG	

	S	P	C	X	F
TRIXIS P. Br. **Asteraceae** *(Compositae)*					
radiale Lag.				SAL	
TROPAEOLUM L. **Tropaeolaceae**					
gracile (Hook. & Arn.) Sparre				CHL	
TSUGA Carr. **Pinaceae**					
canadensis (L.) Carr.				USA	
heterophylla (Rafin.) Sarg.				USA	
mertensiana (Bong.) Carr.				USA	
TUBOCAPSICUM Makino **Solanaceae**					
anomalum (Fr. & Sav.) Makino			JPN		
TULIPA L. **Liliaceae**					
montana Lindl.				PK	
TUNICA Hall. **Caryophyllaceae**					
prolifera (L.) Scop. *(See* **Dianthus prolifer** L.)			POR		AUS
velutina (Guss.) Fisch. & Mey.			POR		
TUPA G. Don **Campanulaceae**					
portoricensis Vatke			PR		
TURGENIA Hoffm. **Apiaceae** *(Umbelliferae)*					
latifolia (L.) Hoffm.			IRA LEB	TUR	
TURNERA L. **Turneraceae**					
ulmifolia L.			PR SAL	BOL BRA DR FIJ JAM PER	
TUSSILAGO L. **Asteraceae** *(Compositae)*					
farfara L.			ENG FIN IRA SOV	FRA GER ICE ITA NET TUR USA	AUS POL
TYLOPHORA R. Br. **Asclepiadaceae**					
indica Merr.				CEY	
laevigata Decne.				MAU	
sylvatica Decne.				GHA	
TYPHA L. **Typhaceae**					
angustata Bory & Chaub.			IRQ ISR	IND PK	EGY POR

TYPHA L. Typhaceae *(Cont'd)*	S	P	C	X	F
angustata Bory & Chaub. *(Cont'd)*				TUR	SUD
angustifolia L. (Syn. *T. domingensis* Pers.)	HUN	AUS IDO IND ITA PER SUD SWE	ARG KEN PR	AFG ARB AST BEL BOR BRA CAB CAN CHL CNK CR CZE DR ECU ENG FIJ FRA GER GRE HAW HON IRA ISR JPN KOR MEL MEX MOR MOZ NET NOR PHI PK POL SUR THI TNZ TUR UGA USA VEN VIE	
australis Schum. & Thonn.			KEN POR		GHA
capensis Rohrb.			KEN		
domingensis Pers. *(See* T. **angustifolia** L.*)*	AUS		ARG	BRA GUA HON IDO URU USA	
elephantina Roxb.				THI	
glauca Godr.				USA	
javanica Schnitzl *ex* Zoll.				CEY	
latifolia L.	HUN	AUS GER ITA RHO SPA TUN	ARG IRA KEN POR USA	AFG AST BEL BOT BRA CAN	ISR

TYPHA L. Typhaceae *(Cont'd)*	S	P	C	X	F
latifolia L. *(Cont'd)*				CHL	
				CZE	
				ECU	
				EGY	
				ENG	
				FRA	
				GHA	
				GRE	
				IND	
				JOR	
				JPN	
				LEB	
				NET	
				NOR	
				PK	
				POL	
				SOV	
				SWE	
				TUR	
				VEN	
muelleri Rohrb.			AUS		
orientalis Presl	AUS			NZ	
TYPHALEA Neck. **Malvaceae**					
fruticosa (Mill.) Britton			PR	DR	
TYPHONIUM Schott **Araceae**					
divaricatum (L.) Decne.			IDO		
trilobatum (L.) Schott			BOR	CEY	
ULEX L. **Papilionaceae** *(Leguminosae)*					
europaeus L.	HAW	AUS	IRA	BRA	
	NZ	CHL	ITA	ENG	
			POL	FRA	
			TAS	GER	
				IND	
				NGI	
				SPA	
				TRI	
				USA	
minor Roth				NZ	
ULMUS L. **Ulmaceae**					
alata Michx.				USA	
americana L.				USA	
crassifolia Nutt.				USA	
parvifolia Jacq.				USA	
procera Salisb.				USA	
pumila L.				USA	
rubra Muhl.				USA	
serotina Sarg.				USA	
thomasi Sarg.				USA	

UMBELLULARIA Nutt. Lauraceae	S	P	C	X	F
californica Nutt.				USA	
URARIA Desv. **Papilionaceae** *(Leguminosae)*					
lagopodioides (L.) Desv. *ex* Dc.			PHI	NGI PLW	
picta (Jacq.) Desv. *ex* Dc.				IVO	GHA
URECHITES Muell.-Arg. **Apocynaceae**					
lutea (L.) Britt.			PR	DR	
URENA Dill. *ex* L. **Malvaceae**					
capitata L.				IDO	
lobata L.	FIJ MEL PLW	AUS IND PK TAI ZAM	BOR HAW IDO PR TRI	BRA CNK DAH GHA HON IVO JAM KEN MAD MAL MAU MIC NGI NIG PHI RHO SEG SUD SUR THI USA VIE	HK
sinuata L.			PR TRI		
trilobata Vell.			PR	DR	
URERA Gaudich. **Urticaceae**					
baccifera (L.) Gaudich.			PR	BOL COL DR VEN	HON PAN SAL
cameroonensis Wedd.				CNK	
hypselodendron Wedd.				CNK	
URGINEA Steinh. **Liliaceae**					
maritima (L.) Baker			MOR		
UROCHLOA Beauv. **Poaceae** *(Gramineae)*					
bolbodes (Steud.) Stapf				ANG	
mosambicensis (Hack.) Dandy		AUS MOZ			BUR SAF
panicoides Beauv.	KEN THI	SWZ UGA	RHO SAF	AUS BOT	HAW IND

	S	P	C	X	F
UROCHLOA Beauv. **Poaceae** *(Gramineae)* *(Cont'd)*					
panicoides Beauv. *(Cont'd)*			SUD	MAU PK	
pullulans Stapf	ZAM				
UROSPERMUM Scop. **Asteraceae** *(Compositae)*					
picroides (L.) F. W. Schmidt			EGY SPA		ISR POR
URSINIA Gaertn. **Asteraceae** *(Compositae)*					
nana Dc.			SAF		
URTICA L. **Urticaceae**					
dioica L.			EGY ENG FIN IRA SAF SOV SPA USA	AUS CHL GER ICE ITA NZ	POL
gracilis Ait.				USA	
incisa Poir.			AUS		
lyallii S. Wats.				USA	
massaica Mildbr.			KEN		
pilulifera L.			ISR	GRE	
procera Muhl.				USA	
urens L.	ENG POL SPA	COL GER NZ SWE	ARG AST AUS EGY FIN HUN IRA ISR ITA SAF SOV TUN	AFG BEL BRA CAN CHL ECU ICE IND IRQ JOR LEB NET NOR PER SAL TUR USA YUG	URU
URVILLEA H. B. K. **Sapindaceae**					
ulmacea H. B. K.				TRI	
UTRICULARIA L. **Lentiburiaceae**					
aurea Lour. (Syn. *U. flexuosa* Vahl)					
bifida L.				THI	
exoleta R. Br. *(See* **U. gibba** L. *ssp.* exoleta (R. Br.) P. Taylor)		IND		CAB	

UTRICULARIA L. Lentiburiaceae *(Cont'd)*	S	P	C	X	F
flexuosa Vahl (*See* **U. aurea** Lour.)		THI	MAL	BND CAB VIE	
foliosa L.				USA	
gibba L. *ssp.* exoleta (R. Br.) P. Taylor (Syn. *U. exoleta* R. Br.)					
inflata Walt.				USA	
inflexa Forsk.		GHA		USA	
odorata Pellegr.				CAB	
purpurea Walt.				USA	
stellaris L. *f.*		IND		SUD	
thonningii Schum. & Thonn.				SUD	
vulgaris L.				ENG TUR USA	
VACCARIA Medic. **Caryophyllaceae**					
parviflora Moench				SOV	
pyramidata Medic.				CAN MOR TUR	AUS EGY IND IRA IRQ LEB NZ POL POR
segetalis (Neck.) Garke				CAN SOV USA	ARG AUS ISR LEB POR
vulgaris Host (*See* **Saponaria vaccaria** L.)				SOV	
VACCINIUM L. **Vacciniaceae**					
angustifolium Ait.				USA	
arboreum Marsh.				USA	
ovalifolium Sm.				USA	
ovatum Pursh				USA	
oxycoccus L.				USA	
parvifolium Smith				USA	
stamineum L.				USA	
vitis-idaea L.				USA	

	S	P	C	X	F
VACHELLIA Wight & Arn. **Mimosaceae** *(Leguminosae)*					
farnesiana Wight & Arn. *(See* **Acacia farnesiana** (L.) Willd.)					COL
VAHLIA Thunb. **Saxifragaceae**					
digyna (Retz.) O. Ktze.				EGY	
VALERIANA L. **Valerianaceae**					
officinalis L.		SPA		BEL	
VALERIANELLA Hall. **Valerianaceae**					
boissieri Krok			LEB		
dentata (L.) Pollich			ENG		POL
discoidea (L.) Loisel.			MOR		POR
locusta Betcke			ENG		
olitoria (L.) Pollich			JPN	USA	
rimosa Bast.			CHL		
vesicaria (L.) Moench			LEB		
VALLARIS Burm. *f.* **Apocynaceae**					
heynei Spreng.				IND	
VALLISNERIA L. **Hydrocharitaceae**					
aethiopica Fenzl				AFG JPN RHO SAF SEG SUD TNZ UGA	
alternifolia Roxb. *(See* **Nechamandra alternifolia** (Roxb.) Thw.)				IND	
americana Michx.		USA		CAN HAW JAM	
gigantea Graebn.		AUS			
spiralis L.		IND PK	AUS POR	AFG ARG CHN CNK GRE HAW IDO IRA IRQ ITA JPN KOR MEL NZ SUD THI TNZ UGA	

VALLISNERIA L. Hydrocharitaceae *(Cont'd)*	S	P	C	X	F
spiralis L. *(Cont'd)*				USA	
VANDELLIA L. Scrophulariaceae					
anagallis (Burm.) Yamazaki			JPN		
anagallis (Burm.) Yamazaki *var.* verbenaefolia (Colsm.) Yamazaki (Syn. *Ilysanthes serrata* (Thunb.) Makino)					
angustifolia Benth.		JPN			
crustacea (L.) Benth.		JPN	CHN	IND	
pedunculata Benth.		IND			
setulosa (Maxim.) Yamazaki			JPN		
VELLA L. Brassicaceae *(Cruciferae)*					
annua L.				AUS	
VERATRUM L. Liliaceae					
album L.				GER	
californicum Durand				USA	
lobelianum Bernh.		SOV			
viride Ait.				USA	
VERBASCUM L. Scrophulariaceae					
blattaria L.			AUS USA	CHL NZ	
lychnitis L.			SOV	USA	
nigrum L.			SOV		
orientale Bieb.			SOV		
phlomoides L.			SOV	USA	
thapsiforme Schrad.			SOV	CHL	
thapsus L.			AUS SOV USA	CHL NZ	
virgatum Stokes			ARG AUS	CHL NZ	
VERBENA L. Verbenaceae					
bonariensis L.			ARG AUS SAF USA	BRA MAU NZ URU	
bracteata Cav. *ex* Lag. & Rodr.			USA		
ciliata Benth.		MEX			
corymbosa Ruiz & Pav.				CHL	
dissecta Willd. *ex* Spreng.				URU	

VERBENA L. Verbenaceae *(Cont'd)*	S	P	C	X	F
gracilescens Cham.			ARG		
hastata L.				USA	
hispida Ruiz & Pav.				ARG COL	
intermedia Gill. & Hook.			ARG		
litoralis H. B. K.			ARG COL HAW	BRA CHL NZ PER URU USA VEN	
officinalis L.	KOR	POL	AUS CHN HUN IRA IRQ JPN LEB MOR POR SAF SOV SPA TAI	EGY IND ITA JOR KEN MAU MEX NZ PK RHO SUD THI TUR USA VIE YUG	HK ISR
peruviana (L.) Britton				URU	
rigida Spreng.			AUS	NZ	
stricta Vent.			USA		
supina L.				MOR	
tenera Spreng.			AUS		
tenuisecta Briq.			SAF		
urticifolia L.				USA	
VERBESINA L. Asteraceae *(Compositae)*					
alata L.		PR			
caracasana Rob. & Greenm.				VEN	
encelioides (Cav.) A. Gray		ARG HAW	AUS IND	USA	SAF
persicifolia Dc.		MEX			
subcordata Dc.			ARG		
VERNONIA Schreb. Asteraceae *(Compositae)*					
altissima Nutt.			USA		
ambigua Kotschy & Peyr.				GHA	
baccharoides H. B. K.				PER	

VERNONIA Schreb. **Asteraceae** *(Compositae) (Cont'd)*	S	P	C	X	F
baldwini Torr.				USA	
brasiliensis Less.				COL VEN	
cainarahiensis Hieron.				PER	
camporum M. E. Jones				GHA	
chinensis Less. *(See* **V. patula** (Dryand.) Merr.*)*				VIE	IDO
cinerea (L.) Less. (Syn. *Senecioides cinerea* (L.) O. Ktze.)	GHA	IND THI	BOR CEY CHN HAW JAM NIG PHI TRI UGA	ANT AUS CAB DAH FIJ IDO IVO MAL MAU MEL MIC MOZ NGI PK PLW PR SEG SUD SUR USA VIE ZAM	HK NZ
colorata Drake				GHA	
divaricata Sw.				JAM	
flexuosa Sims				BRA	
intermedia Dc.				URU	
kotschyana Sch.-Bip.				SUD	
lasiopus O. Hoffm.				KEN	
nigritiana Oliv. & Hiern				GHA	
pallens Sch.-Bip.				MEX	
patens H. B. K.				COL CR HON PAN SAL	
patula (Dryand.) Merr. (Syn. *V. chinensis* Less.)				PHI	
pauciflora Less.				GHA	
polyanthes Less.		BRA			
poskeana Vatke & Hildeb.				ANG	
scabra Pers.				BRA	
scorpioides Pers.				BRA	

VERNONIA Schreb. Asteraceae *(Compositae) (Cont'd)*	S	P	C	X	F
sericea L. Rich.			PR		
undulata Oliv. & Hiern				SUD	
VERONICA L. Scrophulariaceae					
agrestis L.			CHN ENG MOR SPA	GER IND TUR USA	AUS PK POL SOV
americana (Rafin.) Schwein.			JPN	NZ	
anagallis-aquatica L.			IND JPN POR	CHL CHN NZ PK TUR	AUS ISR SAF
anagalloides Guss.				TUR	
arvensis L.	AFG ITA POL	AST CHL KOR SWE TUN	ARG ENG FIN HUN JPN SOV SPA TUR USA	BEL CAN CHN EGY GER IRA JOR LEB MOR NZ PK PLW POR SWT	AUS
beccabunga L.			EGY	BEL TUR	
chamaedrys L.			FIN	TUR USA	POL
chamaepitys Griseb.				TUR	
cymbalaria Bodard				TUR	
didyma Tenore			JPN		
filiformis Sm.				USA	
hederaefolia L.		JPN	ENG SPA	FIN GER ICE ITA USA	
javanica Bl.			IDO TAI		
officinalis L.			USA	CHL	
opaca Fries.				GER	SOV
peregrina L.			ARG JPN USA	CHL COL	
persica Poir.	ENG GER	AST CZE	ARG COL	AFG BEL	AUS KOR

VERONICA L. Scrophulariaceae *(Cont'd)*	S	P	C	X	F
persica Poir. *(Cont'd)*	ITA POL	IRA JPN NZ SWE	HUN NET NOR SAF SPA TUR	CAN CHL CHN FIN ICE JOR NEP PER PK SOV USA	POR
polita Fries			ENG IRQ SOV	GER USA	
serphyllifolia L.			CHN ENG FIN SOV	CHL GER NZ USA	POL
tournefortii C. C. Gmel.				CHL	
triphyllos L.				GER TUR	
undulata Wall. *ex* Roxb.			TAI	JPN	
VETIVERIA Bory Poaceae *(Gramineae)*					
zizanioides (L.) Nash *ex* Small				VIE	
VIBURNUM L. Caprifoliaceae					
acerifolium L.				USA	
alnifolium Marsh.				USA	
dentatum L.				USA	
lentago L.				USA	
prunifolium L.				USA	
rafinesquianum Schultes				USA	
rufidulum Rafin.				USA	
VICIA L. Papilionaceae *(Leguminosae)*					
angustifolia L. *(See* **V. sativa** L. *ssp.* angustifolia (L.) Gaudich.)		JPN	CAN FIN IRQ SOV USA	AUS COL GER NZ	ISR POL
atropurpurea Desf.				CHL	SAF
benghalensis L.			MOR POR		
bithynica L.			POR		
calcarata Desf.			EGY IRQ		AUS IRA
cracca L.		FIN	CAN POR SOV	GER NZ TUR	JPN POL

VICIA L. Papilionaceae *(Leguminosae) (Cont'd)*	S	P	C	X	F
cracca L. *(Cont'd)*			SPA	USA	
ervilia (L.) Willd.			POR		
faba L.		TUN	CHN		
gracilis Loisel.			POR		
graminea Sm.				PER	
hirsuta (L.) S. F. Gray (Syn. *Ervum hirsutum* L.)		IND JPN	ENG FIN POR SOV	BND CHN GER NZ TUR	AUS POL
hybrida L.				MOR	
lathyroides L.				MOR	
lutea L.			MOR POR SPA	GER NZ	ISR
monantha Retz.			MOR	AUS	
monanthos (L.) Desf.			POR		
narbonensis L.		TUN	IRA LEB	TUR	ISR
onobrychioides L.			POR		
peregrina L.			MOR POR		IRA ISR POL
sativa L.	IND ITA POL POR	IRA KOR NZ PK	CHN EGY ENG FIN GER HUN IRQ ISR JPN NEP SOV SPA TUN TUR	AFG ARG BEL BND CAN CHL COL GRE JOR LEB MAU MOR NET USA YUG	AUS HK
sativa L. *ssp.* angustifolia (L.) Gaudich. (Syn. *V. angustifolia* L.)					
sepium L.			SOV	GER	
sibthorpii Boiss.				GRE	
tenuifolia Roth				GER	
tetrasperma (L.) Schreb.			CHN ENG JPN SOV	CAN GER NZ PK SPA USA	AUS ISR POL

VICIA L. Papilionaceae *(Leguminosae) (Cont'd)*	S	P	C	X	F
villosa Roth			IRA POR SOV	CHL COL GER ITA MOR USA	AUS ISR POL
VICOA Cass. Asteraceae *(Compositae)*					
auriculata Cass.				IND	
indica Dc.				GHA IND	
VIGNA Savi Papilionaceae *(Leguminosae)*					
luteola (Jacq.) Benth.			ARG	PER SUR	
marina (Burm. *f.*) Merr.				FIJ PLW	
repens (L.) O. Ktze.			PR	COL	
VIGUIERA H. B. K. Asteraceae *(Compositae)*					
anchusaefolia (Dc.) Baker			ARG	BRA	
annua (Jones) Blake				USA	
dentata Spreng.		MEX			
stenoloba Blake				USA	
VILLEBRUNEA Gaudich. Urticaceae					
frutescens Bl. (Syn. *Boehmeria frutescens* Thunb.)					
VINCA L. Apocynaceae					
herbacea Waldst. & Kit.				TUR	
major L.				CHL NZ USA	AUS
minor L.				TUR USA	
rosea L. (*See* Catharanthus roseus G. Don)				BRA JAM	AUS ISR
VIOLA L. Violaceae					
arvensis Murr.	FIN		ENG POR SOV	ICE NZ USA	
hederacea Labill.				AUS	
japonica Langsd.			JPN		CHN
kitaibeliana Roem. & Schult.				USA	
mandshurica W. Becker			JPN		
ovata-oblonga (Miq.) Makino			JPN		

VIOLA L. Violaceae *(Cont'd)*	S	P	C	X	F
tricolor L.				ENG SOV	CHL COL GER USA	ISR POR
verecunda A. Gray				JPN		
VISCUM L. Loranthaceae						
album L.					ITA TUR	
VISMIA Vand. Clusiaceae *(Guttiferae)*						
cayennensis (Jacq.) Pers.					PER	
VITEX L. Verbenaceae						
agnus-castus L.					TUR	
trifolia L.				HAW	FIJ	
VITIS L. Vitaceae						
aestivalis Michx.					USA	
candicans Engelm.					USA	
hastata Miq.					MAL	
japonica Thunb. *(See* Cayratia japonica (Thunb.) Gagnep.)				CHN JPN		
rotundifolia Michx.					USA	
rupestris Scheele					USA	
tiliaefolia Roem. & Schult.					HON SAL	
trifolia L.					IND	
vulpina L.					USA	
VITTADINIA A. Rich. Asteraceae *(Compositae)*						
triloba Dc.					AUS	
VOGELIA Medic. Brassicaceae *(Cruciferae)*						
apiculata Vierh.					SOV	
paniculata (L.) Hornem.					SOV	LEB
VOLUTARELLA Cass. Asteraceae *(Compositae)*						
divaricata Benth. & Hook. *f.*					IND PK	
ramosa (Roxb.) Santapau					IND	
VOSSIA Wall. & Griff. Poaceae *(Gramineae)*						
cuspidata (Roxb.) W. Griff.					BND SUD VIE	GHA IND KEN

VULPIA C. C. Gmel. **Poaceae** *(Gramineae)*	S	P	C	X	F
bromoides (L.) S. F. Gray			AUS POR SAF	NZ	
dertonensis Volkart			ARG		
geniculata (L.) Link			POR		
hybrida (Brot.) Pau			POR		
megalura (Nutt.) Rybd.			AUS	CHL NZ	HAW IND
myuros (L.) C. C. Gmel.			AUS POR SAF	CHL NZ	HAW USA
WAHLENBERGIA Schrad. **Campanulaceae**					
gracilis Schrad.				AUS	
WALTHERIA L. **Sterculiaceae**					
americana L. *(See* **W. indica** L.*)*		BRA	HAW PR	FIJ JAM PAN SAL USA	
indica L. (Syn. *americana* L.)				COL GHA KEN MAU VIE	
ovata Cav.				PER	
WEDELIA Jacq. **Asteraceae** *(Compositae)*					
asperrima Benth.				AUS	
biflora (L.) Dc.			IDO	FIJ PLW	
calendulacea Less. *(See* **W. chinensis** (Osbeck) Merr.*)*			IND TAI		
chinensis (Osbeck) Merr. (Syn. *W. calendulacea* Less.)			TAI		
glauca Hoffm.	ARG			CHL URU	AUS
gracilis Rich.				ANT JAM	
paludosa Dc.				BRA	
trilobata (L.) Hitchc.		TRI	PR	DR JAM PAN SUR	
WIEDEMANNIA Fisch. & Mey. **Lamiaceae** *(Labiatae)*					
orientalis Fisch. & Mey.				TUR	

WIGANDIA H. B. K. Hydrophyllaceae	S	P	C	X	F
caracasana H. B. K.				GUA HON SAL	
WIKSTROEMIA Endl. Thymelaeaceae					
gampi (Sieb. & Zucc.) Maxim.				JPN	
indica (L.) C. A. Mey. (Syn. *W. viridiflora* Meissn.)					
viridiflora Meissn. (*See* **W. indica** (L.) C. A. Mey.)				MAU	
WISLIZENIA Engelm. Capparidaceae					
refracta Engelm.				USA	
WISSADULA Medic. Malvaceae					
excelsior (Cav.) Presl				PER	
WITHANIA Pauq. Solanaceae					
coagulans Dun.				PK	
somnifera (L.) Dun.			EGY RHO SAF	IND SUD	PK
WOLFFIA Horkel *ex* Schleid. Lemnaceae					
arrhiza (L.) Wimm.	GER IND	POR		CAN CHN DAH GHA HUN IDO ITA JOR KEN MEX NZ PK POL RHO THI TNZ UGA	KOR MAD
columbiana Karst.				USA	
WOODWARDIA Sm. Blechnaceae					
areolata (L.) Moore				USA	
WYETHIA Nutt. Asteraceae (*Compositae*)					
amplexicaulis Nutt.				USA	
XANTHIUM L. Asteraceae (*Compositae*)					
ambrosioides Hook. & Arn.			ARG		
argenteum Widder			CHL		
brasilicum Vell.			EGY LEB MOR		

XANTHIUM L. Asteraceae *(Compositae)* *(Cont'd)*	S	P	C	X	F
californicum Greene		AUS			
cavanillesii Schouw			ARG	CHL URU	
chinense Mill.	AUS		PR	DR USA	FIJ
echinatum Murr.				USA	
italicum Moretti		AUS		BND CHL MEL USA	
macroparpum Dc.				TUR	
occidentale Bertol.				COL	
orientale Bl.			JPN POR	AUS CHL	
pensylvanicum Wallr.		CAN	USA		
pungens Wallr.	FIJ IND	AUS SAF	RHO	MEX	
saccharatum Wallr.			ARG HAW	MEX	
speciosum Kearney				USA	
spinosum L.	AUS	ARG TAS	EGY MOR RHO SAF SPA	BRA CHL ETH GRE ISR NGI NZ POL POR SOV TNZ TUR URU USA YUG	
strumarium L.		IND ISR YUG	HAW IRA IRQ JPN LEB SAF SPA TAI	AUS CHN ETH HUN MEL PHI PK POL SOV THI TRI TUR USA	HK KOR
XANTHORRHOEA J. E. Sm. **Xanthorrhoeaceae**					
preissii Endl.		AUS			
XANTHOSOMA Schott **Araceae**					
helleborifolium (Jacq.) Schott				ANT	

	S	P	C	X	F
XANTHOSOMA Schott **Araceae** *(Cont'd)*					
helleborifolium (Jacq.) Schott *(Cont'd)*				TRI	
XANTHOXALIS Small **Oxalidaceae** *(See* **OXALIS** L.*)*					
corniculata (L.) Small *(See* **Oxalis corniculata** L.*)*				DR	
XANTHOXYLUM J. F. Gmel. **Rutaceae**					
fagara Sarg.			MEX		
XEROPHYLLUM Rich. **Liliaceae**					
tenax (Pursh) Nutt.				USA	
XIMENIA L. **Olacaceae**					
caffra Sond.				RHO	
XYRIS L. **Xyridaceae**					
complanata R. Br.				VIE	
indica L.		IDO		THI VIE	IND
melanocephala Miq.				IDO	
YOUNGIA Cass. **Asteraceae** *(Compositae)*					
japonica (L.) Dc. (Syn. *Crepis japonica* (L.) Benth.)			HAW JPN TAI	USA	AUS MAU
YUCCA L. **Agavaceae**					
elata Engelm.				USA	
glauca Nutt.				USA	
torreyi Schafer				USA	
ZAMIA L. **Cycadaceae**					
loddigesii Miq.				HON	
ZANNICHELLIA L. **Zannichelliaceae**					
palustris L.			POR USA	SUD	ISR
ZANTHOXYLUM L. **Rutaceae**					
americanum Mill.				USA	
clava-herculis L.				USA	
fagara (L.) Sarg.				USA	
martinicense Dc.				JAM	
ZEBRINA Schnizl. **Commelinaceae**					
pendula Schnizl.		PR		COL DR GUA HON JAM	MAU

ZEBRINA Schnizl. Commelinaceae *(Cont'd)*	S	P	C	X	F
pendula Schnizl. *(Cont'd)*				MEX SAL VEN	
ZEHNERIA Endl. **Cucurbitaceae**					
thwaitesii (Schweinf.) C. Jeffrey				NIG	
ZEPHYRANTHES Herb. **Amaryllidaceae**					
andersonii Nichols.				URU	
atamasco (L.) Herb.				USA	
eggersiana Urb.				ANT	
rosea (Spreng.) Lindl.				PLW	
ZIGADENUS Michx. **Liliaceae**					
fremontii Torr. *ex* S. Wats.				USA	
paniculatus (Nutt.) S. Wats.				USA	
ZINNIA L. **Asteraceae** *(Compositae)*					
multiflora L.			SAF		
pauciflora L.			HAW		AUS
peruviana L.			AUS		
ZIZANIA L. **Poaceae** *(Gramineae)*					
aquatica L.				USA	
ZIZANIOPSIS Doell & Aschers. **Poaceae** *(Gramineae)*					
bonariensis Speg.				ARG URU	
miliacea (Michx.) Doell & Aschers.				USA	
ZIZIPHUS Mill. **Rhamnaceae**					
mauritiana Lam.	AUS				
mucronata Willd.		RHO			
nummularia (Burm. *f.*) Wight & Arn.		IND		PK	
rotundifolia Lam.				IND	
rugosa Lam.				IND	
ZORNIA Gmel. **Papilionaceae** *(Leguminosae)*					
diphylla (L.) Pers.				BRA IND	
latifolia Dc.				GHA	
ZOYSIA Willd. **Poaceae** *(Gramineae)*					
japonica Steud.			JPN	THI	HAW IND SAF
matrella (L.) Merr.				THI	HAW

ZOYSIA Willd. Poaceae *(Gramineae)* *(Cont'd)*	S	P	C	X	F
matrella (L.) Merr. *(Cont'd)*					IDO IND SAF
tenuifolia Willd.			TAI	THI	HAW IND SAF
ZYGADENUS Michx. **Liliaceae**					
elegans Pursh.				USA	
gramineus Rybd.				USA	
ZYGOPHYLLUM L. **Zygophyllaceae**					
apiculatum F. Muell.				AUS	
fabago L.				USA	ISR

_____ Date

Send to

Professor Le Roy Holm
Department of Horticulture
University of Wisconsin
Madison, Wisconsin   53706
U.S.A.

_____ Country

I. These additional species are present as weeds in my country:

Species Name	Serious	Principal	Common	Present	Flora

II.   These corrections are in order for A Geographical Atlas of World Weeds:

III.   Additional information offered for A Geographical Atlas of World Weeds:

Name and address of cooperator

_____

_____

_____

_____

We are grateful to you.